OXFORD LOGIC GUIDES: 10

OXFORD LOGIC GUIDES

CANTORIAN SET THEORY
AND
LIMITATION OF SIZE

Michael Hallett

Department of Philosophy, McGill University,
Montreal

CLARENDON PRESS · OXFORD

Oxford University Press, Walton Street, Oxford OX2 6DP
Oxford New York Toronto
Delhi Bombay Calcutta Madras Karachi
Petaling Jaya Singapore Hong Kong Tokyo
Nairobi Dar es Salaam Cape Town
Melbourne Auckland
and associated companies in
Beirut Berlin Ibadan Nicosia

Oxford is a trade mark of Oxford University Press

Published in the United States
by Oxford University Press, New York

British Library Cataloguing in Publication Data
Hallett, Michael, 1930-
Cantorian set theory and limitation of size.
1. Set theory—History 2. Numbers,
Transfinite—History
I. Title
511.3'22   QA248
ISBN 0-19-853283-0

Printed and bound in Great Britain
by Biddles Ltd, Guildford and King's Lynn

To the memory of my father

FOREWORD

Michael Dummett, New College, Oxford

Set theory is that branch of mathematics in which the classical or platonist conception receives its fullest mathematical expression. It also has a foundational character, providing at least one general setting for most other mathematical theories. The work of Georg Cantor, the nineteenth-century founder of set theory, is for these reasons universally acknowledged as of fundamental importance. Michael Hallett's book is the first full-length study of Cantor that does justice both to the mathematical importance of his work and to the philosophical ideas which governed it. Cantor's own writings lay much stress upon the philosophical views which, for him, were indispensable to an understanding of the significance of his work and of the justification for the manner in which he carried it out. Hitherto, these philosophical ideas have been neglected by philosophers and mathematicians alike. Hallett supplies us with a deeply illuminating account of Cantor's development of set theory as informed by his philosophical opinions. He supplements his study of Cantor's own work by tracing the subsequent contributions of his immediate successors, and demonstrates to how great an extent the later history of set theory has depended upon the fundamental principles by which Cantor was guided. In so doing, he corrects a number of prevalent misunderstandings of the early history of set theory. The result is of much more than purely historical interest: by going back to the sources, Hallett gives a stimulus to renewed philosophical reflection on set theory, which has for some time now been virtually becalmed. This stimulus derives both from his patient and sympathetic interpretations of the philosophical ideas of the founders of set theory, and particularly of Cantor, and from his own often perceptive comments on them. Although the book is very well-informed mathematically, a great deal in it will be readily intelligible to non-specialists, and hence should be of great interest to both mathematicians and philosophers generally.

PREFACE

In nature's infinite book of secrecy a little I can read.

Shakespeare, *Antony and Cleopatra*, Act I, Sc. 2.

We shall not weary ourselves with disputes about the infinite, since, finite as we are, it would be perverse to attempt to make any determination of it, turning it, as it were, into something finite that we could conceive . . .

Descartes, *Principia* I, XXVI.

The usefulness of mathematics, the esteem in which it is held, and the honourable name of 'exact science *par excellence*' rightly given it are all due to the clarity of its principles, the rigour of its proofs and the precision of its theorems.

In order to ensure the perpetuation of these precious merits in so beautiful a part of our knowledge, we seek *a clear and precise theory of what is called* Infinite *in mathematics*.

Competition announcement of the Berlin
Academy of Sciences for 1786.

Learning the axioms for set theory one could be forgiven for thinking that the set concept is both elementary and simple. Of course, the axiom systems did make it elementary (no pun intended), since they undertook to base all of mathematics on it, and it on nothing else. But understanding the set concept is anything but simple: there is a mystery surrounding it which easy familiarity with the axioms tends to conceal. The judgement that the set concept is highly problematic is not new; it was the opinion, for varying reasons, of Borel, Lebesgue, Lusin, of Brouwer, Poincaré, Russell, and Weyl, to mention a few. Indeed, the last four thought 'set' too unclear a notion on which to base mathematics. And this challenge to the foundational status of the set concept has resurfaced recently, particularly with the growth of interest in category theory, topos theory, and intuitionism. It is not the purpose of this book to discuss the case for any of these alternatives. But these challenges undoubtedly promote interest in the philosophical foundations of classical set theory itself, and this in turn prompts us to inquire into the philosophical and conceptual ideas behind the creation and development of the theory. This, then, is the primary aim of this book, to contribute to a live and important philosophical debate by tracing the foundations of set theory to their historical and conceptual source.

Set theory, as we know, is the creation of Georg Cantor, and here I am concerned largely with the ideas on which Cantor's theory is founded and their impact on the subsequent evolution of the theory. For anyone interested

in philosophical problems surrounding set theory now (for example, problems
with the concept of set itself, or perhaps with the question of what would or
would not be an acceptable addition to the basic modern axioms) four questions
naturally suggest themselves: How did 'set' become the fundamental notion in
Cantor's theory? What was Cantor's own conception of set? What effect did
Cantor's philosophical and foundational ideas, about which he wrote a great
deal, have on the shape of his own theory and on what came later? And how
much of modern axiomatic set theory is Cantorian set theory? I hope to shed
light on all these questions in the course of this book. Part 1 is concerned more
with the first three questions, and Part 2 more with the fourth, though there is
no completely rigid separation. At the risk of some duplication, I want to give
a brief review here of some of the themes dealt with in what follows.

A major aim in Part 1 is to present an integrated account of both Cantor's
metaphysical and mathematical theories of infinity. There have been accounts
of Cantor's work before (Cavaillès [1962] is by far the best), but none has
taken Cantor's metaphysical writings very seriously either as a way of understanding the detailed content of his mathematical theories or as a way of better
understanding the philosophical foundations of modern set theory. I do not
claim that Cantor was a great or important philosopher; he was not. And actually
this is beside the point; there are even places in his work where he would have
been hampered by sharper philosophical acumen. Nor do I claim, of course,
that modern axiomatic set theory has to stand by all of Cantor's metaphysical
views. But I do claim that these views did contribute enormously to the shape
and development of the key Cantorian concepts, and that *therefore* there is a
direct route from Cantor's metaphysics to the substance and nature of modern
set theory. Historical, critical, and philosophical analysis of both the Cantorian
concepts and the metaphysical ideas underlying them seems to me thus an
essential prerequisite for discussions on the foundations of set theory. My plan
has been as follows. In the Introduction to Part 1 I have summarized the core
of Cantor's foundational doctrines in three basic principles (p. 7), together
with a short sketch of their importance (pp. 7-9). In Chapters 1 and 3 I have
tried to elaborate the philosophical, metaphysical and theological context in
which these principles were set, while in Chapters 2 and 4 I have concentrated
a little more on their mathematical importance, on their involvement in shaping
the detail of Cantor's mathematical theory of infinity.

I stress again that I have attempted to give an integrated account of Cantor's
philosophical and mathematical theories; indeed, in my view the two strands
cannot be treated separately. In any case, to understand fully why the set
concept became so important foundationally one has to understand both the
mathematical aims of Cantor's theory of infinity and its mathematical setting.
The central specific problem which occupied Cantor was the celebrated continuum problem posed by him in 1878, i.e. the problem of whether there are
more than two infinite sizes represented in the real line. Cantor does not say

much about this problem directly in his writings: unfortunately, there are no long, speculative passages setting out reasons why the continuum hypothesis should be true. But this does not mean that he did not regard it as important: on the contrary, virtually all of his later work on the theory of infinity stems from it. As I explain in the Introduction to Part 1, the key element in both of Cantor's lines of attack on the continuum problem was the notion of infinite ordinal number, and especially the ordinal theory of infinite cardinality. This made the extended ordinal number sequence the central pillar of Cantor's theory of infinity, a fact of crucial importance for understanding subsequent developments. But any attempt to extend the natural number sequence 'into the infinite' has to face up to a difficult foundational problem: what is the common basis shared by the natural numbers and the new elements one wants to add? This question was particularly acute for Cantor since he had to counter philosophical and mathematical attitudes which were strongly against 'actual infinities'. His answer was to argue that the notions of ordinal and cardinal number, whether infinite or not, are founded on the notion of set *and* that infinite sets are essential to mathematics, not just at the periphery, but rather at its heart, in analysis itself. The pattern of Cantor's argumentation here is not particularly transparent. But however vague and confused the arguments, the foundational tendency in Cantor's work after 1882 is steadily more and more towards set-theoretic reductionism, eventually to the thesis, not just that sets are indispensable in mathematics, but that the key mathematical objects—numbers—are themselves just sets. I try to make this progression clear, along with Cantor's somewhat hazy argumentation, in §§1.2–1.3 and 2.1, and then in Chapter 3.

This helps to explain how the set concept became fundamental. But what of the set concept itself? I mentioned above that there is a troubling mystery surrounding it, and this mystery in fact goes right back to Cantor's own system. For while the direction of his work was steadily towards making the set concept fundamental in mathematics in place of number, he was never very clear about what sets are. One of the complicating and perhaps confusing factors here is the theological element in Cantor's metaphysics. I have formed no final judgement of how important Cantor's theology was in his foundational work, of whether it actually shaped his realism or whether it was only introduced a little later as a way of justifying the realism *post hoc*. Certainly, Cantor's theology must have some prominence in any detailed discussion of his foundational work, not least because it helps us to understand some aspects of his realism much better. For instance, mixed in with Cantor's prevailing realism are splashes of what could well be called constructivism, and this applies particularly to two crucial elements of his theory, the notion of well-ordering and the set concept itself. (See §§1.3 and 3.5.) It could well be that Cantor was happy to use 'constructivist' or 'creational' metaphors in his explanations of key concepts because they can be backed by assumptions of Deific omnipotence: God can surely conceive of such and such an 'operation' being performed even though we could not possibly carry it out; thus we can assume that there is a result of the operation which exists completed in the Divine intellect. This would fit with Cantor's support for the thesis that 'possibility implies existence' (see §1.1), and also with his view that while we might explain the natural number sequence as being generated by a process of succession, actually the numbers exist completed as a totality, even as a set, as an idea in the Divine intellect. It seems quite plausible that the creational metaphor in the explanation of the set concept generally (where a set is said to be a collection which is put together as a whole in thought) is backed by the same kind of assumption of Divine power. (See §1.3 and Chapter 3.)

But this reliance on theology actually obscures philosophical problems rather than solving them, for it gives the impression that the underlying realism is reasonable without providing any solid arguments in support. Later set theorists like Zermelo and Gödel certainly shunned theological backing for their doctrines, though they certainly took over Cantor's realism. No doubt they were right to dispense with the theology; but surely if this is removed it makes no sense any more to stick to the constructional metaphors. For instance, it is surely nonsense to talk of infinitely many elements being 'put together to a unity' or of an arbitrary set being 'arranged' in a well-ordering as if these were operations which could actually be carried out. In any case, as I show in Chapter 6, it seems clear in retrospect that these constructional metaphors are quite inadequate for supporting the full weight of realist assumptions, especially the assumed exist- ence of uncountable infinite sets (the power-set principle), and this no matter how broadly we construe 'constructivism'. The difficulty is particularly acute since the set concept itself is the fundamental concept. Perhaps it is natural to try to explain sets by metaphors such as 'gathering' or 'completing' or the like. But these metaphors cannot do the job a fully realist set theory such as that created by Cantor requires. There is some evidence that Cantor himself was unhappy with these 'constructional' or 'psychologistic' metaphors; but so many attempted explanations of the set concept still rely heavily on them (see Chapter 6). Surely, is not some other kind of supportive argument needed? If so, how *are* we to understand the set concept?

Part 2 of the book deals primarily with the influence and status of the 'limita- tion of size' idea, an ideal focus for the question of how far modern set theory is Cantorian set theory, for there is no doubt both that limitation of size was of great significance in the development of the axiomatic theory of sets, and that it stems rather directly from Cantor's metaphysical doctrine of Absolute infinity. Nevertheless, it is important to be more precise about the role limita- tion of size played (and plays) and not to exaggerate its strength. Part of my purpose is to correct the impression often given (an impression which stems from Fraenkel) that axiomatic set theory avoids the standard paradoxes by choosing axioms which do not create overly large sets, it being assumed that the collection of ordinals, the Russell 'set' of all self-membered sets and the set-theoretic universe itself are all 'too big' in some sense. In §5.1 I argue that not only is it extremely difficult to specify a notion of 'overly large' which successfully embraces all these collections, but that in any case no form of this limitation of size argument can justify the adoption of the impredicative power-set axiom. This failure is related to the inadequacy of various attempts to explain the acceptance of the usual axioms by their being true in some informally presented, iterative universe of sets. These iterative justifications are linked with one form of Fraenkel's limitation of size argument with 'completability' as the main explanatory metaphor in place of 'limited comprehension'. They also suffer from the difficulties over constructional metaphors outlined above, for in the informal accounts of the set notion 'completability' is usually given a quasi-constructivist sense, and the force and plausibility of the account stems from the intelligibility of this. But this means again that these accounts cannot possibly explain the adoption of the highly *non*-constructive power-set axiom. Despite this, the iterative and limitation of size 'arguments' are still important as partial explanatory frameworks once some additional justification is given for the power-set axiom, or once one assumes, say, that infinite power-sets are relatively 'small'. There are strong parallels between these partial frameworks and Cantor's own conception of the structure of the set theoretic universe, as I explain in §§1.4 and 5.2. The iterative theories and 'completability of sets' are discussed at length in Chapter 6.

Limitation of size was important historically as a heuristic guide in selecting *some* of the axioms, and through this there is a direct connection between Cantor's metaphysical theory of infinity and the shape of the modern axiom systems, as I try to show in Chapters 4, 7, and 8. Cantor may not have said much about the nature of sethood itself, but his theories do imply a good deal about the structure of the universe of sets. Contrary to what one frequently reads, he did not allow that *any* collection is a set. For Cantor, therefore, the task of describing the composition of the mathematical universe is not a trivial one. This view was not a late arrival stemming from the discovery of the set-theoretic contradictions (which Cantor knew in 1895 or 1896); rather it was a conse-quence of his notion of Absolute infinity, a notion introduced as early as 1882 along with the ordinal theory of cardinality. The notion of Absolute infinity was originally presented as a metaphysical doctrine about what is necessarily excluded from mathematical treatment, a doctrine adopted largely for theo-logical reasons, to adapt the new theory of *mathematical* infinity to the tradi-tional Aristotelian-Scholastic teachings on infinity. (See pp. 12–14 and §1.4.) And this is already in effect a limitation of size theory, Absolute infinity being in a sense something incomprehensible. In later correspondence with Dedekind this metaphysical limitation of size conception was cashed out in more precise mathematical form with the ordinal number sequence used to determine which collection are 'absolute' or 'too large' to be sets, in effect, a weaving together of the earlier theory of cardinality and the metaphysical theory of Absolute infinity. This is explained more fully in Chapter 4, along with the work of Russell and Jourdain on similar lines. Chapter 8 describes von Neumann's later version of a very similar theory. What relates this particularly to Cantor is the use of limitation of size as an elaboration of the ordinal theory of cardinality, and the desire to reaffirm that the ordinal numbers are the central pillar of set theory. This dominance of the ordinals is perhaps the most striking feature of Cantor's work and the sway of the ordinal theory of cardinality itself yields a large part of the answer to the question of how far modern set theory is Cantorian. For the main spur to the set-theoretic work of Zermelo and von Neumann was the desire to recreate this ordinal theory axiomatically, and this was a major determinant of the shape of the modern axiom system. Cantor's theory immediately gave rise to the well-ordering problem, for in order for it to work every set must be well-orderable (see p. 64 and §3.6). The problem Zermelo took up was that of providing an axiomatic framework in which the well-ordering condition can be rigorously proved without relying any more on Cantor's rather vague assumptions about 'arrangements' in well-ordered form. (See §3.5.) But while Zermelo succeeded in this, he neglected or tried to dispense with the Cantorian ordinal numbers themselves. Thus, while his system laid the groundwork for the Cantorian theory of cardinality, it is not itself sufficiently Cantorian. The importance of von Neumann's work was precisely that it restored the ordinal numbers to their central position, and in doing so effected a crucial extension of Zermelo's system. A key element in this was von Neumann's axiomatized version of a limitation of size principle similar to Cantor's of 1899, for it was this which enabled him to formulate an adequate version of the replacement principle crucial to the full ordinal theory. Von Neumann's assumptions were later simplified in the so-called 'Zermelo–Fraenkel' system.

Later developments showed that Cantor's ordinal theory of cardinality is inevitable. For one thing, it turned out that if one is to have a theory of infinite cardinal number at all there are very strong theoretical reasons for choosing Cantor's. (See pp. 86-9.) Secondly, investigation of the structure and especially the length of the ordinal number sequence has remained the most promising way to tackle the continuum problem, and this was exactly why Cantor originally regarded the ordinals as so important. There were concerted and partially successful attempts to dispense with transfinite ordinals, particularly in analysis and the theory of point-sets. But they have always reappeared. In particular, Gödel's 'large cardinals' programme was a clear revival of Cantor's plan to solve the continuum problem by investigating what ordinals there are; and although it has not succeeded in its primary aim, some of the 'large' ordinals have proved to be the only way of achieving quite minimal information about relatively simple sets of points of the real line. (See §2.3(c).)

But to understand fully why these later developments of von Neumann and Gödel are essentially Cantorian in spirit, one has to grasp why the ordinals and the ordinal sequence were so important in Cantor's system and why he adopted the ordinal theory of cardinality long before there was any hint of its inevitability. A large part of the answer lies again in his philosophical approach to infinity, through the doctrine that there is no distinction in principle between the finite and the vast realm of infinity that comes within the orbit of mathematics. I call this doctrine, which is again bolstered by appeal to God's supposed intellectual powers, Cantor's 'finitism' (see pp. 7–8). This 'finitism' is explained in §1.3, and its reflection in the theory of number (the 'counting' theory of infinite cardinality which gave rise to the ordinal theory) in §2.2 and then in Chapter 3. What we see here is thus the most important instance of how Cantor's philosophical theory played a key part in determining the shape of his and subsequent mathematical theories.

This gives a somewhat brief explanation of the choice of raw material for this book. A work of this kind is necessarily incomplete, completeness requiring more than an abstract grasp of actual infinity. What it lacks above all, I think, is a detailed account of philosophical theories of infinity before Cantor, and a conceptual analysis of the development of constructive notions of set. For an analysis of Cantorian set theory surely demands, too, an analysis not just of what preceded it but also of modern, non-Cantorian treatments of sets and infinity. This is especially so given Cantor's own reliance on quasi-constructivist ideas, since the conflict between constructivism and fully-fledged Cantorian realism is already to some extent present in his work. In addition, I think some detailed account is required of how much the ideas and development of Hilbert's programme owe to philosophical problems with Cantor's 'programme', aside, that is, from the difficulties of the paradoxes. But there comes a point when one has to call a halt to extensions and improvements and simply let go. If this book arouses interest in the nature of Cantorian set theory and sparks off some argument about it, it will have served a large part of its purpose.

A few more general remarks about the text. Those who shy away from technical exposition may like to skip some of the detail in §§2.2, 2.3(*b*), 2.3(*c*), 4.3, and much of 7.3(*b*). But, as I have said, technical understanding cannot be avoided altogether if one is to properly understand the philosophical and foundational problems Cantor tackled or those he bequeathed. In any case, this is not a textbook on set theory, and quite often results, theorems, concepts and techniques are mentioned without proof or without explicit or formal definition. Complementary material may therefore be required. Fraenkel, Bar-Hillel, and Levy [1973] gives an excellent informal account of axiomatic set theory (modulo my disagreement in Chapter 5 below over limitation of size, and over one or two other historical points). Levy [1979] and Kunen [1980] are both marvellous textbooks, the former better suited for consequences of the theory itself, the latter for metamathematical results about it. For these Chapter 10 of Bell and Machover [1977], Bell [1977] and Dodd [1984] are excellent too. Drake [1974] is the standard source of information on large cardinals; and if there's something you want to know about set theory which you cannot find in any of these texts you probably *will* find it in the encyclopaedic Jech [1978]. As for companion historical reading, Hawkins [1970] is an excellent guide to the development of real analysis in the period embracing the genesis of set and transfinite number theory; Dauben [1979] contains much useful information (including biographical details) on Cantor and for those who cannot read Cantor in German it has an account of the contents of his papers. Moore [1982] appeared too late to be taken into account in the present work, though on a quick look through it seems full of valuable detail. I certainly agree with Moore that it is much too simplistic to say that the paradoxes were the main spur to Zermelo's axiomatization of set theory, and Moore rightly focuses instead on Zermelo's desire to secure his proof of the well-ordering theorem against criticism, including doubts engendered by the paradoxes. (See Moore [1982], §3.2.) A similar focus is taken in §7.3 below, though I have gone slightly further since I claim that the selection of the axioms themselves was guided by the demands of Zermelo's reconstructed proof. But there were surely other pressures too, not least the lack of clarity over the set concept which had been there from the beginning, long before any doubts raised by the contradictions. One thing I would criticise in Moore's approach is the talk of 'implicit' or 'unconscious' uses of the axiom of choice before Zermelo [1904]. Zermelo's intention was to radically and fundamentally change the nature of the mathematical treatment of well-ordering and well-ordering arguments, and the axiom of choice was a key part in this (see §3.5 and the Conclusion below). It was something quite new with Zermelo, and to talk of implicit uses *before* may obscure some of its radicalness, no doubt unintentionally but in any case wrongly.

As the reader no doubt will have guessed, references to the literature are in the style of name followed by a date in square brackets (e.g. Weyl [1949]) or name and date followed by a lower case italic letter (e.g. Cantor [1883*b*]) when various works published by the same author in the same year have to be separated. (Sometimes when the context is clear the name is omitted and only the bracketed date or date plus letter appears.) Full details of the publication are then given in the Bibliography. There is one major exception to this practice, namely reference to the three relevant sections of the Cantor *Nachlass*, three letter books (*Briefbücher*) in which Cantor drafted his letters before (presumably) sending them. Here I use the style 'Cantor *Nachlass*' followed by one of the Roman numerals *VI, VII,* or *VIII,* thus borrowing the classification of the *Nachlass* hitherto used by the *Universitätsbibliothek* in Göttingen; the details are again given in the Bibliography. (The classification has recently been changed, so beware! For more details, see the Bibliography, p. 308.) The Bibliography itself is divided into two sections, *A* containing details of the Cantor works referred to, *B* details of the others. I have normally used the most accessible source whenever quoting or referring to other works. However, when quoting from French or German works I have given not the original *but* an English translation, almost always my own. (My apologies for their inelegance!)

And this has sometimes meant shunning an existing available translation and referring instead to the handiest French or German source, even though the published translation might be more accessible. Thus, to take an example, I have translated passages from Cantor [1895] and [1897] myself, referring to the page numbers of their republication in Cantor [1932] (the most accessible German source) rather than use Jourdain's existing translation (published by Dover Publications under the title *Contributions to the founding of the theory of transfinite numbers*). The reason is the obvious one; I regarded it as much safer to look at what the author actually wrote! (It is surprising how often one disagrees with key passages in existing translations.) Of course, this leaves readers of this work facing a translation with some of my interpretation of the author built into it. But the quoted passage in the original can easily be found, and any disagreement over interpretation will be disagreement with me and not with some third party. Whenever I have used or modified an existing published translation, this is made clear, usually via the Bibliography.

While I had written a little about the influence of Cantor's metaphysical doctrines before (a sketch in my [1979a], a little more in [1979b]) I embarked on and wrote much of the present study during my stay in Germany for the academic year 1981-2 as a visiting member of the Philosophy Seminar of the Georg-August University in Göttingen. My work in Göttingen was made possible by the award of a Research Fellowship from the Alexander von Humboldt Foundation. I would like to express my deep gratitude to the staff of the Foundation for their generosity and kindness, and also to Professors Günther Patzig and Erhard Scheibe for making me welcome in the Göttingen Philosophy Seminar and for providing excellent working facilities. The stay in Göttingen not only gave me the freedom to work and write, but also the opportunity to study the Cantor *Nachlass* which is kept in the University Library there on behalf of the Göttingen Academy of Sciences. It seems appropriate that this *Nachlass* came to rest in Göttingen despite the fact that Cantor never worked there. For although it never became a Cantorian paradise, Göttingen did develop a strong Cantorian tradition. The members of the great Göttingen quartet, Hurwitz, Klein, Hilbert, and Minkowski, were among the first in Germany to recognize the fundamental importance of Cantor's set-theoretic work, and Göttingen subsequently became a centre of active interest in the structure and foundations of set theory. One thinks immediately of Hilbert and Zermelo, and then of Weyl, Nelson and Husserl, and even von Neumann, who, while not a Göttingen mathematician, was a prominent visitor and was enormously influenced by Hilbert and the ideas of Hilbert's programme. This tradition alone (not to mention the marvellous libraries and archives) makes Göttingen an ideal place to reflect on and write about Cantorian ideas. I would like to thank all those whose friendship and care made my stay in Göttingen productive and pleasant, in particular Josef and Ursula Ackerman for their infinite kindness and generosity, Erhard and Maria Scheibe for their warmth and hospitality, Tun for many things, not least buttressing my German and helping me consume vast quantities of jasmine tea, Ulli Majer and Norbert Schappacher (in one sense at least a 'key' man) for their conversation, consolation, and wine. I would also like to thank Herr Dr Klaus Haenel and Herr Heyn of the Handschriftenabteilung of the University Library in Göttingen where the Cantor *Nachlass* is kept. Their helpful cooperation and unflagging good spirits made my afternoon visits a distinct pleasure. I am grateful to the Göttingen Akademie der Wissenschaften and Herr Dr Wilhelm Stahl (Cantor's heir) for permission to publish the many quotations from Cantor's *Nachlass* that I have used below. I am similarly indebted to the University Library of Freiburg in Breisgau in respect of the few quotations from the Zermelo *Nachlass*.

In a time of excision in Higher Education in Britain, I consider myself very fortunate to have had opportunities to carry out this work at all. It would not have been possible but for the financial support of the London School of Economics, two research funds based there (the Imre Lakatos Memorial Fund and the Andrea Mannu Memorial Fund), the Alexander von Humboldt Foundation, and my present College, Wolfson. But behind institutions are people. I would like to express my gratitude to the members of the Philosophy Department at LSE, especially John Worrall, and to the Trustees of the two memorial funds mentioned above, for their backing and encouragement through many years learning, teaching and research. Without their help survival would have been impossible. My contacts with Oxford have been very important too. I am enormously grateful to Daniel Isaacson for his careful criticism of earlier versions of the text, for suggestions for extensions, and above all for his enthusiasm for the project; to Bill Craig for various helpful comments; and to Michael Dummett for his kindness in writing the Foreword. I would like to thank, too, Dana Scott (the Editor of the Oxford Logic Guides) for his encouragement, and the staff of Oxford University Press, who have been not just kind and helpful but patient as well! I am also grateful to the President and Fellows of Wolfson College both for providing a lovely haven in the present gloom, and also for allowing me the year's delay in taking up my Fellowship which enabled me to carry out my research in Göttingen.

But two people deserve very special mention here, for the book owes its existence to them both. John Bell not only kindled my interest in set theory but, more importantly, gave me the confidence to write about it. For more than ten years he has been a constant source of wisdom, stimulus and solace, lightening despondency with marvellous humour and continually infecting me with his own enthusiasm and sense of wonder. Works on the development and nature of mathematics should be a rich mixture of history, philosophy, and technical understanding, a mixture that both Herman Weyl and Imre Lakatos, for example, beautifully achieved in their very different ways. If this book has any richness at all, it is largely because of the influence of John Bell. My debt to him, and to his wife Mimi, is very great, and it is a pleasure to express my thanks and deep affection here. My deepest debt, though, is to Sharon Whyatt. Despite her own massive burden of work (much more valuable than mine), she put up with my variations of mood and confidence with enormous love, care, and patience. For all this, and much more, I thank her.

Wolfson College, Oxford   M. H.
June 1983

CONTENTS

PART 1

THE CANTORIAN ORIGINS OF
SET THEORY

INTRODUCTION: THE BACKGROUND
TO THE THEORY OF ORDINALS

Cantor was the founder of the mathematical theory of the infinite, and so one might with justice call him the founder of modern mathematics. Certainly a large part of his achievement was to help make the notion of set the basic one in mathematics. But in many ways the core of his work was his theory of transfinite number, especially the concept of *ordinal* number. The transfinite ordinal numbers were first introduced in two papers [1880] and [1883b], respectively the second and fifth in a series of six with the title 'On Infinite, Linear Point-Manifolds'. Their introduction involved Cantor not just in novel and fundamentally important mathematical theories but also in crucial *foundational* questions. The philosophical and heuristic framework he developed to tackle these questions had a lasting effect on set theory and modern mathematics, and is, in essence, the recurrent theme of this book. But first to understand *why* the ordinals were so important we have to go back to some of Cantor's earlier work and the problem of powers or infinite sizes that it raised.

The problem of powers in its most general form is to find a calculus of absolute size (power) adequate for describing the sizes of arbitrary infinite sets. This problem has its origin in Cantor's two striking papers [1874] and [1878]. The former demonstrated that the continuum is markedly different from the sequence of natural numbers since no denumerable sequence of real numbers can contain *all* real numbers. (For a description of the proof, see §2.3(a).) Since both the natural and the real numbers are infinite collections, this result showed that there are different kinds of infinity and thus raises the question: in what way precisely are they different? By 1878 Cantor had clearly taken the view that what is shown is that the real-number continuum is a *larger* infinity than the natural numbers. That is to say, he had adopted the principle of using one–one correspondences to measure the *relative* size of sets, a principle which Bolzano ([1851], p. 98), for example, had earlier rejected. The term power is used in a correspondingly relative sense. Thus, Cantor asserted that '*A* and *B* have the same power' if there is a one–one correspondence between the sets (or 'manifolds') *A* and *B*, or '*A* has smaller power than *B*' if there is a one–one correspondence between *A* and a subset (or 'component') of *B*, but not between

A and the whole of *B* (see Cantor [1878], p. 119).[1] But it is clear also that he had gone further, and that he already had firmly in mind the idea that an infinite set has an *absolute* size or power. Thus, to take one instance, Cantor talks of 'the smallest amongst infinite powers' ([1878], p. 120).[2] By the time he wrote his [1882] work he had openly adopted the principle that every 'well-defined manifold' has a definite power. (See [1882], p. 150. For a discussion of Cantor's notion of 'well-defined manifold' see §§1.4 and 7.1.)

This consideration of infinite size in terms of separate and independent powers raises a host of difficult questions. For example, Cantor had already assumed ([1878], p. 119) that the powers of the finite sets are the natural numbers, but what are the infinite powers? Is it possible to set up a scale of size and an effective arithmetic for infinite power just as we have a scale and an arithmetic for finite power? Just how many powers are there altogether? Looked at in this light the discovery of 1874 raises a particular problem. The 1874 work shows that there are at least *two* powers, or infinite sizes, represented within the real line. The *problem of the continuum* which Cantor set himself was then: how many powers in all are represented in the real line? In Cantor's subsequent work the problem of powers and the problem of the continuum cannot really be separated, for it was this latter, apparently more specific, problem which provided the incentive for and the main purpose behind the thorough investigation of powers generally.

In his [1878] Cantor showed that *all* Euclidean intervals of whatever dimension (even of countable dimension) have the same power as any linear interval. This was a somewhat surprising result (see, for example, the correspondence with Dedekind for the period May 1877 to January 1879 in Noether and Cavaillès [1937], pp. 21-50), and it showed that, as regards the sizes they contain, the higher Euclidean spaces are just the same as the real line. It also showed that within the normal Euclidean frameworks with which mathematics then dealt producing variations of size is not at all easy. In the 1874 paper Cantor showed not just that any interval is larger than the natural number series, but just as strikingly that the collection of algebraic (*a fortiori*, rational) numbers is *not*. Thus, study of standard collections of mathematical objects had so far only revealed *two* different sizes. This paucity of power possibly suggested to Cantor that there are very 'few' infinite powers in Euclidean space. In any case, in his [1878] he put forward the strongest possible version of this thesis, for he conjectured that there are *just* two. He conjectured that any infinite linear set

---

[1] In what follows the relations '*A* is cardinally equivalent to *B*', '*A* is strictly smaller than *B*', and '*A* is equivalent to or smaller than *B*' between the sets *A* and *B* will be denoted by '*A* ~ *B*', '*A* < *B*' and '*A* ≲ *B*' respectively.

[2] It is interesting to note that Cantor here ([1878], p. 119) *assumes* that < must be a linear ordering, that is that all sets *must be comparable*. With the introduction of a (linearly ordered) scale of size, the comparability question becomes just that of the adequacy and completeness of the scale. This plunged Cantor into the well-ordering problem (see §2.2).

must either have the power of the natural numbers or the power of the whole line—the first version of Cantor's celebrated *continuum hypothesis*.

The desire to prove this conjecture was the main creative spur to Cantor's work from this time on. For example, in his famous report on set theory of 1899 Schoenflies remarked of the continuum problem that '. . . we have this to thank for a large part of his [i.e. Cantor's] set-theoretic investigations' (Schoenflies [1899], p. 49). He later commented that Cantor struggled with the problem on and off throughout his life, and put his 'highest ability' into trying to solve it (see Schoenflies [1927], p. 16). (The effects of this struggle are dramatically presented in the excerpts from Cantor's letters to Mittag-Leiffler quoted by Schoenflies [1927] and in Schoenflies's comments on them.) Certainly, as I have said, the continuum problem was the main reason for tackling the general problem of powers. Cantor admitted the possibility that his hypothesis might be false. Despite his belief in its truth, his aim as a mathematician was to decide the issue, first by looking for a proof, and then, if a proof was not forthcoming, by looking for a disproof. From Cantor's viewpoint in 1878 there were two possible ways of approaching the continuum problem. One was to try to prove (or disprove) directly that any infinite linear point-set is either denumerable or has the power of the continuum. The second was via a solution to the general problem of powers, by defining an arithmetical scale of infinite size, showing that all sets must be represented in the scale, and then discovering at what place the continuum is represented. If it is represented by the power in second place, the continuum hypothesis must be correct; if not, then it must be incorrect. Cantor tried both lines of attack, and both in different ways involved the transfinite ordinals. The basis of the second line of attack, which will be explained at length in §2.2, was to define a scale of infinite powers in terms of the ordinal numbers; in short, to reduce cardinality in the infinite realm to ordinality. For the moment I want to say a little about Cantor's line of direct attack which, although it avoided the general problem of powers, still involved the transfinite ordinals. Not only did this raise fundamental questions about the ordinals themselves, but it eventually led to Cantor's best result concerning the continuum problem (see §2.3(b)).

Cantor's direct approach to solving the continuum problem was to set up a further method of classifying point-sets alongside the classification by one–one correspondence and to try to obtain information about the latter via the former. His further method of classification involved certain decomposition properties using the idea of *point-set derivation*, which goes back to his remarkable [1872]. The central topic of this paper is the problem of the uniqueness of trigonometric series expansions of functions. (The history of this problem is described by Hawkins [1970] and Dauben [1971] or [1979], chapter 2.) However, it was in this paper that Cantor first began to use the idea of arbitrary infinite *sets* of points, which are not just collections of objects *but are themselves subject to mathematical operations* (see §1.1). (Cantor's classical definition of real numbers via Cauchy sequences of rationals is also given in this paper.) The basic operation Cantor introduces is that of the derived set of a point-set P denoted by $P^{(1)}$ and containing exactly the 'limit points' (i.e. accumulation points) of P. If P is bounded and infinite then according to the Bolzano-Weierstrass theorem $P^{(1)}$ will be non-empty. Correspondingly, if $P^{(1)}$ is bounded and infinite it too will have a non-empty derived set $(P^{(1)})^{(1)}$, which Cantor denoted by $P^{(2)}$. Cantor extended this to $P^{(n)}$ for any finite n. This is as far as he went at this stage: the central theorem of his [1872] uses only sets P for which $P^{(n)} = \emptyset$ for some n.[3] However, Cantor recognized even this early that there is a much more general heuristic method here, namely to discover properties of P by investigating the sequence of derived sets. Indeed he wrote ([1872], p. 97; the italics are mine):

If a point-set is given in a finite interval, then in general a second point-set is given with it, and with this second a third, and so on [Cantor means $P^{(1)}$, $P^{(2)}$, etc.]. *These point-sets are essential in order to be able to understand the nature of the first point-set.*

However, there are clearly point sets for which $P^{(n)} \neq \emptyset$ for any n, for example an interval or the rationals in an interval. As Cantor commented later ([1884a], p. 218):

We also saw that the construction of the concept of derivatives [or derived sets] of different orders, *which is so essential for the investigation of the nature of a point-set P,* is *is no way* completed with the construction of derivatives with finite ordinal number ν, . . .

Thus, in order to sub-classify these 'second-species' sets ('first species' sets being those with $P^{(n)} = \emptyset$ for some n; these terms were introduced in [1879], p. 140), Cantor required some way of analysing the derivatives beyond the $P^{(n)}$. To this end he introduced in his 1880 paper new 'symbols of infinity', which were *contextually* defined via the derivation process. Thus,

$$P^{(\infty)} = \bigcap_{n=1}^{\infty} P^{(n)}, \qquad P^{(\infty+1)} = (P^{(\infty)})^{(1)}$$

and so on (see Cantor [1880], pp. 357-8).

Cantor's hope was that this more general idea of derived set would provide an exhaustive classification of point-sets discriminating enough to give information about the power of any set by reducing it to components whose power is known or easily calculable. The introduction and use of the symbols in his [1880] and [1883a] constituted the first stage in the creation of Cantor's new theory, but for

---

[3] The theorem is:

If $f(x)$ is a function which is represented by a trigonometric series at all points of the set $[0, 2\pi] - P$, where $P \subseteq [0, 2\pi]$ such that $P^{(n)} = \emptyset$ for some n, then the representation is unique.

(See Cantor [1872], pp. 99–101, or Hawkins [1970], pp. 21–8, or Dauben [1979], chapter 2.)

both logical and mathematical reasons he was forced to shift to a second stage in which the 'symbols' were reintroduced as transfinite ordinal *numbers*.

In the first place, Cantor had already begun instinctively to use an arithmetic of symbols, combining them both with themselves and with the natural numbers. To take an example from his [1880] (pp. 147–8) he introduces $P^{(\infty^2)}$, and more generally $P^{(n_0 \infty^\nu + n_1 \infty^{\nu-1} + \ldots + n_\nu)}$ for natural numbers $\nu, n_0, \ldots, n_\nu$, followed by $P^{(\infty^\infty)}$, and so on. This in effect presupposes that the 'symbols' and the natural numbers are objects of the same kind, subject to the same arithmetical operations and obeying generalized arithmetical rules. But this meant that the 'symbols' ought to be presented as *numbers* and the generalized arithmetic spelt out. Secondly, some classification of the symbols (or numbers) was required. As Russell later remarked ([1903], p. 324), the derived sets of a given set 'measure' or 'gauge' its degree of concentration:

Popularly speaking, the first derivative consists of all points in whose neighbourhood an infinite number of terms of the collection are heaped up; and subsequent derivatives give, as it were, different degrees of concentration in any neighbourhood. Thus, it is easy to see why derivatives are relevant to continuity; to be continuous, a collection must be as concentrated as possible in every neighbourhood containing any terms of the collection.

Hence, it was crucial for Cantor's project to find a precise way of framing the statement that a collection is 'as concentrated as possible'. This in particular required an investigation of such questions as how many new symbols are required to cover all cases, how 'far' derivation can usefully be taken, and so on. This meant that Cantor could no longer rely on *ad hoc* insertion of 'symbols', but required principles for their introduction and classification.

Already, then, we see that the direct attack on the mathematical problem of the continuum led Cantor to the idea of a general theory of finite and infinite ordinal *numbers*, treated as real objects existing in their own right away from the context of point-set derivation, and satisfying a generalized ordinal arithmetic. Without this the theory of derivation lacks coherent structure. Cantor himself noted in his [1883b] (p. 166): 'I was led to these infinite, real whole numbers already many years ago, without becoming fully conscious that they are concrete *numbers* with real meaning.' In a footnote (not reproduced in Cantor [1932] but on p. 547 of the original publication) Cantor adds: 'Hitherto I have called them "definitely defined symbols of infinity".' (See also the passage from Cantor's letter to Mittag-Leffler, below, p. 51.) One central aim of the [1883b] paper was to explain why the transfinite ordinals are not mere 'symbols of infinity' but 'concrete numbers with real meaning'.

Cantor's *indirect* attack on the continuum problem, which was much more important for the development of pure set theory, also depended on a clear theory of transfinite ordinal numbers. For it was Cantor's idea, also put forward in [1883b] (see §2.2), to use the sequence of finite and transfinite ordinals to characterize *all* powers (through division into the various ordinal number classes), and thus to provide a framework for solving the continuum problem. Obviously one of the necessary conditions for the fulfilment of this project is again an analysis of the key properties of the ordinals themselves. These two lines of development make it quite clear *mathematically* why Cantor begins his remarkable [1883*b*] with the following comment on the ordinals:

The presentation hitherto of my investigations in the theory of manifolds has reached a point where continuation depends upon an extension of the concept of real, whole number [*i.e.* natural number] out beyond its known limits and in a direction which, to my knowledge, no-one previously has pursued.

My dependence on this extension of the number concept is so great that without it it would be scarcely possible for me to take freely the smallest step forward in the theory of sets. These circumstances therefore should justify, or, if necessary, constitute an apology for, my introducing seemingly strange ideas into my work. For what is involved here is an extension resp. a continuation of the real whole number series out beyond the infinite; . . . (Cantor [1883*b*], p. 165)

However, it is now *also* clear that Cantor was facing not just technical mathematical problems, but philosophical and foundational problems too. Is a theory of infinite number (this 'strange idea') metaphysically and philosophically legitimate? If so, then what exactly are these numbers? What, if any, is the common basis they share with the ordinary familiar finite numbers? How do they merge with these? Cantor was not a particularly careful or systematic philosopher, unlike Frege who was concerned with related problems at roughly the same time, but he did tackle these fundamental questions in a courageous way. His philosophy was in many points vague or shaky, but the core of the answers he produced was nevertheless of fundamental importance, not perhaps for (pure) philosophy (again in contrast to Frege) but certainly for mathematics, and particularly for the *foundations* of mathematics. For the ideas contained here led clearly and directly to modern set theory.

As will be obvious from the above remarks, the starting point in considering Cantor's foundational work is the paper [1883*b*]. While the other papers in the series 'On Infinite, Linear Point-Manifolds' are by and large mathematical (though, to be sure, occupied with new mathematics), this fifth paper is much more speculative, probing, and openly philosophical. Cantor indeed also had it published separately under the title *The Foundations of a General Theory of Manifolds* ([1883*d*]) which has an interesting additional *Foreword*, unfortunately *not* included in Zermelo's edition of Cantor's works [1932]. Referring to the four previous papers in the series, as well as [1874] and [1878], Cantor remarks:

Since the present paper in many respects takes the subject much further than the earlier papers and in essentials is independent of these, I have decided to allow it to appear as a separate work and to supply it with a title which corresponds to its content.

He goes on:

In publishing these pages, I wish to point out that I have written them with two main groups of reader in mind—philosophers who have followed the development of mathematics up to the present time, and mathematicians who are familiar with the most important older and newer publications in philosophy.

I am well aware that the theme I deal with here has at all times encountered the most widely varying opinions and interpretations, and that neither mathematicians nor philosophers have achieved comprehensive agreement. Consequently I do not delude myself that I am in a position to say the last word on such a difficult, involved and far ranging subject as the Infinite. However, through many years research in this subject I arrived at definite convictions, which were not swayed in the course of wider studies, but on the contrary have become more fixed. I believe myself, therefore, to have a certain duty to bring order to these convictions and to make them known.

I hope that my endeavour to find out and give expression to the objective truth is hereby achieved.

Unfortunately, despite what Cantor says here about 'bringing order' to his ideas, he never gave the philosophical and foundational side of them a systematic presentation. He published three largely philosophical works later ([1886a], [1886b], [1887-8]—the latter being the most important) which help to fill out the picture given in [1883b]. However, none of these four papers taken separately gives a unified treatment of his ideas. In attempting to understand what he writes in one place one often has to interpolate consequences of his writings elsewhere, or refer to unpublished correspondence, or even sometimes just conjecture how a certain opinion would best fit with other views he utters. The nature of [1887-8] does not help in all this, it being largely a collection of passages from letters Cantor wrote between 1884 and 1886, put together with some introductory commentary and footnotes which are often very long. However, despite the rather rambling way in which Cantor presented his philosophy, there is a unity to his ideas; at least, this is what I suggest and attempt to capture in Chapter 1. The importance of this core is then traced in the rest of the book—its effect on Cantor's own theory is discussed in Chapters 2–4, and the effect on modern axiomatic set theory is discussed in Part 2.

The aim of Cantor's philosophical work was to provide arguments to support the legitimate mathematical employment of actual infinities, especially infinite numbers. What emerges from this can be summarized in three key principles:

principle (a): *Cantor's principle of actual infinity, or the domain principle*
Any potential infinity presupposes a corresponding actual infinity

principle (b): *Cantor's principle of finitism*
The transfinite is on a par with the finite and mathematically is to be treated as far as possible like the finite

principle (c): *Cantor's principle of Absolute infinity*
The Absolute infinite cannot be mathematically determined

(*a*) is not quite the thesis of set-theoretic reductionism, i.e. the thesis that mathematics just *is* set theory, but it goes quite a long way towards it. For it means with Cantor that completed domains are the fundamental subjects of mathematical study, and since one wants to treat most of these as genuine objects (a move which is largely legitimated by (*b*)) (*a*) is transformed into the thesis that mathematics is occupied at the most fundamental level with the study of collection-objects, i.e. *sets.*

Cantor's reductionism, or at least his tendency towards reductionism, is discussed in §§ 1.2 and 2.1 and taken up again in Chapter 3. The reductionist attitude towards numbers is already present in [1883*b*], though here the numbers are not explicitly 'reduced' to sets in anything like the modern sense. The tendency towards full reductionism becomes clearer a little later, with the determination that numbers should be explained as sets. Cantor certainly did not create a pure set theory, not least because his own reductionist explanations of number are vague. Nevertheless, there is a clear sense in Cantor that problems of infinity reduce to problems about sets. This does not mean that we can forget about the existence of numbers, that the step taken in 1878 of shifting from comparing sets to comparing *numbers* is now abolished. Not at all. Indeed for Cantor the numbers were and remained the central mathematical objects. Rather the reductionist element was an essential part of his argument that infinite numbers exist, it being part of his foundational framework that the existence of infinite sets is essential for mathematics. It follows from Cantor's position that a necessary condition for reductionism (or the reductionist tendency) is that it *must* be able to explain the structure and existence of the numbers. I stress this here because this condition was later *abandoned* by Zermelo and Fraenkel, who tried to reconstruct set theory without taking the demands of a reductionist theory of number into account. In a sense, this *was* an attempt to return to the pre-1878 theory, where 'number theory', not just ontologically but in spirit too, is nothing but the comparison of sets. Von Neumann subsequently restored Cantor's condition and this necessitated a considerable extension of Zermelo's system. In essence, von Neumann recreated modern set theory—along more strictly Cantorian lines. (Zermelo's and von Neumann's different versions of reductionism are discussed in Chapters 7 and 8.)

Principle (*b*) (which I have called, for reasons which will become clear in § 1.3, Cantor's *finitism*) is the thesis that the mathematical treatment of infinity is accompanied by a decision to treat mathematical infinities in the same way as finite objects, at least so far as fundamental properties are concerned. The main heuristic consequence of this is discussed in §§ 2.2 and 3.5, namely the beautiful counting (ordinal) theory of cardinality—Cantor's answer to the problem of the numerical scale. Here the ordinals appear clearly as the most important mathematical objects, a position which is strengthened later (in 1899) by their structural application in Cantor's 'limitation of size' theory, described in

Chapter 4. Other aspects of Cantor's 'finitism' are discussed in Chapter 1 and §2.1. Particularly important here is the treatment of infinities (infinite collections) as *single objects*, or *wholes* (§1.3), and also the pressure that 'finitism' exerts to find a unified treatment of the mathematical finite and the mathematical infinite (Cantor's transfinite). This pressure was one of the things which led Cantor in the direction of reductionism.

Cantor's doctrine of the Absolute (principle (*c*)) was designed to put bounds on the mathematization of the infinite, to effect a kind of type distinction. While in its original form this doctrine helps to give a certain shape to the mathematical universe, and, for example, indicates that 'very extensive' collections cannot be taken as sets (see §1.4), it is rather imprecise. Faced with the need to prove the aleph theorem as well as avoiding what became known as Burali-Forti's paradox, Cantor transformed the doctrine into a theory of 'absolute collections', and attempted thus to put it to precise mathematical use. This involved the ordinal numbers, as representative of the Absolute, as the central part of a 'limitation of size' theory (Chapter 4), an extremely powerful idea in the later development of set theory as we shall see in Part 2 of this book. Here again what finally dominated, despite the 'abolitionist' moves of Zermelo, Fraenkel, and others, was the Cantorian idea that the ordinals are the key structural objects, whether reduced to sets or not.

The ideas contained in this (foundational) core derive in part from Cantor's theology. *In the end*, it is important to divorce the main content of Cantor's ideas (e.g. the three key elements sketched above) from their theological origins. This is partly because to appeal to God in the way Cantor often does is ultimately no great help. For example, to claim that certain infinite sets or certain infinite numbers exist because it is possible for God to conceive, and therefore to create, them, or that they exist because they are 'ideas in the divine intellect', tells us no more than the bald claim that they exist. Moreover, while Cantor's foundational ideas were of profound significance for the later development they were taken over apparently stripped of their theological overtones. Indeed, sometimes the later abandoning of reliance on God helped to make Cantor's position much *clearer*. (To take one example, Zermelo adopted the axiom of choice in place of the Cantorian assumption that every set can be counted in some more or less literal sense, a doctrine which with Cantor is most naturally associated with his theological ontology. This removed considerable confusion and tension from Cantor's system. See §3.5.) Nevertheless, we cannot ignore the theological side of Cantor's ideas altogether, for in many cases it at least helps us to understand what position Cantor takes (e.g. about the existence of infinities) and perhaps a little of why he took it. This is particularly so with his 'finitism' and the doctrine of the Absolute (see Chapter 1).

Before approaching Cantor's theory of infinity and his foundational ideas in detail, it is wise to stress here how important theology was to Cantor. The mathematician Gerhard Kowalewski, who had quite regular personal and

academic contact with Cantor at the end of the 1890s, remarked in his autobiography ([1950], p. 108): 'Cantor was a deeply religious man. What he
witnessed with the construction of his *Mengenlehre* moved his innermost soul.'
And the close connection *for Cantor* between achieving the correct picture
of the foundations of mathematics and understanding something of the nature
of God is particularly clear in his correspondence. In a letter to Pater Thomas
Esser of 1 and 15 February 1896 (Cantor *Nachlass VIII,* p. 135; Meschkowski
[1965], p. 513) Cantor wrote:

> The general *Mengenlehre* . . . belongs thoroughly to metaphysics. You can easily
> convince yourself of this by examining the basic concepts of *Mengenlehre*, the
> categories of cardinal number and ordinal type, and noticing not only the degree
> of their generality, but also how thinking [*Denken*] with them is fully pure,
> so that there is not the slightest room for fantasy.

A large part of what Cantor means by 'metaphysics' here would today come
under the heading of 'foundations'. That is, as became clear in the background
to [1883*b*], he is occupied with providing in a quite general way a framework
within which mathematics (and he would also say natural science) can be
developed. Questions about the nature, structure, and existence of numbers and
sets are questions which have to be asked and coherently answered *before* we
can proceed to develop specific branches of mathematics with any confidence
that we know what we are doing, or what it is that we are talking about. The
foundational element of this 'metaphysics' was recognized by Cantor himself.
In the same letter, he notes

> The grounding of the principles of mathematics and natural science is a matter
> for metaphysics. Metaphysics has therefore to look upon these two sciences not
> only as its servants and helpers but also as its children which it should not let
> out of its sight, but must watch over and control . . . (Meschkowski [1965],
> p. 512)

In his unpublished paper of 1884 Cantor distinguished explicitly between pure
set theory ('pure mathematics being, in my view, nothing other than pure
Mengenlehre' (Grattan-Guinness [1970], p. 84)) and applied set theory. This
latter for Cantor included 'point set theory, function theory and mathematical
physics'. More broadly

> Under *applied set theory* I understand that which one cares to call the *theory
> of nature* or *cosmology*, to which belong all so-called natural sciences, those
> relating both to the inorganic and to the organic world. (Grattan-Guinness
> [1970], p. 85)

It is clear that Cantor understands *pure* set theory as a quite general *foundational* theory which prepares the way for any theory which uses or relies on sets
or numbers. But now we come back to theology and God, for this foundation,
this understanding of what numbers are, or what sets etc. exist, is for Cantor
intimately connected with the attempt to understand God's whole abstract
creation and the nature of God himself. Cantor states this indirectly in the same

letter to Esser quoted above, for he speaks there of the 'inseverable bond' between theology and metaphysics. But it is also clear in a letter to Vatson of 31 January 1886 (Cantor *Nachlass VI*, pp. 43–4). Cantor disparages the 'materialism and positivism of the present time, which has developed into a kind of monster' which he says derives from Newton, indeed from 'the great metaphysical deficiencies and perversities of his whole system'. Cantor is not explicit about these 'deficiencies' but it becomes clear that what is lacking according to Cantor is a proper theology. First Cantor notes that Newton's achievements are what results when 'the greatest work of a genius, despite his personal religiosity, is not united with the true philosophical and historical spirit'. Then later he adds that his *own* work represents 'a quite different and new ordering of ideas'. He goes on:

It is not that they relate to something beyond nature; rather that they aim at a more exact, more complete, finer knowledge of *nature itself* than can be achieved through Newtonian principles, certainly not without contact with Him who stands above nature, since it is His own free creation. (Cantor *Nachlass VI*, p. 44)

In short, for Cantor classical mathematics and Newtonian-style mathematical physics need transfinite *Mengenlehre*, and for the theory of the transfinite we need God:

I entertain no doubts about the truth of the Transfinite, which with God's help I have recognized and studied in its diversity, multiformity and unity more than twenty years. Every year and almost every day brings me further in this science. (Letter to Fr. Ignatius Jeiler, Whitsun 1888; Cantor *Nachlass VI*, p. 169)

1

CANTOR'S THEORY OF INFINITY

The most convenient place to begin discussion of Cantor's theory of infinity is with his schema or categorization of the inifinite. The schema is based on Aristotle's distinction between

(i) the potential infinite

and

(ii) the actual infinite.

But Cantor insists on a crucial modification, namely that (ii) must be broken down further into

(ii) (*a*) the increasable actual infinite, or *transfinite*
(ii) (*b*) the unincreasable or *Absolute* actual infinite.[1]

(i) is what is involved when the mathematician says 'let *n* be an arbitrarily large natural number' or 'given an arbitrarily small rational number ϵ . . .'. In either case, any specific value given to *n* or ϵ will be *finite* but can be assumed to be greater or smaller respectively than any pre-assigned finite bound. The potential infinite is therefore really the idea of unlimited variability. Cantor expresses it thus:

The potential-infinite is mostly witnessed where one has an undetermined, *variable finite* quantity which either increases beyond all limits (here we can take as an example that so-called time which is counted from a definite initial moment), or which decreases beneath any finite small limit (as, for example, in the correct presentation of a so-called differential). More generally, I speak of a Potential-Infinite whenever it is a question of an *undetermined* quantity which is capable of innumerably many determinations. ([1887–8], p. 40; see also [1886*b*], p. 371)

In contrast to this:

By an Actual-Infinite is to be understood a quantum which on the one hand is *not variable*, but rather is fixed and determined in all its parts—a genuine constant—but which at the same time surpasses in magnitude *every finite quantity of the same kind*. ([1887–8], p. 40)

[1] This full schema is presented as such in [1887–8], pp. 401 and 405. (The letter from which this passage comes, however, is dated 1886; see [1887–8], p. 400, and also [1886*a*].) Although not presented in exactly this way, the elements of this schema are already present in [1883*b*] (see in particular pp. 165–6 and p. 205, n. 2). Cantor calls (i) the *improper* (*uneigentlich*), (ii) the *proper* (*eigentlich*) infinite, instead of the later 'potential' and 'actual'.

Before Cantor, it was generally thought that science and mathematics is and can only be concerned with *potential* infinity; the actual infinite, if admitted at all, was usually identified with what Cantor calls here the Absolute infinite. (A notable exception here is Bolzano; see his [1851], *passim*.) This was certainly the case with many of the important Christian philosophers whose writings were of the greatest importance to Cantor. For example, Aquinas claimed that God represents an actual (absolute) infinity, but that there is no other existing actual infinity. Hence, since neither science nor mathematics has God for its object of study, and since it is impossible to investigate His essence scientifically, it must follow that mathematics (science) can concern itself only with the unlimited, i.e. potential infinity (see Aquinas: *Summa Theologica*, Part I, Question 7). In a certain sense, this doctrine conveniently expresses the magnitude, the greatness of God, and thus with it the qualitative gap between God and His created subjects. God is infinite; man, however, is inherently finite, and cannot share in, or understand, or rationally subjugate the infinite. Actual infinity is an Absolute beyond human grasp.

For Cantor too this notion of a transcendent Absolute was of the highest importance, not least mathematically, as we shall see. He certainly accepted the view that this notion of the Absolute is an appropriate symbol of the power and transcendence of God, and also that it is *not* a possible subject for precise scientific investigation. For example: 'The Absolute can only be acknowledged and admitted [*anerkannt*], never known [*erkannt*], not even approximately.' ([1883b], p. 205). The Absolute, says Cantor, is 'the veritable infinity' whose magnitude is such that it

. . . cannot in any way be added to or diminished, and it is therefore to be looked upon quantitatively as an absolute maximum. In a certain sense it transcends the human power of comprehension, and in particular is beyond mathematical determination. (Cantor [1887–8], p. 405; the passage stems from 1886.)

In a letter of 28 November 1885 to Carbonelle, Cantor refers to: '. . . the actual-infinite in God, which in my "Grundlagen" ([1883b]), as you will have seen, I call the Absolute, . . .' (Cantor *Nachlass VI*, p. 34). Much later (in a letter to G. C. Young, June 20, 1908; Dauben [1979], p. 290) he wrote:

I have never proceeded from any "Genus Supremum" of the actual infinite. Quite the contrary, I have rigorously proven that there is absolutely no "Genus Supremum" of the actual infinite. What surpasses all that is finite and transfinite is no "Genus"—it is the single completely individual unity in which everything is included, which includes the "Absolute" incomprehensible to the human understanding. This is the "Actus Durissimus" which by many is called "God".

Where Cantor diverged, from the scholastic philosophers, however, was over the claim that the Absolute is all there is to the actual infinite. With reference to the schema above, traditional doctrine held that (ii) (*b*) belongs to God, (i) to mathematics, and (ii) (*a*) is empty. Cantor disagreed:

It is my conviction that the domain of definable quantities is not closed off with the finite quantities, and that the limits of our knowledge may be extended accordingly without this necessarily doing violence to our nature. I therefore replace the Aristotelian–Scholastic proposition [*infinitum actu non datur*] mentioned in §4 by the following:

Omnia seu finita seu infinita *definita* sunt et excepto Deo ab intellectu determinari possunt. [All forms whether finite or infinite are definite, and with the exception of God, are capable of being intellectually [mathematically] determined.] ([1883*b*], p. 176)

This realm of infinities which *can* be rationally subjugated, which *are* mathematically determinable, is what Cantor calls the *transfinite*. Again, referring to the Aristotelian–Scholastic theory of infinity he writes (in a letter to Aloÿs Schmid, 26 March 1887) that these 'definitions' of the infinite

. . . only fit to the potential infinite, or to the absolute completeness, perfection in God, but not to that infinite which I call the transfinite. This is in itself constant, and larger than any finite, but nevertheless unrestricted, increasable, and in this respect thus bounded. Such an infinite is in its way just as capable of being grasped by our restricted understanding as is the finite in its way. (Cantor *Nachlass VI*, p. 99)

So far what I have cited here are only statements of the *conviction* that category (ii) (*a*) is populated. Cantor's foundational task was to provide some argument which supports such conviction. Cantor in fact provided two lines of argument. The first expounds a realist theory of concepts and of the existence of mathematical objects. At first this was presented independently of any reference to God. But later, as we shall see, Cantor brings in God in a kind of Berkeleyian way, by claiming that mathematical objects exist as ideas in the Divine intellect. While this is not much help philosophically, it is nevertheless important since it helps to clarify Cantor's notion of the Absolute. The second line of argument stems from an analysis of the foundations of contemporary mathematics and his own contribution to it. This leads Cantor to adopt a quite different view of (i) from (most of) his predecessors, and is closely tied to his shift in the direction of reductionism. I shall suggest, though, that for Cantor even this line of argument must ultimately rely on the first.

1.1 Free mathematics

The most useful point with which to begin tracing Cantor's line of argument is his stress on science as a 'free' conceptual construction. First, he says, in order to achieve 'certain knowledge' we must concentrate not on the *senses*, and thus not on the physical world as we can grasp it through our senses, but on concepts and ideas and the relations and connections between them. This part of the strategy is clear in the following passage. After quoting Sponoza's *Ethics*: 'The order and connection of ideas is the same as the order and connection of things.' (part II, prop. VII, Elwes translation), he goes on:

The same epistemological principle is hinted at even in Leibniz's philosophy. Only since the new empiricism, sensualism and scepticism, and the Kantian criticism that emerged from it, has it been believed that the source of knowledge and certainty is located in the senses or in the so-called form of pure intuition of the world of ideas [i.e. the Kantian *Vorstellungswelt*] and must be restricted to these. According to my conviction, however, these elements do not at all furnish certain knowledge. This can only be obtained through concepts and ideas [*Ideen*], which are at best only stimulated by outer experience, but which are principally formed through inner induction and deduction, like something which, so to speak, already lay within us and is only awakened and brought to consciousness. ([1883*b*], p. 207, n. 6)

This perhaps also makes a little clearer Cantor's foundational remarks about Newtonianism and mathematical physics which I quoted in the Introduction. For Cantor here is saying very clearly that reading directly from the book of nature will not tell us very much about the fine structure. Rather we must approach nature armed with a framework of concepts or ideas. This applies not just to the physical sciences, but especially to mathematics, even to those parts of mathematics which find application in the physical world or which might claim to be a more or less direct description of basic physical structures. It is no surprise that Cantor makes special mention of the continuum. His view was that, far from constructing or justifying the mathematical continuum (and the physics founded on it) on the basis of perception of space and time, the procedure is reversed. According to Cantor, we form first an *abstract conceptual* mathematical theory of space or continua and explicate and investigate the notions of real time and physical space with the help of this. For example, on time Cantor writes:

. . . I have to declare that in my opinion reliance on *the concept of time* or *the intuition of time* [*Zeitanschauung*] in the much more basic and more general concept of the continuum is quite wrong. In my opinion, time is an idea [*Vorstellung*] whose clear explication presupposes the independent concept of continuity, and which even with the help of this latter can be conceived neither objectively as a substance, nor subjectively as a necessary a *priori* form of intuition [*Anschauungsform*]. Rather, it is nothing other than a relational concept, by whose aid the relation between various motions we perceive in nature is determined. ([1883*b*], pp. 191–2)

And on space:

Likewise it is my conviction that one cannot attempt to gain knowledge of the continuum by starting with the so-called *form of spatial intuition* [*Anschauungsform des Raumes*], since space and the structure attributed to it achieve that substance by which they can become the object not merely of aesthetic reflections or philosophical scrutiny or imprecise comparisons, but of careful and exact mathematical investigations, *only* with the help of a continuum conceptually *fashioned and already available*. ([1883*b*], p. 192. Cf. also Cantor [1882], pp. 156–7)

In other words we *begin* our description of the physical world pre-equipped with an abstract, and (Cantor would say) Platonic arithmetic or set-theoretic theory of the continuum, and *conjecture* that physical space fits that theory. A similar position was taken by Dedekind in his [1888] (*Vorwort* to the first edition):

Only by means of the purely logical construction of the science of numbers, and the continuous number-domain achieved with it, are we in a position to investigate precisely our conceptions of space and time, by tying these to just that number domain created in our intellect. ([1888], p. iii, my translation)

Cantor and Dedekind are perfectly right—mathematics and natural science must be based very largely on conceptual frameworks provided by us not simply dictated by Nature, though, as Cantor says, Nature may 'stimulate' us. (Cantor's criticism of 'Newtonianism' is to some extent misplaced, certainly as far as the physics is concerned. For, regardless of any strong 'empiricist' claims made for it, the physics *is* based on a conceptual framework provided by man, and cannot have been dictated by Nature alone.) But the key questions now are the following. What kind of conceptual frameworks are permitted? How free are we in proposing frameworks? In other words, how do we justify the introduction of new concepts? It is interesting at this point to contrast Cantor's views with those of the constructivist Weyl, whose views sometimes coincide with Cantor's (most strikingly in believing that 'true infinity' belongs to God and is outside man's grasp (see Weyl [1932])). Weyl like Cantor stresses the conceptual independence of mathematics; it may often start with considerations of direct physical import (thus like Cantor's 'stimulation'), but its characteristic feature is precisely that it abstracts from the particularity of the physical (see Weyl [1940]). But Weyl (at least in much of his philosophical writing) had quite definite views about the nature of mathematics which strongly diverged from Cantor's. He took the view that mathematics is wholly and genuinely 'constructed' by man, not discovered, and that it must be subject to the demands of a coherent and genuine constructivism. Cantor's position was quite different. For although he held that we do not obtain our concepts directly from physical nature, that we in some sense 'form' them ourselves, this formation process is not that of the later constructivists. Rather this formation is actually a process of discovery (one is tempted to say even revelation) of a world of ideas in something like the Platonic sense. Ultimately, Cantor's position is supported by appeal to God. But let us first trace the line of argument as it is presented without theology in [1883b].

The route to Cantor's Platonism begins with his various statements about the introduction of new concepts.

Mathematics is completely free in its development and only bound by the self-evident consideration that its concepts must be both consistent in themselves and stand in an orderly relation fixed through definitions to the previously formed concepts already present and tested. ([1883b], p. 182)[2]

---

[2] On the same page we find Cantor's famous remarks that 'the nature of mathematics lies precisely in its freedom', and that pure mathematics would have been better termed *free* mathematics.

At first sight it seems that Cantor is adopting a somewhat formalist position: *all* that matters is the consistent integration of a notion with notions already accepted. But this is simply misleading, as is readily seen from the footnote which continues the above passage:

The process with the correct formation of concepts is in my view always the same. One lays down a thing without properties [*ein eigenschaftsloses Ding*] which at first is nothing other than a name or a sign *A*. One then duly gives to it different even infinitely many predicates, whose significance is known through already present ideas, and which must not contradict one another. In this way the relations of *A* to concepts already present and especially to similar concepts are determined.

These are, loosely, the 'formal' aspects, but then he goes on:

With this one has then completely finished; all conditions for the awakening of the concept *A* which slumbered within us are present, and it comes completed into being, furnished with that intrasubjective reality which is generally all that can be demanded from concepts. To establish its transient significance is then a matter for metaphysics. ([1883*b*], p. 207, n. 7, 8)

What Cantor is getting at, then, despite the 'formalist' gloss, is nothing more or less than the reality of the concepts or objects considered. He says explicitly of numbers:

In particular one is only obliged with the introduction of new numbers to give definitions of them through which they achieve such a definiteness and possibly such a relation to the older numbers that in given cases they can be distinguished from one another. As soon as a number fulfills all these conditions, it can and must be considered in mathematics as existent and real. ([1883*b*], p. 182)

Let us first be quite clear that it *is* genuine *existence* that Cantor is getting at, and that he is not *just* saying that consideration of existence can be dispensed with in favour of this idea of consistent integration. We can see this quite clearly in what Cantor says further about the reality of numbers.

Cantor in his [1883*b*] distinguished two senses in which mathematical objects, and he speaks specifically of numbers, can be said to exist. The first he called *intrasubjective* or *immanent reality*:

In the first place, integers may be considered real in so far as they occupy an entirely definite place in our understanding on the basis of definitions and can be precisely differentiated from all other parts of our thought and stand in determinate relationships to those parts, and accordingly modify the substance of our thought [*Geistes*] in a determinate manner; I propose to call this kind of reality their *intrasubjective* or *immanent reality*. ([1883*b*], p. 181)

Already the talk here of numbers or concepts 'modifying the substance of our thought' strongly suggests that Cantor is being anything but formalist. Indeed he remarks here in a footnote:

What I call here 'intrasubjective' or 'immanent' reality of concepts or ideas could be said to correspond with the designation 'adequate' in the sense that this word is used by Spinoza when he says, *Ethics*, part II, def. IV: 'Per ideam adaequatam intelligo ideam, quae, quatenus in se sine relatione ad objectum consideratur, omnes verae ideae proprietates sive denominationes intrinsecas habet.' ['By *an adequate idea*, I mean an idea which, in so far as it is considered in itself, without relation to the object, has all the properties or intrinsic marks of a true idea.'] ([1883*b*], p. 206, n. 5)

(The English translation of Spinoza is from Elwes's translation of 1883, p. 82 of the Dover reprint.) This strongly suggests already that by inventing or discovering a concept which is successfully integrated into the existing schema one is on the track of something much more important than the mere self-consistent. And indeed this is confirmed by Cantor's introduction of the second kind of reality. Numbers, he says, can also be *transubjectively* or *transiently* real:

Secondly, reality can be ascribed to numbers in so far as they must be taken as an expression or image of the events and relationships of that outer world which is exterior to the intellect. So for instance, the various number-classes (I), (II), (III) etc. are representatives of powers which are actually found in corporeal and intellectual nature. This second species of reality I call the transsubjective or transient reality of the integers. ([1883*b*], p. 181)

And crucially, immanent and transient reality are intimately connected:

With this thoroughly realistic, but none the less idealistic, foundation of my reflections, there is no doubt in my mind that these two forms of reality are always connected with one another. For, a concept said to exist in the first sense also always possesses in certain, even infinitely many, respects a transient reality . . . ([1883*b*], p. 181)

As Cantor himself says ([1883*b*], p. 206, n. 6), what he proposes is a Platonic principle: the 'creation' of a consistent coherent concept in the human mind is actually the uncovering or discovering of a permanently and independently existing real abstract idea.

It is now clear why Cantor considered mathematics as so free. It does concern itself with objective truth and an independent (Platonic) realm of existents in so far as its objects of study are transiently real. But it need not attempt to investigate this transient reality directly, or even worry about the precise transient 'significance' of a concept. All that mathematics need worry itself with is 'intrasubjective' reality, and once this is established it is *guaranteed* that the concepts are also transiently real. There may be all kinds of ways in which transient reality is manifested; in particular concepts might be represented or instantiated in the physical world. Cantor for example firmly believed this of the transfinite numbers, that there is a range of different infinities in 'corporeal nature'. But he also accepted that for mathematics itself this is largely irrelevant, that mathematics can proceed without such reassurance of physical application.

There is undoubtedly a certain similarity between Cantor's position here and that of later formalism, according to which mathematics in the last resort need only concern itself with consistency. Cantor remarked (see the passage quoted on p. 17 above) that establishing the precise 'transient significance' of a concept was a matter for metaphysics and not mathematics. (At this stage it seems he was not inclined to call mathematics a branch of metaphysics.) For Hilbert, too, mathematics is free in a similar sense, that is to say, free in principle (according to the aims of Hilbert's programme) from serious 'extra-mathematical' philosophical investigation of its concepts, particularly of the infinite. (As Bernays puts it the doctrine that 'existence in the mathematical sense denotes nothing other than freedom from contradiction ... means that for mathematics no philosophical existence question arises' ([1950], p. 92).) But, by virtue of this similarity, it is important to stress again the *differences* between Cantor's position and that of the formalists. Cantor was not just or indeed very seriously occupied with the non-contradictoriness of concepts. (Certainly he was not concerned with formal consistency, since he had no idea of a formally presented theory.) Rather, Cantor's stress on a new concept's having 'orderly relations to existing concepts' is much better described as a notion of *coherence*, of the coherent integration into the existing conceptual framework rather than of *mere* non-contradiction. Moreover, while he mentions non-contradiction, Cantor deals with it only in so far as he attempts to rebut suggestions that the very notion of infinite number is self-contradictory.[3] Outside of these negative rebuttals, which are, of course, a long way from a consistency *programme*, Cantor concerns himself entirely with what I call here coherence. We know from the development of Hilbert's programme that to make an explicit demonstration of consistency a necessary requirement for the acceptance of a mathematical theory may be a very strong demand indeed, certainly if one insists on absolute consistency. But in practice mathematics has concentrated much more on something like Cantor's coherence, the plausible extension and generalization of theories already reasonably well established. And in this sense it has been rather free.

In addition, as will be clear from the above discussion, for Cantor concentration on coherence was not a means of avoiding discussion of existence but rather a means of guaranteeing it. 'Coherence' may be looked upon as a kind of minimal condition that mathematics has to respect but according to Cantor's doctrines this minimal condition is an existential maximal principle: as many things as possible exist. Thus in the case of infinite numbers, which occupied him above all, coherent integration of the concept 'transfinite ordinal number' guarantees that there must be objects which fall under it. (Cantor's argument that this concept *is* successfully and coherently integrated is based on his

---

[3] For example, Cantor ([1883*b*], p. 178) deals with the argument that since an infinite ordinal number would be both even *and* odd, and no number can be both, there cannot possibly be such infinite numbers. He points out that just because there are no odd and even natural numbers it does not follow that no numbers can be even and odd. (The smallest infinite ordinal ω is both even and odd, since it equals $1 + \omega$ and $2 + \omega$.) In his later writings Cantor also considered the Scholastic arguments for the impossibility of infinite numbers (see below).

attempt to explain all numbers in terms of sets. This will occupy us in §2.1 and
Chapter 3.)

As Dummett makes clear in his [1976] (particularly p. 234), Frege rejected
the position that showing the non-contradictory nature of a concept is enough
to establish that an object falls under it. Indeed Frege noted, no doubt with
damning intent:

> This theory imagines that all we need do is make postulates; that these are
> satisfied then goes without saying. It conducts itself like a God, who can create
> by his mere word whatever he wants. ([1884], p. 119)

Now Cantor stresses something other than non-contradiction, though it is not
clear precisely what this consists in. (In the case of the transfinite numbers,
establishing 'coherence' involves a 'reduction' of the concept of number to the
concept of set. Thus, here Cantor invokes a form of relative coherence, and
therefore relative consistency, argument.) Nevertheless, one might still apply
Frege's caution here: it is not at all obvious that showing 'coherence' (or even
relative coherence) should be a sufficient condition for existence. Perhaps
Cantor himself realized this, for in his later writings he devoted more attention
to existence. In one place in his [1883*b*] Cantor states rather mysteriously that
the connection between immanent and transient reality is founded in the 'unity
of the All to which we ourselves belong' ([1883*b*], p. 182). It may be that with
this somewhat mystical comment Cantor is already appealing to God as a main
support of his philosophical position. But in any case, this is what he does in his
later treatment of existence.

What Cantor suggests first is that freedom from contradiction (or if we follow
the [1883*b*] presentation, coherence) of a concept makes it *possible* that objects
fall under it. Then, in this case, says Cantor, such objects are creatable by God.
For example, in a letter to Pater Ignatius Jeiler of 13 October 1895, he writes
concerning the transfinite:

> Such a transfinite imagined both *in concreto* and *in abstracto*, is free from
> contradiction, thus is possible, and is therefore just as much creatable by God as
> a finite form. (Cantor *Nachlass VII*, p. 195 or Meschkowski [1967], p. 258)

Cantor had already elaborated this idea nine years earlier in a letter to Eberhard
Illigens, 21 May 1886:

> I would like to point out above all [*zuvorderst*] . . . that I do *not* make the
> distinction you point to between 'pure possible being' and 'existing possibly'
> [*'Existieren können'*] or 'conceivably existent being'. If I have recognized the
> inner consistency of a concept which points to a being, then the idea of God's
> omnipotence impels me to think of the being expressed by the concept as in
> some way actually realizable. Consequently I call the being concerned a 'possible'
> being. By this is not meant that the being somewhere, somehow and sometime
> exists in actuality [*Wirklichkeit*], since that depends on further factors, but only
> that it can exist. Thus for me the two concepts 'suited for existence i.e. for being
> created' and 'possibility' coincide. (Cantor *Nachlass V1*, pp. 52–3)

But (certainly in the case of the transfinite) Cantor concludes that, whether or not the 'possible' object in question *is* 'created' and exists in the concrete world, nevertheless, since it is 'realizable' to God, it does have abstract existence as an idea *in God's intellect*. In the case of the transfinite at least, one concludes this from Cantor's statement ([1887-8], p. 405) that the transfinite 'expresses the extensive domain of the possible in God's knowledge [*Erkenntnis*]'. This view of the nature of the existence of the transfinite forms is confirmed in the 1895 letter to Jeiler:

The transfinite is capable of manifold formations, specifications, and individuations.

In particular, there are transfinite cardinal numbers and transfinite ordinal types which, just as much as the finite numbers and forms, possess a definite mathematical uniformity, discoverable by men.

All these particular modes of the transfinite have existed from eternity as ideas in the Divine intellect.

I take it that by 'capable of manifold formations, specifications, and individuations' Cantor means that we can articulate a coherent theory of the transfinite. This would mean that transfinite 'beings' are possible 'beings', and this is what must, according to Cantor, entail that they already have real and independent existence as ideas in 'the Divine intellect'. This kind of appeal to the Divine intellect, and no doubt to the perfection of God, finally guarantees the correspondence between immanent and transient reality. It is rather ironic that Frege rather damningly refers to God in his condemnation of the thesis that consistency entails existence. For Cantor's later advocation of 'coherent integration entails existence' depends more or less on God creating not so much 'whatever He wants', as Frege put it, but 'everything He can'.

Given this, the mathematician does indeed appear quite free. Moreover, the appeal to God here, given that God's intellectual capacity should be immeasurably more powerful than man's, suggests both that much more exists than has so far been conceived by man and that the conditions on the 'formation' or 'construction' of concepts by *us* need be only rather weak. Thus, from this position one could argue that we can quite well proceed with the development of a theory of objects of a certain kind without necessarily having a very precise idea of what the objects are (for example, without precise definitions). All that is required is *some* reasonably coherent conception of them. God does the rest. Cantor nowhere clearly expounds this position, but one can, I think, see a tendency to adopt it both in his treatment of sets and in his treatment of ordinal numbers. For example, although we cannot directly grasp the collection of natural numbers as a whole, Cantor assumed that God can; thus this unity, this *set* of all natural numbers exists as an idea in God's mind, and can therefore be taken as an object by mathematics. God is thus used to bridge a gap that we cannot bridge by ourselves and without it being clearly explained what the nature of this unity (this sethood) is. (I come back to this in my discussion of Cantor's finitism.) Similarly with the ordinal numbers, Cantor clearly regarded it as enough to *indicate* that they are 'conceptual representatives' of well-ordered sets in order to conclude that they must exist without attempting (at least not at first) a precise explanation of what they are or what their structure is. (See §2.1.)

The central passages quoted here were written as long as 13 years after Cantor's [1883*b*], although it should be noted that the key passage from [1887–8] stems from 1886. But a similar position (reliance on possibility, indeed coherence, and the power of the Creator) is also clearly evident much earlier in Cantor's reply to Aquinas's argument against an actual created infinity. Aquinas begins his argument with a quote from the *Book of Wisdom* (Chapter 11, verse 21): 'Thou hast ordered all things in measure, and number and weight.' He then concludes that there can be no actually infinite multitude:

. . . since every kind of multitude must belong to a species of multitude. Now the species of multitude are to be reckoned by the species of numbers. But no species of number is infinite, for every number is multitude measured by one. Hence it is impossible that there be an actually infinite multitude, either absolutely or accidentally. Furthermore, multitude in the world is created, and everything created is comprehended under some definite intention of the Creator; for no agent acts aimlessly. Hence everything created must be comprehended under a certain number. Therefore it is impossible for an actually infinite multitude to exist, even accidentally.[4]

Cantor certainly took Aquinas's argument seriously. Indeed, in a letter to Schlottman of 9 April 1887, he remarks, after pointing out the above passage from *Summa Theologica*:

The thoughtful arguments which one finds here and with the Church Fathers (Augustine should be excepted) *against* the standpoint represented by me were thoroughly justified, and it was therefore right and correct to oppose actually infinite numbers, so long as a principle of *individuation*, *specification* and *ordination* of that actual infinite which I call the *transfinite* had not been found. However, all these arguments crumble as soon as one set up such a principle and it had been demonstrated as *true*. This however one finds in my works. (Cantor *Nachlass VI*, p. 110)

Thus without a coherent *theory* of the transfinite one is right to dismiss the notion of a non-Absolute actual infinite, as Aquinas does. But, Cantor says, this changes once one introduces such a coherent (and here he goes much further and says 'true') theory. This he claims he has done, based on the concepts of infinite power and transfinite ordinal number (and one might add on the crucial result of 1874 which shows that in so far as the completed infinite does enter mathematics, there must already be a 'differentiation'). Thus, says Cantor

---

[4] St. Thomas Aquinas, *Summa Theologica*, part I, question 7, article 4, quoted from Anton C. Pegis (ed.), *Introduction to St. Thomas Aquinas*, Random House, New York, 1948, p. 60. Cantor quotes this passage in Latin ([1887–8], pp. 403–4, footnote).

. . . I believe that the passage from the Holy Scripture 'Thou has ordered all things in measure and number and weight' . . . in which was seen a contradiction of actually infinite numbers does not in fact contradict them. For suppose there are, as I believe I have shown, actually infinite 'powers' *i.e.* cardinal numbers, and actually infinite 'enumerals of well-ordered sets', *i.e.* ordinal numbers, . . . then quite certainly these transfinite numbers would also be embraced by the cited passage from the holy scripture. Thus, in my opinion, if one wishes to avoid circularity this passage cannot be taken as an argument against actually infinite numbers. (Cantor [1887–8], pp. 399–400; the passage stems from 1886.)

For Cantor then, the introduction of a coherent theory of the transfinite with 'individuation', 'specification', and so on is enough to counter the argument that a created actual infinite is automatically ruled out. It is possible because 'coherence' implies possibility. But it is clear that Cantor again goes further. This possibility means for Cantor, as he says, precisely that the transfinite is available to God (presumably the ideas are 'in the Divine intellect') and therefore that He *could have* created the concrete world in such a way as to instantiate various 'modes' of the transfinite. Thus, as he says in another part of his reply to Aquinas:

. . . in the transfinite a vastly greater abundance of forms and of *species numerorum* is available, and in a certain sense stored up, than there is in the correspondingly small field of the unbounded finite. Consequently, these transfinite species were at the disposal of the intention of the Creator and his absolutely inestimable will power just as were the finite numbers. ([1887–8], p. 404, footnote)

Of course, as far as mathematics is concerned establishing possibility and abstract existence (via coherence or inner consistency) is all that matters. And this is the main thrust of his 'justification' of the transfinite. But it is no surprise that Cantor went further and claimed that much of this abstract transfinite world is represented in the concrete physical world of God's creation. For example, Cantor writes:

That an 'infinite creation' must be assumed to exist can be proved in many ways . . .
    One proof stems from the concept of God. Since God is of the highest perfection one can conclude that is is possible for Him to create a *transfinitum ordinatum*. Therefore, in virtue of His pure goodness and majesty [*Allgüte und Herrlichkeit*] we can conclude that there actually is a created *transfinitum*. ([1887–8], p. 400; this is also from 1886)

Thus, we seem to have here the principle of maximum possibility applied also to the created world. (In his [1885*b*] (pp. 275–6) Cantor indicated with some remarks about 'aether monads' and 'monads of matter' how he believed the first two powers to be represented in created nature.) The connections between 'possibility', 'abstract existence', 'creatability', and 'concrete existence' are then all summed up in a further remarkable passage:

. . . the transfinite not only expresses the vast domain of the possible in God's knowledge, but also presents a rich and continually increasing field of ideal discovery. Moreover, I am convinced that it also achieves reality [*Wirklich-keit*] and existence in the world of the created, so as to express more strongly than could have been the case with a mere 'finite world' the majesty of the Creator following his own free decree. ([1887–8], pp. 405–6; see also p. 400)

There is a firm suggestion throughout these discussions that Cantor believed not only that the idea that man can coherently grasp the infinite does *not* demean God or necessitate rejecting the very special position that He has, but on the contrary the 'availability' of the transfinite serves to glorify God. Certainly the rigid traditional dichotomy between finite man and infinite God has been weakened, since man is permitted to some extent to share the infinite. But the recognition of the transfinite really serves to enhance our conception of God's grandeur, since the expanse both of possibility available to Him and of His actual creation is admitted to be much greater. And the distance or difference in capacity between God and man is not profoundly disturbed. Indeed, the traditional dichotomy reappears (re-upholstered) in the careful (if not precise) distinction between the transfinite and the Absolute infinite. Thus, rather than Christian philosophy being opposed to the actual infinite, Cantor claimed to have '. . . made acceptable to Christian philosophy for the first time the true theory of infinity in its beginnings'. (Cantor letter to Pater Thomas Esser, 1 and 15 February 1896, Cantor *Nachlass VIII*, p. 137, or Meschkowski [1965], p. 513.)

As I mentioned in the Introduction, Cantor's appeal to God is often of no great philosophical help. This is certainly the case here. We are not much wiser about the nature of mathematical objects having followed Cantor's arguments through. Nevertheless, clarification of Cantor's appeal to God does at least explain some of the strength of Cantor's realism, specifically his belief in an independent world of ideal mathematical objects. Indeed, we are by no means finished with Cantor's appeal to God, particularly not with its effect on the conception of the Absolute nor its connection with realism, as we shall see shortly. Now I turn to Cantor's mathematical argument for the non-emptiness of category (ii) (*a*).

1.2 The potential infinite and reductionism

The argument is based upon a simple application of principle (*a*) mentioned in the Introduction, namely that every potential infinity presupposes a corresponding actual infinity. Let us begin, therefore, with Cantor's statement of this.

No one would deny that the potential-infinite is of fundamental importance in mathematics, for example, and most importantly in the notion of endless addition of 1 (the natural numbers) or variability or continuity in the real line, here particularly after the introduction of ϵ–δ methods in the nineteenth century. What Cantor says, though, is that once one accepts this self-evident statement, one has already admitted the *actual*-infinite into mathematics:

There is no doubt that we cannot do without *variable* quantities in the sense of the potential infinite; and from this can be demonstrated the necessity of the actual-infinite. In order for there to be a variable quantity in some mathematical study, the 'domain' of its variability must strictly speaking be known beforehand through a definition. However, this domain cannot itself be something variable, since otherwise each fixed support for the study would collapse. Thus, this 'domain' is a definite, actually infinite set of values.

Thus, each potential infinite, if it is rigorously applicable mathematically, presupposes an actual infinite. (Cantor [1886*a*], p. 9)

(This passage also appears in a section of *Nachlass VI*, pp. 63-6, dated 1886, and which is published in Meschkowski [1967], pp. 249-51, and again in [1887-8], pp. 410-1.) Or, again speaking specifically of classical analysis,

. . . the potential infinite is only an auxiliary or relative (or relational) concept, and always indicates an underlying transfinite without which it can neither be nor be thought. (Cantor [1887-8], p. 391)

In other words, it does not make sense to speak of variability without speaking of variability over a completed domain. And as soon as one is concerned with the *unlimited* variability characteristic of mathematics, one must be concerned with variability over completed *infinite* domains. Cantor's argument is now that not only does this show that (ii) (*a*) (the transfinite actual infinite) is non-empty but also, because these domains (or many of them) are basic in mathematics, mathematics should and must study them.

Let us fill out the argument a little. Of course, it could be the case that the only completed domain which, say, variable numbers point to is the domain of *all* objects, which would in effect be the Cantorian Absolute. But Cantor assumes that each number concept gives rise to a homogeneous domain of all and only the objects falling under it; thus for example the three concepts natural number, rational number, and real number point to three completed homogeneous number domains. Then, or so the argument would go, these domains cannot be Absolute because they are increasable, or extendable, and thus they must for Cantor fall under the *transfinite*, i.e. under (ii) (*a*) and not (ii) (*b*). Cantor does not say much about increasability in his statements on the categorization of infinity. There are clear informal, intuitive senses in which the natural, rational, and real numbers form extendable domains. The natural numbers form part of the rational numbers, and these form part of the real numbers. Thus each successive domain seems richer in elements. Dedekind notes in his [1872], p. 9: 'The straight line *L* is infinitely richer in point-individuals than the domain *R* of rational numbers in number-individuals.' Bolzano, who also accepted that the actual infinite is capable of increase, also conceived increase in this intuitive sense of adding more elements. (See [1851], pp. 82 and 95.) Moreover, looked at geometrically, rational points are embeddable in the real points, real points of 1-space in higher spaces, and so on. But by the time Cantor introduced the distinction between the transfinite and the Absolute it is quite clear that he

understands increasability in the light of the notion of cardinality introduced in [1878]. He says this explicitly in [1887-8], but one even obtains a very strong sense of it in [1883b] where he describes the Absolute as 'symbolized' by the sequence of number-classes (I), (II), . . . , etc. and their cardinalities (see §1.4, especially p. 42). Although Cantor's results showed that some intuitions of size are not borne out (the rationals are not bigger than the natural numbers, Euclidean n-space is not bigger than Euclidean m-space even when $n > m$), some of them crucially are substantiated: the domain of real numbers *is* strictly bigger than the domain of natural numbers ([1874], [1891]), the domain of real functions *is* bigger than the domain of real numbers ([1891]), and so on, apparently indefinitely. Thus, these domains *are* cardinally increasable infinities, and thus fall under (ii) (*a*). (ii) (*a*) is therefore not only populated, but populated by perhaps the most important of mathematical domains. Such basic domains must, of course, be studied. But in particular, since the cardinality results show that there is a fundamental structural difference between the two most important domains (natural numbers and real numbers), the difference must be explained and analysed. This, of course, is what the theory of transfinite numbers is designed to do. As Cantor himself states:

These 'domains of variability' are the correct foundation for both analysis and arithmetic. Thus they themselves deserve in the strongest possible way to be taken as the subject of investigation as is done by me in my 'Mengenlehre'. (Cantor [1886a], p. 9; also *Nachlass VI*, p. 66, or Meschkowski [1967], p. 250, or Cantor [1887-8], p. 411)

It seems at first sight that Cantor has here a strong argument for (ii) (*a*) and indeed for the whole project of transfinite *Mengenlehre*, an argument which is moreover independent of appeal to the 'theological ontology'. Its strength is somewhat increased by Cantor's further claim that there is no way to develop a theory of real number other than by using completed infinite domains in some way. Thus

. . . the provision of a foundation for the theory of irrational numerical quantities [real number] cannot be effected without the use of the actual infinite in some form. (Cantor *Nachlass VI*, p. 64 or Meschkowski [1967], p. 250)

Here Cantor is, of course, referring to the so-called 'arithmetization of analysis' which both he and Dedekind had independently effected. I shall return to this a little later, since this indeed does lend Cantor's position considerable strength.

Let us look at the argument now a little more critically, beginning with the domain principle itself. The principle is a realist one as can be seen as soon as we understand its conceptual content. We are incapable of directly perceiving or intuiting the whole of an infinite domain either as a totality or indeed one by one. Thus, to take one example, our grasp, such as it is, of natural number must come through some intensional means, through a concept or perhaps in this case it is better to say through a process or operation—the operation

of addition of 1. But Cantor's principle asserts that in addition to this intensional understanding, the grasp of a concept or process, the *extension* of the
concept or the process *already exists in totality*. This is clearly and unashamedly
a realist position. To take the example of the natural numbers again, the most
basic way to attempt to grasp the notion is through the process of successive
generation, and it is quite beyond our capacity to bring this generation to
completion. But principle (*a*) asserts that, regardless of our limited ability to
construct or generate natural numbers, *there is* a complete sequence of these
numbers. Our constructive frailty is irrelevant.

One way Cantor would support this thesis, particularly with respect to the
natural numbers, is to say that if generation or construction is all we have to
rely on then we would have to abandon even large finite numbers for these are
also 'beyond' our immediate grasp. As he says specifically with reference to the
transfinite numbers (and thus transfinite totalities, since for Cantor transfinite
numbers are derived from infinite totalities), one might argue against them

. . . on the ground that *we* with our restricted being are not in a position to
actually conceive the infinitely many individuals (ν) belonging to the set (ν)
[i.e. $\{n : n \in \omega\}$ in modern notation] in one intuition. But I would like to see
that man who, for instance, can form the idea distinctly and precisely in one
intuition of all the unities in the *finite* number 'thousand million', or some
even smaller numbers. No one alive today has this ability. And yet we have the
right to recognize the finite numbers, however great, as objects of discursive
human knowledge, and to investigate their properties scientifically. *We have the
same right also with respect to the transfinite numbers.* ([1887–8], p. 402)

Thus, Cantor would say, since we deal quite happily with *all* finite numbers, no
matter how large, complete construction or generation by us must be irrelevant.
Again, for Cantor, mathematical practice does not rely on a direct intuition of
numbers, but just on their assumed independent existence.

So principle (*a*) is a realist one, legitimized or bolstered *post hoc*, Cantor
seems to be saying, by the necessity of building classical mathematics upon it.
The development of twentieth century mathematics has to some extent
challenged this appeal to completed domains, and the claim of necessity
advanced for it. One finds a good illustration of this challenge in the writings of
Hermann Weyl, for example. Weyl, like Cantor, believed that mathematics is
fundamentally the 'science of the infinite' and that this must be clearly grasped
(as it possibly was not, in modern times, before Cantor). But Weyl maintained
that the potential not the actual infinite is basic in mathematics, and indeed that
we have no right whatever to operate with completed infinities. Mathematics as
far as the infinite is concerned must be the science of *becoming* and not the
science of *being*. One has access to processes or intensions, but *only* to these.
He held that it is a fundamental error to leap from a 'field of constructive
possibilities' which an intension opens up to a 'closed aggregate of objects
existing in itself' (see Weyl [1949], particularly Chapter 2, §8, or [1931]).

Also as far as God is concerned, and in contrast to Cantor, Weyl would say: God *is* the [only?] completed infinite 'and cannot and will not be comprehended by the human mind' ([1932], p. 84). (In other words, there are potential infinities and an Absolute infinite, but nothing in between.)

The challenge thrown down by Weyl here is not merely a philosophical one. Brouwer's development of intuitionism (as Weyl well knew) shows that one can build a perfectly coherent and quite extensive mathematics in which there are no extensionally definite, actually infinite sets. Therefore Cantor's adoption of his principle (*a*) is not a necessary condition for the development of an extensive analysis. So, it transpires, there is no clear 'pragmatic' justification for the domain principle. This appears to focus attention back onto the principle itself, away from its consequences. So, why should we adopt it? Cantor's answer would of course be quite simple: the principle is just a consequence of his realist theory of concepts and objects. Formulation (or grasp) of a coherent concept like natural number is enough to show that all natural numbers already exist, that the collection of these numbers is already completed. Thus, a concept or a process only really points to something which already exists—its domain. And ultimately Cantor's justification for this must be simply that the objects exist as ideas in God's mind. This is nicely illustrated by Cantor in a letter to Veronese of 7 September 1890. Speaking of the numbers Cantor says:

. . . one must distinguish between them as they are *in and for themselves, and in and for the Divine intelligence* and how these same numbers appear in our restricted, discursive comprehension and are differently defined by us for systematic or pedagogical purposes. (*Nachlass VII*, p. 3)

Cantor goes on to say that *we* may grasp or define the numbers through some process of succession, but that nevertheless they already exist independently of such a process, and, as he says in the passage quoted, 'in the Divine intelligence'. Cantor was also firm that however *we* might conceive of or grasp actual infinities (transfinites) these properly speaking have nothing to do with processes:

My conceptual grasp of the transfinite excludes properly and from the beginning 'process', since this denotes a 'change'. According to me transfinite = definite, greater than anything finite of the same kind, but nevertheless still capable of increase. Because of this last property change is still nevertheless possible in the domain of the transfinite; so, for example, the ordinal number ω can become $\omega + 2$, and so on. Thus even transfinite processes would be conceivable in so far as one understands by this processes in the domain of the transfinite. But a transfinite process properly speaking seems impossible according to my understanding of the 'transfinite', because here the two mutually exlusive predicates 'definite = constant' and 'variable' would be joined. (Cantor, letter to Harnack, 3 November 1886, *Nachlass VI*, p. 86)

What Cantor is saying is again the old point expressed by principle (*a*): we can certainly conceive of variability in the transfinite, but it is *just* that, i.e. variability over the completed domain of the transfinite. And I take the last comment about not equating 'definite' and 'variable' to mean that one cannot properly speak of a transfinite number, say, as 'coming into being' through some process; for Cantor it is already something existent and fixed ('definite', 'constant'). Thus, the position Cantor adopts is in stark contrast to that taken later by the constructivism of Weyl and Brouwer where there is no 'definite' transfinite but *only* processes (see in particular Weyl [1931]). But the point I wish to make here is that clearly Cantor's mathematical argument for the non-emptiness of (ii) (*a*) (the transfinite) must ultimately rest, like the argument of §1.1, on his realist theory of mathematical objects. And this for Cantor depends in its turn on the goodness of God, on the assumption of the independent existence of transfinite objects as ideas in the Divine intellect. Thus the argument is neither as strong nor as independent as it seemed at first glance.

But having said all this, we must begin to be a little careful. It may be right as Weyl did to question the intuitive correctness of basing the notion of varying quantity on that of fixed totality. But one cannot deny either the power of Cantor's conception or its heuristic fecundity, particularly when associated with principle (*b*), his finitism (see §1.3). Neither can one deny its profound effect on subsequent mathematics. Indeed we must be very wary of criticizing Cantor on the basis of developments which came only much later. Certainly by 1872 Cantor (among others) had already demonstrated its power. In the nineteenth century many of the earlier difficulties over infinitesimals had been clarified by resort to the notion of finite limit and the so-called ϵ-δ method. But the notion of limit itself stood in need of clarification, particularly the question of under what circumstances sequences possess limits. Various important theorems discovered and proved by Weierstrass assumed that any bounded infinite sequence has at least one sub-sequence which converges to an actual finite limit value. The crucial problem was then to justify this assumption, that is to say, to provide a definition or characterization of real number which yields exactly this result (or equivalently, that every bounded infinite set has a least upper bound). This was stated with crystal clarity by Dedekind in his [1872] (*Vorwort*). The problem was solved by Cantor and Dedekind (independently) through the misnamed 'arithmetization of analysis'.[5] For here while the solution is arithmetic in the sense that the natural numbers are involved—and in the sense that it is intended to be separated from *geometrical* ideas—the crucial new element in both Cantor's and Dedekind's solutions is domain-theoretic. In the first place, the definition of the real numbers must be based on the assumed presence of the completed domain of rational numbers, and hence the completed domain of the natural numbers. Secondly, whichever definition one follows, each real number

[5] For Cantor's solution (via the use of Cauchy sequences) see Cantor [1872] or [1883*b*], or Dauben [1979], p. 37 ff. For Dedekind's solution, see his [1872]. For a modern treatment, see Feferman [1964], particularly Theorem 7.9, p. 233, and Theorem 7.26 (especially step 6), p. 244. Weierstrass also had a solution: see van Dantscher [1908] or Jourdain [1909]. For an account of how the problem of completeness of the real numbers arose in Weierstrass's work, see again Jourdain [1909].

is either *itself* a completed infinite domain of a certain kind (an equivalence class of sequences of rational numbers or a segment of the rational numbers), or is a primitive term defined by reference to such a domain.

For both Cantor and Dedekind, the whole purpose of the exercise was to free analysis from formal reliance on geometrical or spatial intuition (see Dedekind [1872], *Vorwort*, and the letter to Lipschitz of 27 July 1876 in Becker [1964], p. 241, and Cantor [1882], p. 156, particularly n. 1). This they did by producing purely abstract definitions in terms of concepts which are quite divorced from empirical or spatial considerations. Indeed, for both, however strong our spatial intuitions may be, in the end we have to clarify the notion of geometrical line by appeal to the arithmetic or domain theoretic notion of real number. Of course, as we have seen, this abstract solution brings with it rather strong ontological commitments—namely, the independent existence of completed domains. But it was nevertheless extraordinarily successful. Until the work of Weyl and Brouwer (one might also add Russell), the Cantor-Dedekind theory of real number was the only coherent and rigorous theory. Indeed one can go further. It remains the only theory which yields the sought after Bolzano-Weierstrass theorem. For the proof of this depends crucially on the acceptance of impredicative definitions, which of course, no constructive theory can accept. The theory and the domain principle which underlies it put analysis on a new and clearer footing, and this was perhaps the crucial step in making domain-theoretic (and, as we shall see, set-theoretic) methods fundamental in modern mathematics. The 'domains of variability' may not be finally justified as 'the correct foundation for both analysis and arithmetic' as Cantor had claimed, but they do provide an impressively clear and powerful foundation. And, in view of this, Cantor was certainly right to claim that they 'deserve in the strongest possible way' to be investigated. In a way, then, *Mengenlehre* began with the successful application of the domain principle to the foundations of analysis. As Russell remarked:

The mathematical theory of infinity may almost be said to begin with Cantor. The Infinitesimal Calculus, though it cannot wholly dispense with infinity, has as few dealings with it as possible, and contrives to hide it away before facing the world. Cantor has abandoned this cowardly policy, and has brought the skeleton out of its cupboard. He has been emboldened in this course by denying that it is a skeleton. Indeed, like many skeletons, it was wholly dependent on its cupboard, and vanished in the light of day. (Russell [1903], p. 304)

And Frege also explicitly supported Cantor's domain principle (see his [1890-2], pp. 68-9, and also [1892], p. 163). (It should be noted that if an axiomatic rather than an explicitly set theoretical approach to the real numbers is adopted, then the crucial 'completeness' property must be adopted as an axiom see e.g. Dieudonné [1969, p. 17, axiom IV.)

Cantor's approach to real numbers is a good point at which to switch discussion away from the arguments for the existence of transfinites to the nature of the transfinite itself, and the relation between the transfinite and the Absolute. There are three more aspects of this approach to real number which should be mentioned, and which will occupy us a good deal in what follows. The first is that there is a clear tendency to *reductionism*. Set-theoretic reductionism, that is the desire, the intention, to explain all mathematical objects as sets, to allow *Mengenlehre* to embrace all of mathematics, became more important and gradually stronger throughout Cantor's work. Set reductionism is not yet present in the 1872 treatment of real number, but certain important elements of it are. For one thing while for Dedekind and Cantor (at least at first) real numbers are still primitive objects, each such number is defined by reference to a complex structure, a collection of sequences of rationals or a cut, i.e. it is something defined in terms of rationals by appeal to the notion of collection. This is already reductionsim of a sort. Of course, the reduction is effected by bringing in a new element of a rather general sweeping kind. And the success of this procedure compared with the muddle surrounding continuity and real numbers previously suggests that the new notion of collection is indispensable. Since this new element apparently cannot be done away with, it may be quite natural to ask whether it is possible to get rid of some of the other elements that remain, for example, the claim that real numbers are new *primitives* tied to the defining collections. There is no evidence that Cantor put such a question to himself at a very early stage of his work, or even explicitly at all, but I shall attempt to show in Chapters 2 and 3 that this tendency lies behind his whole treatment of ordinal and cardinal numbers, and was in fact the keystone of his justification of transfinite numbers. If one shows that infinite domains (collections) are indispensable in mathematics *and* that (transfinite) numbers 'arise out of' or are indissolubly linked with these domains or, better, are such domains themselves, then one has indeed a strong argument for their acceptance. They too are 'necessary', not in the sense (yet) that they have any mathematical importance, but in the sense that they cannot be denied. This was just the path Cantor took from his [1883*b*] on. Each step in the chain of reasoning adds to the importance of the domains. And finally Cantor accepted, and tried to show, that numbers must be the same kind of thing as the domains they number, i.e. domains themselves.

So far this is only reduction to domains (collections). But there is a second important aspect of the treatment of real number here which hitherto we have suppressed. The real numbers defined by the methods of 1872 are certainly taken as single objects (individuals). And they are either themselves collections or are co-ordinated with collections or a domain of collections, and these themselves therefore should be subject to predication, be the arguments of functions, etc. But this means accepting either way that collections (multiplicities) can also at the same time be single objects (individuals). Moreover, in his [1872] Cantor immediately takes real numbers (i.e. these complex collections) as elements of other collections (point-sets). So it seems that we have a recursion (an iteration): collections form objects which are again taken as elements of collections and so on. This iterative element is the third important feature and will be of great importance when we come to consider the relationship between the increasable transfinite and the Absolute in §1.4.

This treatment of collections as objects, which is essential here, marks the first genuine intrusion of *set theory*. For collections which are also objects (individuals) are just what Cantor called sets (*Mengen*). Thus the reductionist tenor of the domain principle and its application to the definition of real number marks the beginning of set-reductionism. This prominence of sets, and therefore of the transfinite, brings us to what I call Cantor's 'finitism'.

1.3 Cantorian finitism and the concept of set

In his [1949], p. 47, Weyl remarks: '. . . , for set theory, there is no difference in principle between the finite and the infinite' (see also [1931], p. 14). Weyl is perfectly justified in emphasizing the unified nature of set theory, for as we shall see, it goes right to the very heart of the Cantorian theory. Indeed, the unity which Weyl points to is so much a fundamental part of Cantorianism (at least when we substitue 'transfinite' for 'infinite') that I have called it Cantor's principle of finitism (why I use this term will become clear in what follows) and designated it as one of the three basic Cantorian principles (see Introduction, p. 7). Moreover, it is this doctrine which gives principle (*a*), which we discussed in §1.2, its force. There are two senses in which we should take notice of Cantor's 'finitism'. First, it expresses a certain 'finistic' attitude to sets (mathematical objects) and which is what gives the theory its unity. Namely, sets are treated as simple objects, regardless of whether they are finite or infinite. Secondly, all sets have the same basic properties as finite sets. I shall try to make clear later what is meant by this. But for the moment I want to observe that this gives rise to a certain heuristic attitude which guided the construction of Cantor's theory in certain key aspects, above all with regard to the theory of number. Recognition of this helps greatly in understanding the detailed construction of Cantor's theory.

I shall come to numbers a little later. I begin with Cantor's attitude to sets, and the role that his 'finitism' plays here.

We saw that the Cantor and Dedekind definitions of real number force us to admit certain infinite collections as single objects, as individuals. Thus, while in construction (i.e. in terms of their definition) these real numbers are highly complex, the theory says that once we have defined them we can treat them as simple objects—forget the complexity. What Cantor does is to extend this doctrine to *all* collections that one wishes to subject to mathematical examination. They can all be treated as single objects. Indeed this is turned eventually into a declaration of reductionist principle: mathematics just is the study of such collections, namely of sets. This is in the first place the foundation for a prin-

ciple of iteration, and at the same time for a principle of type collapse. Principle
(*a*) will yield a completed domain from a concept. But now, if there is good
reason to accept that this domain is not an Absolute domain, then it too must
form a simple object (a set), and (the unity doctrine says) of the same fundamental kind as the objects which make it up. In particular it will be available as
a member of further collections, and this ability to iterate (collection to a
domain, collapse to a single object, further collection) is ultimately what gives
the domain principle (*a*) its strength. (Indeed it seems immediately to give rise
to the notion of power domains. Suppose principle (*a*) yields a domain A which
is not Absolute. Then any property or concept ϕ will yield a sub-domain A, of
A which, since it cannot be Absolute either, will be an object. But the concept
'subset of A', by the domain principle, will yield a new completed domain,
Pow(A). If one reassures oneself that this is not Absolute, the formation process
can be iterated, and so on. See the further remarks on this in §5.3.) Cantor
himself beautifully summed up the combined nature of these two principles
working together in a remark on the natural number sequence (letter to Aloÿs
Schmid, undated but very likely from April 1887):

The first, simplest fact, accessible to everyone, on which the theory of the transfinite is based is the simultaneous boundlessness and yet the definiteness in itself
of the series of all finite cardinal numbers $1, 2, 3, \ldots, \nu \ldots$ viewed as a constant
set of clearly differentiated things. (Cantor *Nachlass V1*, p. 113)

(After this draft Cantor has written 'N.B. letter not sent' (Cantor *Nachlass V1*,
p. 114).) To repeat, the unity principle is best viewed as a principle of type
collapse: there is no fundamental difference in type between finite and transfinite collections. They are subject to the same mathematical operations, and
crucially, as we shall see, should all be capable of numerical determination.

   But if mathematics must concern itself with this unified realm of sets, what
does Cantor have to say about such things? Cantor makes four main statements
about the concept of set, in [1882], in [1883*b*], in [1895] and in his famous
1899 letter to Dedekind, the third being the best known. I shall examine the
[1882] statement a little later in §1.4. First then from [1883*b*]:

Theory of manifolds. By this I mean a very extensive theoretical concept which I
have only attempted to develop hitherto in the special form of an arithmetic
or geometric *Mengenlehre*. By a 'manifold' or 'set' I understand in general any
many [*Viele*] which can be thought of as one [*Eines*], that is, every totality of
definite elements which can be united to a whole through a law. By this I
believe I have defined something related to the Platonic εἶδος or ἰδέα. (Cantor
[1883*b*], p. 204, n. 1)

And in [1895] we find:

By a 'set' we understand every collection to a whole M of definite, well-
differentiated objects m of our intuition or our thought. (We call these objects the
'elements' of M.) (Cantor [1895], p. 282)

In the 1899 letter we have, somewhat differently:

When . . . the totality of elements of a multiplicity can be thought without contradiction as 'being together', so that their collection into *one* thing' is possible, I call it a *consistent multiplicity* or a *set*. (Cantor [1932], p. 443)

(The translation here is mine; the published translation is given in van Heijenoort [1967], p. 114. The letter is dated 28 July 1899 in both these publications, although according to Grattan-Guinness ([1974], p. 128) this section actually comes from a letter of 3 August 1899.)

What do these statements tell us? The special involvement of consistency in the last 'definition' will be discussed in its context of the theory of absolute collections in Chapter 4. But what all three statements stress is that a set is a collection made up from definite elements but which itself is a whole, that is *one thing* (see also Dedekind [1888], p. 1, quoted in the Conclusion, p. 300). This is precisely the characteristic of sets that I have emphasized above. But Cantor certainly does not say what this unity, this oneness, consists in, what there is to a set which makes it one thing, and not just a multiplicity. The difficulty was certainly recognized. In his [1892], Frege remarks that Cantor is quite 'unclear about what we should understand by "set" ' (p. 164). And in his reply to Cantor's letter of 3 August 1899 Dedekind remarked: 'I don't know what you mean either by the "being together of all elements of a multiplicity" or its opposite' (Grattan-Guinness [1974], p. 129). In the first passage Cantor states that a set is a 'totality united to a whole *through a law*'. It could be therefore that he regards the unity as arising through this intension, this law. However, Cantor seems to have adopted an extensional view of sets. For example, in [1887-8] (in a passage which stems from 1884) he writes of a set '. . . consisting of clearly differentiated, conceptually separated elements $m, m', . . .$ and which is thereby determined and delimited . . .' (p. 387). And again in [1887-8] (another passage which apparently dates from 1884) he speaks of a set as something 'thought of as a thing for itself [*Ding für sich*]' yet '*consisting of* clearly differentiated concrete things or abstract concepts' (p. 411). Indeed, extensionalism is exactly what the domain principle (*a*) leads us to; while an intension is what pointed out the set to us, the set itself is something quite separate from the intension. Extensionalism in general might also be regarded as an aspect of 'finitism'. Small finite collections (and all finite collections in principle) can be treated extensionally, that is, simply by enumerating their elements. Thus, only the elements are important, and not any intension which first brought them to our attention. The unity principle extends this conclusion to the transfinite despite the fact that we cannot enumerate the elements. (See also §3.3.)

In the [1883*b*] passage Cantor speaks of a set as a many 'which *can be thought of* as a one', with a similar emphasis in [1899]. (See also Dedekind [1888], p. 1.) This suggests not only that the unity is something intellectual, but that it is something that we as thinking subjects impose on the collection, that we 'create' the set (unity) from the elements. This 'being thought of as a one' is also in the second of the passages from [1887–8] mentioned above and is hinted at in the [1895] passage. This reference to our capacity to create or arrange or whatever also occurs very frequently in Cantor's discussion of number, particularly with his idea of abstraction (see §§3.2, 3.3 and 3.5). But on the face of it this is of no help at all (though see pp. 301–3). No amount of mental gymnastics can turn a collection of objects into 'one thing', at least not without appeal to intensional means of holding the elements together. And this inability to 'see' or conceive unity has nothing to do with the infinite, or inability to survey the given collection (though introduction of impredicative infinite totalities makes things worse (see Chapter 6 and the Conclusion)). For example, two apples on my table remain two apples no matter how I try to conceive of them as 'one thing'. The only way I can impose unity upon them is by subsuming them under some concept which only they fall under—for instance, being the only two apples on my table at this time. But the unity cannot then be dissociated from this concept: remove it and the apples become a multiplicity again. (See Chihara [1982], p. 223, where he makes a similar point. He too only sees apples on his table and not sets. The point here is, of course, reminiscent of Frege's attack on certain subjectivist accounts of number (see §3.2 below)).

There is evidence to suggest that for Cantor much more important than our ability to conceive of a collection as 'one' was God's ability to do so, and this of course fits well with his theological justification of his ontology. For example, in [1887–8], p. 401, Cantor explains that an actual-infinity is 'not variable' but a 'true constant, fixed and determined in all its parts'. He goes on:

As an example, I give the collection, the totality of *all* finite positive, whole numbers. This set is *a thing for itself* and forms, quite apart from the natural order of the numbers belonging to it, a definite quantum fixed in all its parts . . .

Then in a footnote he cites a stirring and beautiful passage from St. Augustine's *De Civitate Dei* in which he sees an early affirmation of the existence of transfinite forms. Augustine alludes to the incompletability (for us) of the natural numbers, but he goes on to emphasize that God can not only grasp the whole sequence of numbers but grasp them as a *finite* form. I quote only a part of the relevant section, from Book XII, Chapter 19 (Chapter XVIII in some editions):

Every number is defined by its own unique character, so that no number is equal to any other. They are all unequal to one another and different, and the individual numbers are finite but as a class they are infinite. Does that mean that God does not know all numbers, because of their infinity? Does God's knowledge extend as far as a certain sum, and end there? No one could be insane enough to say that.
    Never let us doubt, then, that every number is known to him 'whose understanding cannot be numbered'. Although the infinite series of numbers cannot be numbered, this infinity of numbers is not outside the comprehension of him 'whose understanding cannot be numbered'. And so, if what is comprehended in knowledge is bounded within the embrace of that knowledge, and thus is finite,

it must follow that every infinity is, in a way we cannot express, made finite to God, because it cannot be beyond the embrace of his knowledge.

Therefore, if the infinity of numbers cannot be infinite to the knowledge of God, in which it is embraced, who are we men to presume to set limits to his knowledge, by saying that if temporal things and events are not repeated in periodic cycles, God cannot foreknow all things which he makes, with a view to creating them, or know them all after he has created them? In fact his wisdom is multiple in its simplicity, and multiform in its uniformity. It comprehends all incomprehensible things with such incomprehensible comprehension that if he wished to create new things of every possible kind, each of them unlike its predecessor, none of them could be for him undesigned and unforeseen, nor would it be that he foresaw each just before it came into being; God's wisdom would contain each and all of them in his eternal prescience.

(I have used the English translation by Henry Bettenson, *City of God*, Penguin, Harmondsworth, 1972, pp. 496-7.) Cantor quotes the whole of Chapter 19 in Latin in his [1887-8], pp. 401-2, and was clearly deeply impressed by this passage, and in particular I would suggest by the claim that 'every infinite is made finite to God'. (Recall that Cantor called the Absolute 'the true infinite', thus implying that the transfinite is a quasi-finite.) While in his published writings he stresses only that the natural number sequence should be 'considered' or 'thought of' as a 'thing for itself', in his correspondence he is quite willing to bring in God. For example, in a letter to Pater Ignatius Jeiler, Whitsun 1888, we find:

Each individual *finite* cardinal number is in God's intellect both a representative idea and a unified form for the knowledge of innumerably many compound things, that is, those which possess the cardinal number in question. All *finite* cardinal numbers are thus distinct and simultaneously present in God's intellect. They form in their *totality a manifold, unified thing for itself* [*Ding für sich*], *delimited* from the remaining content of God's intellect, and this thing is itself again an object [*Gegenstand*] of God's knowledge. (Cantor *Nachlass VI*, p. 170)

(Cantor then goes on to stress that since this infinite 'form' is present with God, so must a cardinal number also be present. This is discussed further in §2.1.) The similarity with the passage from *De Civitate Dei* is obvious and remarkable. Of course, it re-emphasizes Cantor's theological realism, specifically that the sequence of natural numbers is complete with God. More importantly Augustine's idea that in some way infinity is 'made finite to God' is here expressed in the very strong form (at least with respect to the natural numbers) that this collection forms a *single object* for God. This means, of course, just that the natural numbers exist as a set, and that it is God, not us, who has conceived the collection with the necessary unity. Since for Cantor the natural numbers form no privileged transfinite collection, the same must hold for any such collection. (This is expressed in a later letter to Jeiler from 13 October 1895 which I quoted in §1.1. Here Cantor refers to each 'individual transfinite and generally every thing which corresponds to a divine idea'. See Meschkowski [1967], pp. 258-9, where the letter is published.)

Thus finally, it seems, sethood is a property of a collection which God is able to perceive. Our ability is presumably now irrelevant, since God can undoubtedly encompass all that we can and much more besides. Indeed, it seems that God is brought in essentially to bridge a gap (between a collection and its unity as a set) that we ourselves cannot bridge. It should be noted that Cantor's clear adoption of the Augustinian idea that 'infinity is made finite to God' is another reason why I call Cantor's unity principle 'finitism'.

Some comments are in order here. First we see that as with principle (*a*) (the completed domain principle), Cantor's finitism ultimately has a theological origin, or at least is given a theological justification. But, again while this helps us to understand what position Cantor took, the involvement of God is of no philosophical help. Certainly the notion of the unity or oneness of a collection remains a mystery. Moreover, with the exception of the natural numbers, we obtain no idea at all from this which among all the possible collections form sets. Indeed, it may be that invoking the Almighty poses a positive danger. For cannot the Almighty conceive of the collection of all things (that is, the universal collection, or something corresponding to Cantor's Absolute) as a unity, thus as a set? This, of course, would appear to go strongly against Cantor's conception of the Absolute.

The unity doctrine, then, leaves us with the following: mathematics is concerned with a unified realm of finite and transfinite forms, basically the 'finite' sets. (One should add numbers here as well, though according to Cantor these ultimately reduce to sets.) But while there is here a strong reductionist tendency (sets are at the very least indispensable according to Cantor's conception) the nature of sethood is left unexplained. Indeed it seems that it is not just unexplained but inexplicable. If one wants to explain the notion of set, or rather do away with the mystery of the 'oneness' of sets, then one must bring in some intensional element, as is done say in intuitionistic set theory, or by Russell originally (before he introduced the axiom of reducibility). Indeed, one of the main reasons why Russell built his system on propositional functions and not classes was that he could not understand how an extensional class could be both 'many' and 'one'. As he remarked in the *Principia*:

It is an old dispute whether formal logic should concern itself mainly with intensions or extensions. In general, logicians whose training was mainly philosophical have decided for intensions, while those whose training was mainly mathematical have decided for extensions. The facts seem to be that, while mathematical logic requires extensions, philosophical logic refuses to supply anything except intensions.

In a footnote he goes on:

If there is such an object as a class, it must be in some sense *one* object. Yet it is only of classes that *many* can be predicated. Hence if we admit classes as objects, we must suppose that the same object can be both one and many which is absurd. (Whitehead and Russell [1910–12], Vol. 1, p. 72)

(For an account of the evolution of Russell's thinking here, see Hallett [1984].) Modern set theory, however, takes a stand at exactly the point Cantor's theory had reached: the concept of set is taken as primitive and is left unexplained. Each set a allowed by the axioms is a collection $\{x : \phi(x)\}$ of objects satisfying some predicate ϕ, but the axiom of extensionality allows us to forget the particular predicate, while at the same time the theory then proceeds to treat a as a single object which can be meaningfully taken as a member of any other set and so on. Looking with hindsight Cantorian set theory was an ideal candidate for being axiomatized even *without* the discovery of contradictions based as it is on a notion which *cannot* or *will not* be further explained. Thus von Neumann, referring to axiomatizations of set theory, remarks:

Here (in the spirit of the axiomatic method) one understands by "set" nothing but an object of which one knows no more and wants to know no more than what follows about it from the postulates. (Von Neumann [1925], p. 395)

There is another sense in which Cantor's theory was a prime candidate for axiomatization. While the theory indicates that the notion of set is fundamental, Cantor never explicitly says which collections are sets. (Though the qualifying phrase 'without contradiction' in the 1899 statement together with his ontological doctrines of [1883b] suggest that Cantor's roundabout answer would be 'as many as consistently possible'.) I have noted that the appeal to God does not help us at all here, and neither do Cantor's 'explanations' or 'definitions' of set. Thus there is some need for specifications, 'axioms'. It is sometimes assumed that Cantor's 'definitions', particularly that from [1895], allow virtually any collection to be a set, and therefore that Cantor's system clearly gives rise to the famous paradoxes, say those of Burali-Forti or Russell. This was strongly hinted at by Zermelo in his [1908b], p. 200, and is found quite clearly in Fraenkel, Bar-Hillel, and Levy [1973]. On p. 15 these authors say that the paradoxes refute Cantor's 'naive concept of set' as it appears in his [1895] statement. Then on pp. 30–1 they say that given the framework of a first-order logical calculus, Cantor's 'definition' translates into the so-called 'comprehension principle', that is the axiom schema according to which each predicate yields an axiom $\exists x \forall y [y \in x \leftrightarrow \phi(x)]$. But this view is quite mistaken. Cantor's 'definitions' only allow as sets those collections which are *wholes* and this does not at all imply that any collection can be a set. Nothing like the comprehension principle of so-called 'naive set theory' follows from Cantor's statements. If 'naive set theory' is characterized as set theory based on the comprehension principle, then this goes back, not to Cantor, but to Russell [1903]. Indeed, Cantor's doctrine of the Absolute explicitly denies that every collection can be a set. And one can argue at least that the universal collection and the collection of all ordinals cannot be Cantorian sets, and this even before there was any suggestion that they are contradictory. Cantor *is* quite vague about what collections should be sets, but when one takes his remarks about the transfinite seriously there is good reason to see in them something more like the iterative universe of the modern axiomatic theory than the indiscriminate 'blanket' universe of naive set theory. I take up this topic again in §1.4. For the moment, we return briefly to Cantor's finitism.

What we have discussed so far is the thesis that the finite and the transfinite form a unified realm where there is no essential difference between finite sets and transfinite sets: both are single objects, which can be members of other collections, and there is in principle epistemological parity in respect to both realms. That is, according to Cantor it is in principle no more difficult for us to acquire knowledge of the transfinite than it is to acquire knowledge of the finite. (See the passage quoted on p. 14, above.) But for Cantor this unity or parity is matched by a structural unity or parity, for he firmly believed that there should be no essential mathematical differences between the objects of the finite and transfinite realms. This does not tell us very much until we know which properties are 'essential'. But there is, of course, one clear candidate that goes back to the prehistory of mathematics—namely, the property of being numerable. The original purpose of the finite numbers was to represent properties (size, or perhaps sometimes the arrangement) of collections or groups of objects. Cantor's insistence on unity implies that transfinite collections must also be subject to numerical determination, something which sharply distinguishes Cantor from Bolzano (see Bolzano [1851], p. 107). Indeed, he insisted that the numerical analysis of transfinite forms constitutes the main body of pure *Mengenlehre*, and the development of a theory of transfinite number is therefore its main task. (This fits perfectly the aim of solving the main specific mathematical problem of set theory, the continuum problem.) Thus in his [1883*b*] he writes:

The assumption that, besides the Absolute, which is unreachable by any determination, and the finite, there are no modifications which I call actually-infinite, that is to say which are determinable through numbers—this assumption I find to be quite unjustified . . . What I assert and believe to have demonstrated in this and earlier works is that following the finite there is a *transfinite* (which one could also call the *supra-finite*), that is an unbounded ascending ladder of definite modes, which by their nature are not finite but infinite, but which just like the finite can be determined by definite well-defined and distinguishable *numbers*. (Cantor [1883*b*], p. 176)

So far what I have described is only a declaration of principle about the mathematical accessibility of the transfinite. But Cantor also took the heuristic power of the unity principle seriously.

First, we should recall that the idea that transfinite sets (or manifolds) must all possess a size (a cardinal size) was adopted long before any extensive theory of infinite size was worked out, and thus before one had any right to accept that it would be possible to develop such a theory. Indeed, concern with infinite size goes back, properly speaking, to [1878]. Now given that one has the development of such a theory in view, the finitism principle puts strong heuristic constraints on the shape the theory can take. Because of the unity with the finite, transfinite sets should be quantitatively measured and compared by numerical means which are clearly seen to be natural generalizations of the finite numbers. This means that the infinite sizes should be *numbers*, that is to say, if possible they should form a numerical scale. And Cantorian finitism was above all important here, for Cantor assumed that transfinite sets should be countable, enumerable, just as finite sets are *and* that in these cases too this counting should give rise to a cardinal size. This was the most important heuristic application of the finitism principle, that is, of assuming that infinite sets will have properties which we know already hold of finite sets. It led Cantor to postulate the existence of a counting (ordinal) number for every set, thus to the well-ordering assumption, and to his ordinal theory of cardinality and a solution to the scale problem. And this ordinal theory is, in effect, the theory of cardinal number which modern set theory adopted. The development of this theory is described in §2.2. It is hard to overestimate the importance of this theory with respect to subsequent developments. Indeed, one can argue that the desire to shape the axiomatic theory of sets in such a way as to allow the reconstruction of this ordinal theory of cardinality gave the modern axiomatic system its strength and form. (Here I am thinking not just of the axiom of choice and the power set axiom, but also of the so-called axiom of replacement which is necessary for the provision of enough ordinals to 'count' every set. See the chapters on Zermelo and von Neumann in Part 2.)

But the unity principle was also important with respect to ordinal numbers. For since finite and infinite sets are the same kind of thing, one must find a uniform account of their ordinal numbers. This Cantor sought and found by relating ordinals to the general theory of well-ordered sets. Using this he not only attempted to explain in general what ordinal numbers are, but could develop a generalized and unified ordinal arithmetic too. Only later, as it were, do we differentiate between the finite and the infinite ordinal numbers. Unification was important here, too, in that it provided Cantor with an argument for the coherence of the concept 'transfinite ordinal number', which (referring back to §1.1) formed the basis of his argument for their existence. This topic will be taken up again in §2.1.

1.4 Cantor's Absolute

So far I have said very little about the transfinite, except that the dominant forms here are object-collections (sets), and nothing at all about the relationship between the transfinite and the Absolute (categories (ii) (*a*) and (ii) (*b*) in Cantor's categorization). This is partly because Cantor says little about which collections are sets. But nevertheless he does have more to say than I have so far indicated, and it is possible to obtain a reasonably clear picture of how the transfinite and the Absolute fit together in Cantor's conception. Of course, he himself did not turn this into a watertight theory of sets. But it might neverthe-less be regarded as a programmatic picture, and one which foreshadows the structure of modern axiomatic set theory.

What then is the structure of the transfinite? What Cantor stresses when introducing his categorization of infinity is that the transfinite, as opposed to the Absolute infinite, is *increasable*. Thus having 'defined' an actual-infinite to be something 'fixed and determined in all its parts', not variable, he states:

If we look more closely at [this definition] then we see that we certainly *cannot* conclude that an actual infinite is *unincreasable* in magnitude. This false assump-tion one finds not only in the old philosophy, following on from the *Scholastics*, but also in *recent* and *most recent* philosophy. Indeed one can almost say that it is generally widespread. Rather we must make a *fundamental* distinction here between:

IIa Increasable actual-infinite or *transfinite*
IIb Unincreasable actual-infinite or *Absolute*
(Cantor [1887–8], p. 405; the passage stems from 1886)

And here we find another aspect of Cantorian finitism, since he also clearly states:

. . . I find in various places in Gutberlet's work [1878] (p. 45 for example) the completely untenable thesis that 'the concept of infinite magnitude excludes all possibility of increase'. This can be admitted only in the case of the Absolute-infinite: the transfinite, although conceived as definite and greater than every finite, shares with the finite the character of unbounded increasability. (Cantor [1887–8], p. 394)

And again in a letter to Eulenberg of 28 February 1886:

. . . the transfinite has two things in common with the finite, for example with the number 3: (1) a fixedness and definiteness in relation to all other numbers; (2) increasability in just that sense that 3 can be increased to larger numbers through addition of new unities. (Cantor *Nachlass VI*, pp. 47–8)[6]

This 'sharing' of increasability is tantamount, of course, to the sharing of numerical determination. I have already pointed out that for Cantor the trans-finite (together with the finite) is the realm of the 'rationally subjugable', the mathematically determinable, and this means above all subject to numerical determination. And when Cantor mentions magnitude here he certainly includes cardinality or cardinal size. For example, in continuation of the first passage quoted above we find:

Thus, the smallest transfinite ordinal number, which I denote ω, belongs [to the transfinite], for it can be increased, enlarged to the next greater ordinal number $\omega + 1$, this again to $\omega + 2$, and so on. But *the smallest actually infinite power or cardinal number* is also a transfinite, and the same holds of the next greater cardinal number, and so on. ([1887–8], p. 405; the italics are Cantor's)

[6] This letter is published with some alterations and deletions in Cantor [1887–8], though the passage quoted above is not included.

But it is not just, as Cantor rather weakly says here, that the infinite numbers belong to the transfinite. They are its most important elements since in the end numerical determination is a criterion of transfinitehood, and therefore of sethood, and moreover they give it its overall structure. The numbers in effect map out or mark the transfinite realm. This idea goes back to Cantor's earliest discussion of the transfinite, namely his [1883b]. To understand completely what Cantor says there, one must first be aware of his classification of the ordinal numbers into the increasing number-classes (I), (II), etc., and his claim that these same classes designate or form *the* scale of cardinal numbers. This theory is discussed in detail in §2.2. Here it is important only to know that the ordinal numbers and the number-classes form two equivalent numerical scales (Cantor assumes that to every ordinal number α there must be a corresponding number-class and therefore power (α)), *and* that the transfinite is marked out or represented by either or both of these two scales. This gives not only a picture of the shape of the transfinite, but also of how the transfinite fits together with the Absolute, as the following long passage from [1883b] shows:

I fix conceptually once and for all the various levels of the proper [i.e. actual] infinite through the number-classes (I), (II), (III), etc., and now consider the only problem to be to investigate the relations of these supra-finite numbers, not just mathematically but also quite generally in tracking down and demonstrating where they appear in nature. I have no doubt at all that in this way we extend ever further, never reaching an insuperable barrier, but also never reaching any even approximate comprehension of the Absolute. The Absolute can only be recognized, never known, not even approximately. For just as inside the first number-class (I) given any finite number, no matter how large, the power of the finite numbers following it is *always* the same, so following each supra-finite number of any of the higher number-classes (II) or (III), and so on, there is a totality of numbers and number-classes which has lost nothing as regards power with respect to the whole of the absolutely infinite totality of numbers beginning with 1. This is quite similar to what Albrecht von Haller says of eternity: "I subtract it (the enormous number) and Thou (eternity) liest intact before me". The absolutely infinite sequence of numbers therefore seems to me in a certain sense a suitable symbol of the Absolute. Whereas, hitherto, the infinity of the first number-class (I) alone has served as such a symbol, for me, precisely because I regarded that infinity as a tangible or comprehensible idea, it appeared as an utterly vanishing nothing in comparison with the absolutely infinite sequence of numbers. It seems to me remarkable that each of the number-classes, and therefore each of the powers, is co-ordinated to a quite definite number of the absolutely-infinite totality of numbers in such a way that to every supra-infinite number γ there is a power which is called the $\gamma^{th.}$ Thus, the different powers form an absolutely-infinite sequence. This is all the more remarkable since the number γ which specifies the order of the power (in the case where the number γ has an immediate predecessor) is ludicrously minute in comparison to the numbers of the number-class which has the power (γ), and this the more so the larger the specified γ. (Cantor [1883b], p. 205, n. 2)

Of course, at this early stage not all of the details of the number-class sequence were worked out. This came only much later. But the programme, the vision of the transfinite, is clear—the transfinite is whatever is representable in these number-scales. The number scales thus form a kind of backbone to the transfinite. But with this comes the first idea of the mathematical status of the Absolute. For while the levels of the transfinite are marked by the number sequences the Absolute itself is necessarily excluded from representation *in* them. Cantor takes the Absolute as symbolized by the whole of either sequence, and this makes it impossible that the Absolute be marked in them. This follows for example from Cantor's remark that if we subtract any 'amount' from the whole sequence we are still left with the same 'amount'. Since the whole sequence represents the Absolute, this would not be true if it was also represented by one level in it. This makes it partly clear why Cantor later calls the Absolute an unincreasable infinity. For if 'increasable' means numerical (cardinal) increasability, and being numerable means being representable in the transfinite number sequence, then the Absolute must be both 'beyond' mathematical numeration *and* unincreasable. What is also interesting about this is that *if* the Absolute is in some way a collection then it is not a numerable collection, in contradistinction to the transfinite *sets*.

Let us grant that the Absolute is not counted in the scale of transfinite numbers. But why should numerability mean just numerability in the transfinite scale? Why does the Absolute not give rise to a further domain of mathematical activity, to super-transfinite numbers, Absolute numbers, or whatever? Why is it as Cantor says an Absolute maximum? One answer that Cantor would give is that to try to mathematize the Absolute would be simply a category mistake: everything mathematizable (or numerable) is already *in* the realm of the finite and transfinite, and the Absolute is simply that which embraces all these. There are no numbers *beyond* all transfinite numbers waiting to enumerate the Absolute. This is not to say that we may not discover *new* types of number, perhaps with surprising properties. For example, Hausdorff later discovered numbers ω_α such that $\alpha = \omega_\alpha$, and since then much larger ordinals have been defined or isolated. But if—to take one example—'the smallest uncountable measurable cardinal' is a genuine number (i.e. if this concept is self-consistent or coherent) then it is not a new Absolute number, but a normal increasable *transfinite* number. We have discovered it *within* the realm of the transfinite. The same would hold of *all* numbers we might define or hope to introduce.

We now have a clear sense in which the Absolute is beyond mathematical determination, and one which stems from the beginning of Cantor's detailed theory of the transfinite. But how does this tie up with Cantor's later Scholastic conception of the Absolute as being in some sense an expression or representation of the magnitude of God? The most natural way to interpret the Absolute as it is presented in [1883*b*] is not as something mathematizable itself but as the category of everything mathematizable. The reason is quite clear: each transfinite number represents all transfinite forms which have that number, and thus either sequence of numbers represents simultaneously *all sets* and (Cantor says)

the Absolute. Thus it is natural to assume that the collection of all sets represents the Absolute, and thus the Absolute is itself an expression for the realm of
all sets, or the realm of everything mathematizable. To put it in modern terms,
the Absolute acts as a kind of *universe* for set theory, though there is no clear
evidence that Cantor conceived of the Absolute as itself a collection. The nearest
Cantor comes to saying this clearly is with his statement:

The *transfinite* with its plenitude of formations and forms necessarily indicates
an Absolute, a 'true infinite' whose magnitude is capable of no increase or
diminution, and is therefore to be looked upon quantitatively as an absolute
maximum. (Cantor [1887–8], p. 405)

This looks somewhat like a final application of the domain principle (*a*): the
variable transfinite must 'point to' or rely on a completed domain of all transfinites which Cantor is here calling the Absolute. If this reading is correct there is
now a clear connection with the theological conception of the Absolute once
one recalls Cantor's ultimate justification for the existence of transfinite (or any)
mathematical forms, namely that they exist as ideas in God's intellect. For if the
Absolute is taken to represent the category of all mathematical forms, it represents then all possible creatable 'multitudes', to go back to Aquinas (p. 22), or
as Cantor says 'the vast domain of the possible in God's knowledge' (p. 23).
Thus, the Cantorian Absolute must represent a substantial part of God's understanding, and God's capability, just as the sequence of natural numbers did previously in Aquinas's conception ('everything creatable must be comprehended
under a certain number'). Thus it is not too surprising that Cantor took over the
Scholastic conception of the Absolute, nor that as Kowalewski described it:

. . . these powers, the Cantorian alephs, were for Cantor something holy, in a
certain sense the steps which led up to the throne of the infinite, to the throne
of God. (Kowalewski [1950], p. 201)

This association of the Absolute with God makes it finally clear why the
Absolute cannot be subjected to any attempt at rational (or in particular, mathematical) understanding. This is a permanent and ineradicable imperfection or
gap in our understanding. We may have a feeling or an inkling about Absoluteness or God, but these feelings can never be explicated in any clear intellectual
sense. This is what Cantor means when he talks of 'the true Absolute, which is
God, and which permits no determination' ([1883b], p. 175). This is just what
principle (*c*) captures (Introduction, p. 7), and why it must be a fundamental or
core principle of Cantorian theory.

[There is one difficulty here, already rasied on p. 37, and connected with
Cantor's idea that sets have a unity because God can conceive of them as
being 'united together'. But then why cannot God conceive of all sets or all
ordinal numbers as 'united together' in one set? In this case, these collections,
which for Cantor represent the Absolute, would be sets, and thus subject to
mathematization. Therefore, seemingly, the Absolute should be mathematically

examinable. Cantor has no easy answer to this. Indeed, the only answer would be to attempt to divorce his ontological assumptions from reliance on the Almighty, and so simply insist (as principle (c) baldly stated does) that there is a rigid type distinction between the transfinite and the Absolute, that the latter is nothing more than the category of the mathematizable and is not mathematizable itself.]

The conception of how the Absolute fits in with the domain of the transfinite is not only interesting in itself, but it later had important mathematical consequences when Cantor (in the 1899 letters to Dedekind) transformed the notion of Absoluteness into a theory of absolute collections and of limitation of size. This subject I take up fully in Chapter 4. But it is already clear from what we have seen so far that since the number sequences represent the Absolute they themselves must be outside mathematical determination, i.e. they can themselves have no number. This already represents a considerable prior protection against, for example, the Burali-Forti paradox.

Let us come back finally to Cantor's conception of set, and how this fits in with the picture given above of the transfinite (which is predominantly made up of sets). Since the Cantorian transfinite is marked by the two (ordinal and cardinal) sequences of numbers the Cantorian universe of sets is thus in some sense divided into layers or levels. There is no indication that Cantor conceived this layering as anything like that of the cumulative hierarchy of axiomatic set theory. Rather the Cantorian levels mark only size accretion, and do not at all indicate the way in which sets are generated. Nevertheless, I have indicated that the Cantorian principles (a) and (b) working together, as in the Cantorian definiton of real number, point to a kind of iterative conception of set. This iterative conception can perhaps be extracted from a 'definition' of set which Cantor gave in his [1882]. Let me quote Cantor first and then explain after why I think there is an iterative element hidden here:

Also the concept of *power*, which embraces the concept of whole number (the groundwork of the theory of magnitudes) as a special case, and which should be looked at as the most general genuine moment with manifolds, is not at all restricted to the linear point sets. Rather it can be considered an attribute of any well-defined manifold, whatever conceptual properties its elements may have.

I call a manifold (a totality, a set) of elements which belong to some conceptual sphere well-defined, if on the basis of its definition and as a consequence of the logical principle of excluded middle it must be seen as *internally determined both* whether some object belonging to the same conceptual sphere belongs to the imagined manifold as an object or not, *as well as* whether two objects belonging to the set are equal to one another or not, despite formal differences in the way they are given. (Cantor [1882], p. 150)

Clearly a conceptual sphere is the extension of a concept, and so conceptual spheres are apparently quite general. Nevertheless, it seems from the way Cantor uses this notion that he has in mind a restricted notion of conceptual sphere. For example, two pages later he writes:

The theory of manifolds, according to the interpretation given to it here, includes the domains of arithmetic, function theory, and geometry, if we leave aside for the time being other conceptual spheres and consider only the mathematical. (Cantor [1882], p. 152)

Cantor seems to suggest with this that in mathematics at least we are concerned with homogeneous domains of definition of a rather limited kind, the natural numbers, the real numbers, points in 3-space, or whatever. According to Cantor's 'definition' of sets, these domains (or 'spheres') give rise to sets separated off from the whole sphere. But more interestingly all the domains which one normally uses in mathematics are naturally restricted, or bounded; indeed each is intuitively embeddable in a higher domain, the natural numbers in the rational numbers, the rational numbers in the real numbers, the real line in the real plane, and so on. Moreover, according to Cantor's view, these are all also *sets*, and thus for him cardinally bounded. That is, not only can they act as domains of definition, but they also appear themselves later as sets. It is not exactly clear from Cantor's writings what set-theoretic mechanisms we are allowed to use in proceeding from one domain to another, for example from the natural numbers to the real numbers. (One forms the domain of rational numbers from the natural numbers, then the domain of all sequences of rational numbers, and then separates the Cauchy sequences from these. But Cantor does not make clear *set-theoretically* how the rational numbers arise from the natural numbers, or how the collection of rational sequences arises.) None the less, it is clear that he allows himself procedures which increase not only the extent of the conceptual spheres, but also their *power*.

Of course, there is nothing in these rather vague statements themselves to prevent the whole universe from being taken as a conceptual sphere and thus from being itself a set, along with any other collection. And certainly, some nineteenth-century logicians and mathematicians operated either with arbitrary extensions (Boole, Frege) or with the whole universe as a conceptual sphere. Thus Dedekind says:

. . . a system S (an aggregate, a manifold, a totality) . . . is completely determined when with respect to every thing it is determined whether it is an element of S or not. (Dedekind [1888], p. 45)

Nevertheless, both possibilities are ruled out for Cantor. First, from the way he elucidates his 1882 'definition' of set, it seems that as far as mathematics is concerned only limited conceptual spheres are used. But secondly, and much more important, use of the universal sphere is ruled but by the doctrine of the Absolute. For, if the 'collection of all things' were a sphere it must also be a set (since we could 'separate' from it the set of all things which satisfy identity, thus yielding the whole collection as a set). But this universe must at least be a 'symbol' of the Absolute, and thus cannot be a set. Thus, for Cantor sets cannot be collections separated off from the universe by defining conditions.

Rather it looks much more as if they are built up in some more piecemeal way. (His critical comments on Frege's [1884] in his [1885a] are best interpreted in this light (§3.4).)

There is certainly some tension here. Despite the heuristic benefits of the doctrine of the Absolute, there has always been a problem deciding exactly how the Absolute fits into the mathematical framework. (Indeed the problem remains—witness the contemporary dispute about how category theory should mesh with set theory (see e.g. Bell [1981]).) Cantor's ontological doctrines (the coherence of concepts etc.), as well as the domain principle based on them, lead us to conclude that the collection of all abstract objects (or all finite and transfinite forms) does exist as a completed legitimate domain. Why *not* then operate with this as a conceptual sphere? The immediate Cantorian answer is theological, as we have seen: this domain, as a reflection of the Absolute, or of God, cannot be rationally subjugated by man, should not therefore be taken by us as *one* object, or ascribed a mathematically usable number. But in a way this just puts off the question, or rather brings back one we asked earlier: what then *are* the sets, and what is the difference between these and non-set domains? After all, if we cannot treat the domain of all things as a set, there are surely many collections contained in it which *are* perfectly good sets. What of these: are they simply lost? Cantor presents in embryo two possible ways of relieving some of this tension; both were elaborated by various of his successors and they were to some extent combined in the system of von Neumann, as Chapter 8 should make clear.

The first is that which I have just discussed, the suggestion of an iterative element in the 1882 treatment of set. While the Absolute still forms part of a useful guiding doctrine, it remains *outside* the system, while within, an effort is made to ignore altogether anything which might be regarded as representing it, i.e. the universal domain and anything equally grand or extensive. Sets are regarded as built up in some way from some initial sets (e.g. the set of natural numbers) with enough strength in the defining mechanisms to guarantee increasing power, and with some kind of guarantee that all the sets which might be actually useful in mathematics will be generated. Certainly Cantor had a mechanism for increasing the power of any set he generated, in effect by relying on something like the power-set operation (strictly speaking the set of all functions from a set to $\{0, 1\}$ (see Cantor [1891] and §2.2)). According to this picture, then, the actual 'construction' of sets in practising mathematics fits passably well with the idea of a scale of increasing power which we find with Cantor's standard portrayal of the transfinite. Thus it seems that the scale reflects *something* of the complexity of generation of the sets. (It was not (still is not) clear *exactly* how this 'natural mathematical' formation of sets fits with the scale of number-classes (alephs). All we know (using the well-ordering assumption) is that the cardinalities of these power-sets fall somewhere in the Cantorian number-scale (see §2.2(*b*)).) The Absolute itself can never be approached by any of these processes of set generation, just as it cannot be approached by any of the cardinal or ordinal numbers in the two number sequences. Each number, each set, is bounded because we can proceed beyond it to a further strictly larger set, a strictly larger number. The 'Absolute maximum' is never approached.

The problem left here was to isolate and articulate an adequate collection of set formation principles, a problem which Zermelo later took up. While there is no clear expression of a fully iterative concept of set in Cantor, the combination of the way the set principle is introduced in [1882] and of the way the transfinite is conceived somewhat suggest the picture later associated with Zermelian axiomatic set theory—'limited' set formation in which none of the 'large' or Absolute collections are approached. This is discussed in more detail in Chapter 5 after the 'limitation of size' idea has been thoroughly examined.

The second approach is to admit the universal domain and any others which might be regarded as reflecting the Absolute as perfectly legitimate collections and, while not allowing them as sets, to use them to determine which collections *are* sets. In other words, instead of saying that the 'Absolute maximum' cannot be reached, one allows the 'Absolute maximum' in from the start and uses this as a kind of 'limit of size' to determine what is permissible below. One finds the beginnings of this approach in Cantor's late work of 1899 in the Dedekind letters which I discuss in Chapter 4. Unfortunately, though, it does not seem to work on its own, that is, without some further specification from *below* of what sets are (see §4.3 and Chapter 5). To *some* extent this second approach demands the demystification of the original Cantorian doctrine of the Absolute, for it demands some kind of mathematical operation (though perhaps quite limited) with collections which reflect the Absolute. Cantor himself certainly made this transition in his 1899 letters but, and this is why the doctrine of the Absolute remained important, in a way directly inspired or guided by the original doctrine. As regards the tension I pointed to above, one might say that this solution adapts somewhat more easily to the pressure from Cantor's ontological doctrines, while the 'iterative' solution sits rather better with the theological aspect of the notion of the Absolute.

Some of these twists and turns in the story of the mathematical assimilation of Cantor's Absolute are dealt with below. But now I want to examine how Cantor's theory of infinity was applied in his conception of the most important elements in the transfinite, the ordinal and cardinal numbers. Here we shall begin to see the strength of Cantor's principles (*a*) and (*b*).

2

THE ORDINAL THEORY OF POWERS

In Cantor's later work there is an attempt to *define* the ordinal and cardinal numbers as sets. Their existence as such is assumed to be demonstrated by *abstracting* from already given sets. Unravelling Cantor's views on abstraction is a difficult matter, and one which I leave for Chapter 3. But before that, I want to discuss Cantor's earlier treatment of number ([1883b]). Here, there was no explicit attempt at reduction to sets. But the earlier treatment is of great importance. For one thing, it develops Cantor's mathematical theory of cardinal number in a way which had a profound effect on the subsequent development of set theory. For another, an understanding of the 1883 treatment itself makes Cantor's later (rather obscure) philosophical explanation easier to understand. For, although there is no explicit reduction, numbers are very closely *tied* to sets since they are taken to represent them, and the way the representation is effected, particularly in the case of powers, foreshadows Cantor's explanation of abstraction. Furthermore, the 1883 treatment of number illustrates just how the heuristic principles (*a*) and (*b*) we have discussed were put to mathematical work, and how Cantor's views on legitimacy of concepts and existence of objects are exploited.

2.1 The generating principles

Cantor's introduction of transfinite ordinal numbers in his [1883b] appears to centre on the two so-called generating principles which state:

(1) if α is an ordinal number (whether finite or transfinite) then there is a new ordinal number $\alpha + 1$ which is the immediate successor of α;
(2) given any unending sequence of increasing ordinal numbers there is a new ordinal number following them all as their 'limit' (that is to say, no ordinal number smaller than this limit can be strictly greater than all ordinals in the given sequence).

(See Cantor [1883b], pp. 166–7 and 195–7, as well as Cantor's letter to Dedekind of 5 November 1882; Noether and Cavaillès [1937], p. 55.) As they stand these are pure existence principles which simply postulate new numbers stretching the natural number sequence into the realm of the infinite. It seems from this presentation that the new objects exist quite independently of the realm of sets.

There is, to say the least, a good deal of arbitrariness here, certainly if we take these principles at face value with Cantor's assertion ([1883b], p. 166) that they are part of the *definition* of the transfinite numbers. Cantor could invoke his claim that mathematics is free: why should it not push on boldly and be judged later by the success of its theories? But we must recall that 'mathematical freedom' is based on some (albeit vague) notion of coherent (consistent) conceptual integration. And in any case, the major philosophical point of Cantor's [1883*b*] was to establish the 'symbols of infinity' as 'concrete numbers with real meaning' ([1883*b*], p. 166). The way to do this, according to the ontological principles outlined in [1883*b*], is to show that the concept 'transfinite ordinal number' is coherent or legitimate. This Cantor attempts by closely tying the concept of number to sets. So the impression of separation of numbers from sets which we obtain from the generating principles is misleading. Rather, the generating principles, introduced *after* the legitimization of the ordinals, are best seen as a summary statement of which ordinals exist (a first, a successor, a successor to this, one after all these successors, and so on). One should note that even as existence statements they rely on *set* existence principles. One gets the scent of this already in generating principle (1). For this is an extension of the successor operation of ordinary arithmetic to arbitrary numbers, and Cantor characterizes this as the 'setting down of a unity', that is to say the addition of a new element to a set. (See [1883*b*], p. 195, or the letter to Dedekind of 5 November 1882 referred to above. For Cantor's theory of numbers as sets of 'unities', see Chapter 3.) This is already a nod in the direction of reductionism, since it ties the existence of successor ordinals to set existence. But the true force of the existence assumptions is really only revealed in what lies behind generating principle (2). I shall come back to this later. Let us first look at legitimacy.

Cantor's argument for legitimacy is somewhat garbled in his [1883*b*]. But its basic structure was outlined with radiant clarity later in a letter to Kronecker of 24 August 1884. Cantor writes:

My conviction that these concepts [the transfinite ordinals] are to be interpreted as numbers is based on the concrete determinacy of their relations with one another, and because they can be subsumed under the same viewpoint as the ordinary whole numbers. I have already for some time had a foundation for these numbers which is somewhat different from that given in my written works, and this will certainly suit you better.

I start from the concept of a 'well-ordered set', and I call well-ordered sets of the same type [*Typus*] (or same enumeral [*Anzahl*]), those which can be related to one another one to one and uniquely in such a way that the *sequence of elements is reciprocally preserved*.

I understand by number the symbol or concept for a definite type of well-ordered set. If one restricts oneself to *finite* sets, one obtains in this way the finite whole numbers. If one goes beyond this, however, and looks at all types of well-ordered sets of the *first* power, then one comes of necessity to the transfinite numbers of the second power.

(Quoted from Meschkowski [1967], p. 240; the whole letter appears there on pp. 240–1. For transfinite numbers of the first and higher powers see §2.2(*a*).)

There are similar remarks on the transition from mere 'symbols of infinity' to numbers in a letter to Mittag-Leffler of 20 October 1884. Concerning some review of his work Cantor writes:

Among other things the reviewer commits a gross error if he believes that those things which I earlier called 'symbols of infinity' and now call transfinite or super-finite numbers can only be used as they historically first appeared and were originally discovered by me, namely with the various derivatives

$$P^{(1)}, \ldots, P^{(\nu)}, \ldots, P^{(\omega)}, \ldots, P^{(\omega^\omega)}, \ldots, P^{(\omega^{\omega^\omega})}, \ldots$$

of an arbitrary point-set P. They also appear on many other occasions, as I will soon show, and *must be defined quite independently of all applications*. This I do in my *transfinite number theory* (which I will soon send you for the *Acta*) with a definition based on the general concept of well-ordered set. Here I will show, in the clearest possible way, that we are here concerned with *real* [*wirklichen*], *concrete numbers* in the same sense as $1, 2, 3, 4, 5, \ldots$ have been designated and looked upon as *numbers* from ancient times. If my memory serves me well, I sent you this foundation of *transfinite* numbers more precisely in a letter some years ago. (Cantor *Nachlass VI*, p. 3)

(Although dated 20 October, Cantor notes 'sent on the 28 October 1884'. Mittag-Leffler was editor of *Acta Mathematica* and the work Cantor says he is 'sending in' to *Acta* is very likely the unpublished paper of 1884 (see Grattan-Guinness [1970]).) From the dates of these letters and from Cantor's remarks that they transcend hitherto published work it is fairly clear that he is pointing partly to his abstractionist account of number which makes its first appearance in the unpublished paper of 1884 (though there is no attempt to *explain* abstraction there; see Chapter 3). But nevertheless the principle of the argument is already found in [1883*b*]. Abstraction was added and refined later to try to fill a gap which the [1883*b*] account leaves open.

The basis of the argument can be explained as follows. First, the concept 'transfinite ordinal' is integrated with the concept of natural number. This is done by taking them both under one *general* notion of 'ordinal number'. This in its turn is explained as 'enumeral of a well-ordered set'. Thus legitimacy depends on two new concepts, that of *well-ordered set* and that of an *enumeral* [*Anzahl*] of a well-ordered set.[1] The argument then rests on the coherence of these. This much at least is clear in [1883*b*].

Cantor's [1883*b*] characterization of well-ordering is not quite the same as the now standard one. For Cantor M is well-ordered by a relation $<$ if:

---

[1] In Frege [1884], to take one example, there is no distinction made between *Anzahl* and *Zahl*; both mean simply 'number'. (See n. * to p. 2 of the English translation by J. L. Austin.) Indeed normally *Anzahl* means number more in the sense of quantity, i.e. cardinal number. But Cantor's use of *Anzahl* is clearly quite different; indeed it is important for his argument that '*Anzahl* of a well-ordered set' *should* point to something conceptually new. Thus to point up this difference I have used 'enumeral' to translate *Anzahl*. Cantor's new usage of *Anzahl* may have been one reason why Frege mistook Cantor's transfinite ordinal numbers for cardinals and why he thus criticized him accordingly (Frege [1884], pp. 97–9).

(i) M is linearly ordered by $<$;

(ii) M has a $<$-first element, m_0;

(iii) whenever $N \subseteq M$ and $\exists m \in M - N \, \forall n \in N \, [n < m]$ then there is a $<$-smallest $m \in M - N$ such that $\forall n \in N \, [n < m]$.

(See [1883b], p. 168. This definition (as one might expect!) is equivalent to the standard definition, a definition first given, as far as I know, by Schönflies in his [1899], p. 36. Schoenflies also shows the equivalence of the two. For let $\emptyset \neq N \subseteq M$. Then if $m_0 \in N$, m_0 is the least element of N. If $m_0 \notin N$, consider $N' = \{m \in M : \forall n \in N, m < n\}$. Then by (iii) $M - N'$ must have a first element and this will be the least element of N. Thus, either way, every non-empty subset of M has a least element; Cantor's well-ordered sets are therefore well-ordered. The converse is immediate. This equivalence was proved, in effect, by Cantor in his [1897], pp. 312-3, theorems A and B.) Through this definition it is now quite clear what well-ordered sets are, and that the notion is sound. For after all there are many finite examples, and, according to Cantor, quite natural infinite examples, for instance the sequence of natural numbers. (Though already with this we begin to see how much Cantor's procedure depends on set existence, and certainly the argument is dependent on the presumed legitimacy of the set concept itself.)

Thus, it seems, the argument must turn on the concept of *enumeral* of a well-ordered set. What of this? The key characteristic of an enumeral is that it should be a 'picture' or 'representational image' of a well-ordered set, thus closely tied to this set. This follows from what Cantor says on [1883b], pp. 168-9, where an enumeral is taken to be a canonical representative of a well-ordered set (or of a class of isomorphic well-ordered sets, [1883b], p. 168). It is not explicitly stated that these representatives must themselves actually be sets, though they were later construed as such by Cantor and this is already hinted at in [1883b], p. 195. Rather, it is only stated that they must stand in a unique well-ordered sequence and that enumeral e represents well-ordered set $(E, <)$ just in case the set of predecessors of e is isomorphic to $(E, <)$. Enumerals then simply measure the length or 'essential nature' of a well-ordering, or rather what the members of a class of isomorphic well-ordered sets have in common. But while Cantor is not explicitly reductionist in that he does not explicitly demand that enumerals *themselves* be sets, his notion of enumeral smacks strongly of reductionism. For what actually canonically 'pictures' a given well-ordered set is the set of predecessors of its enumeral. This is much like choosing a representative set from within the given equivalence class (see §2.2 and Chapter 3). (Incidentally, there is a little messiness with the idea that the predecessors of an enumeral e represent a given set. Since Cantor excluded 0 from the sequence of enumerals, the definition of enumeral representation in the finite case has to be modified to read: e represents $(M, <)$ just in case the set of enumerals from 1 up to and including e is isomoprhic to M.)

The crucial point in all this is that enumerals in Cantor's conception are tied closely to sets, and that ordinal numbers are to be explained as enumerals. Cantor clearly saw himself as returning to a traditional notion of number as based on collections. (This is hinted at in the passage from the letter to Mittag-Leffler recently quoted; see also the remarks on the Greek notion of number in Chapter 3.) According to this, numbering must be just the numbering of collections of objects, and since *sets* are the most important collections in mathematics, strictly speaking numbering should be numbering of sets. So no doubt Cantor saw the invocation of the set concept as a move towards a clarification of numbering and therefore of number. Certainly he saw it as a way to make conceptual legitimacy (of transfinite ordinals via transfinite enumerals) easier to establish. The generality of the notion of enumeral is at least a step in this direction. For if one already accepts the existence of the natural numbers (though Frege's [1884] makes it clear that the natural numbers did not rest on very secure conceptual foundations), then since they are enumerals of finite well-ordered sets, *and* since the concept of well-ordered set and enumeral is framed quite generally, surely transfinite enumerals are legitimate (exist) too. After all, Cantor would say, infinite well-ordered sets certainly exist. This would be, then, a kind of argument by analogy, analogy supported by Cantor's principle of 'finitism'. The concept of transfinite enumeral extends a perfectly coherent basic concept, and since there are finite enumerals (numbers) corresponding to finite sets, why not also transfinite enumerals (numbers) corresponding to the infinite well-ordered sets?

This is no doubt one way Cantor wants to argue for the existence of transfinite numbers. But I suspect that he wants more. I suspect that, relying on the existence of well-ordered sets, Cantor wants to argue directly for the legitimacy of enumerals in a way which does not seem to justify the concept of enumeral by appeal to already existing (natural) numbers. I base this view on a rather obscure and admittedly confusing passage from [1883b]. This passage is important for it helps to explain why Cantor later shifts to an abstractionist account of number. The passage is as follows: referring to the 'new concept of enumeral of the elements of an infinite well-ordered manifold, a concept which has not been known until now' Cantor says

... this concept is always expressed by a quite definite number of our extended number domain, provided only that the order of the elements is determined; ... and since on the other hand the concept of enumeral obtains an immediate objective representation in our inner intuition, then by this connection between enumeral and number [*Anzahl und Zahl*] the reality of the latter stressed by me is proved even in the case where it is definite infinite. (Cantor [1883b], p. 168.)

At first sight it is not at all clear what is going on here: enumerals are first explained as 'numbers of the extended domain'; but then these numbers are said to be real because they are enumerals. However, one can make sense of the passage as follows. We invent the concept of (ordinal) number and even prostulate that such numbers exist; but the concept obtains legitimacy and significance and we realize that the postulated objects actually *do* exist by recognizing that they correspond to enumerals of well-ordered sets. This is the explanation of the perhaps hazily understood concept of number in terms of, for Cantor, the clear concept of enumeral. And, Cantor goes on, the concept of enumeral is in its turn clear (legitimate) because it 'obtains an immediate objective representation in our inner intuition'. This remark is quite vague, and this vagueness constitutes the weak point of the chain of reasoning. Nevertheless, what I take Cantor to be saying is the following. Given a well-ordered set we are able to form (or have access to one preformed) an abstract or conceptual image or picture of the set, an image which represents it and presumably all sets like it. And such an image (or at least an ideal copy which the image points to) is the enumeral of the set. What this means is not fully clear, in particular the nature of the abstract representatives is not clear, nor is it clear how we form them, or how we can imagine them always preformed. But this interpretation of enumerals as 'pictures' of sets at least accords with Cantor's remark that numbers are 'an expression or image of the events and relationships of that outer world which is exterior to the intellect' ([1883*b*], p. 181; see p. 18). (By 'outer world' here Cantor presumably does not mean just the physical world, but rather also the ideal world, the world which sets must inhabit.) There is no special appeal to finite sets or finite numbers now, only to well-ordered sets generally. And crucially, if I interpret Cantor correctly (or plausibly), a certain conceptual priority has been laid down: numbers derive from sets.

This legitimacy argument so far presented is rather weak. Indeed it is best seen as a *suggestion* for an argument rather than a fully worked out cohesive case. What it *wants* to claim or would like to show is that given well-ordered sets (finite or transfinite) it must necessarily or immediately follow that enumerals for these sets exist, or (and for Cantor this amounts to the same) that the concept of enumeral must immediately make sense. Something of this expectation is present in Cantor's remark that the set of natural numbers has an enumeral, a 'new number'

. . . which we call ω and which shall be the expression for the idea that the whole totality (I) [of natural numbers] is given in this natural lawlike succession. (In the same way, ν is an expression for the fact that a certain finite number [*Anzahl*] of unities is united to a whole.) (Cantor [1883*b*], p. 195)

(For an explanation of the reference to 'unities united to a whole' see §§3.2 and 3.3.) Thus Cantor believes and asserts that the step from the well-ordered set to the enumeral is short and immediate. But he does not say why; we are left with the vagueness of 'obtaining immediate representation in our inner intuition'.

Nevertheless, we have here in [1883*b*] the framework of the approach Cantor outlined in the Kronecker letter. By this time (1884) he was trying to bolster the [1883*b*] argument at its weakest point by relying on the idea of abstraction to explain enumerals. This was (in part at least) an attempt to make clear what is not clear in [1883b], namely just what a representational image or picture of a well-ordered set is (an ideal copy of the set's structure) and how we obtain it. Indeed, what Cantor attempts is to show what is just asserted in the passage above, namely that given a well-ordered set, a canonical numerical representative for this set must also exist. In this later elaboration, Cantor also makes the connection to the realm of sets even tighter, for he claims that the enumeral or number is also a certain kind of well-ordered set whose existence is established by 'abstracting' from the nature of the elements of the given set. Cantor's abstrationist theory of number brings with it severe difficulties of its own which I examine in Chapter 3. But his appeal to it does at least suggest that he was building on the 1883 argument, was aware of its weak point, and wanted to strengthen it.

Despite the weakness of the 1883 argument as an attempt to show that the transfinite ordinals must exist, there is another purpose behind it which is to a large extent achieved. In §1.3 we saw how important the unity of the finite and transfinite realms is for Cantor. This is achieved for the transfinite and finite ordinal numbers by a kind of *set* theoretic reductionsim via just these notions of well-ordered set and enumeral that we have been considering. Of course, the reductionsim is not fully fledged, and as we have seen we do not even have a proof of the existence of enumerals (numbers) given the existence of sets. But the tendency is plain: everything depends on the existence of a base of underlying sets, and even the concept of enumeral is a set-related concept. What we see here might be called a kind of *weak* reductionism: the explanation and legitimization of the objects of one realm (numbers) in terms of the objects of another (sets). And this yields the unity for numbers that Cantor was seeking. According to this new presentation ordinal numbering is just enumeration of well-ordered sets, and ordinal numbers are just the resultant enumerals. But the explanations of well-ordering and enumeral are quite general, that is to say without any reference to whether an enumerated set is finite or infinite. One way Cantor expresses the conceptual unity achieved here is to say that, given the existence of infinite well-ordered sets, denial of enumeration in the transfinite realm is tantamount to denial of enumeration in the finite, that is to the unacceptable denial of the existence of finite ordinal numbers. Thus, Cantor insists '. . . if one of these two is dropped, the other must be done away with also; and what would become of us then?' ([1883b], p. 174.) (See also the various similar passages quoted in §3.1.) For Cantor finite and transfinite ordinals are thus indissolubly linked:

. . . the new whole numbers [the transfinite ordinals], even though they are distinguished by more intensive substantial determination than the customary numbers (the finite numbers), *nevertheless as enumerals have just the same kind of reality as these*. ([1883b], p. 178: my italics)

But the unity obtained is not just objectual. Treated as enumerals the finite and transfinite ordinals are easily shown to be united in a common arithmetic. This again is given in terms of well-ordered sets. Thus, for example, if α represents the well-ordered set $(A, <_1)$ and β the well-ordered set $(B, <_2)$, $\alpha + \beta$ is taken to be the unique ordinal representing $(A \cup B, <_3)$ where $<_3$ is the order generated by allowing the A elements to come first in their $<_1$ order and the B elements after in their $<_2$ order. (See [1883b], p. 170. Cantor also characterizes products here, but not exponents, though he does use them both.) Again the arithmetic is quite general; there are not separate realms of transfinite and finite arithmetic, but simply one realm of ordinals and *one* ordinal arithmetic. By this argument the theory of transfinite ordinal numbers therefore now completely coalesces with the theory of finite ordinals. According to Cantor, in the letter to Dedekind of 5 November 1882 (Noether and Cavaillès [1937], p. 57), it is the existence of this generalized arithmetic which finally gives us the right to call the transfinite ordinals *numbers*. They are no longer mere 'symbols of infinity', and Cantor has gone some way towards showing that they are as much 'concrete numbers with real meaning' as are the finite ordinals. Thus, although the notion of enumeral is not fully explained, at least the 1883 argument outlines a programme for the integration of the new numbers into the existing mathematical corpus, an integration on which Cantor laid so much stress. And with this, some of the arbitrariness which surrounds the generating principles is dispelled.

One should note that in this tying of the transfinite numbers to well-ordered sets there is perhaps a curious reversal of the process of discovery. For there would seem to be no doubt that Cantor *first* discovered the possibility of extending the number sequences along the lines of the generating principles (1) and (2), and then only *later* arrived at the notion of well-ordered set. This development route at least explains what now appears as a rather clumsy definition of well-ordering (p. 52). For clearly the clauses of the definition simply mirror the recursive way in which the sequence of ordinals is built up by (1) and (2). Looked at in this way, it would be automatic that any well-ordered set can be represented by an ordinal in Cantor's sense: the recursive 'generation' of the set is simply matched step by step by the recursive generation of the number sequence.

Cantor's early treatment of ordinal numbers if not fully reductionist is, as explained, a strong move in the direction of reductionism. Whether or not ordinals are construed as being themselves *sets*, what ordinals there are *must* depend on what sets there are. Thus, to show the existence of a number one must refer to some appropriate set. For example, the number ω is not, as the second generating principle would imply, simply 'created' as the first number after all natural numbers. Rather, as we saw in the passage from p. 195 of [1883b] on p. 54 above, ω exists because we can exhibit the well-ordered *set* of all natural numbers (though any other similar set would do). And to establish the existence of further ordinals Cantor has to appeal implicitly or explicitly to the existence of further well-ordered sets. For this reason, Cantor's assumptions about ordinal existence are a good guide to his assumptions about *set* existence. And in this respect the generating principles are of particular interest.

Once ω is available, one can use it to form $\{0, 1, 2, \ldots, n, \ldots, \omega\}$ and thus obtain $\omega + 1$ and so on. Strictly speaking, one does not need to use the 'new' ordinal ω here to obtain $\omega + 1$; one could instead use the ordering $1, 2, 3, \ldots, n, \ldots, 0$ of the natural numbers. However, as Cantor implicitly recognized in his letter to Dedekind of 5 November 1882 (Noether and Cavaillès [1937], p. 57), to break out of the so-called ordinal number-classes (say from the finite numbers to ω, as above, or from countable ordinals to ω_1) one does need to gather together *all* 'new' ordinals to a set before going on to 'create' (or, rather, to prove the existence of) an ordinal of higher power. For this purpose, rather strong principles of set existence are required.

The clearest way to explain these principles and their strength is by reformulating generating principle (2) as:

(2') given any unending sequence of increasing ordinal numbers there is a smallest actually infinite domain containing them all; moreover (providing it is not absolutely infinite and is therefore a set) there is a unique ordinal number representing this domain.

(2') is partially sustained by appeal to principle (*a*) but it goes much further as can already be seen with the proof of the existence of ω. The domain principle (*a*) as such says only that there must be some actually infinite domain containing all values of some varying quantity, for example some actually infinite domain containing all natural numbers. But we saw that in this case (and also with the rational and real numbers) Cantor interpreted it as saying that there is a domain which contains *just* the natural numbers, that is, which is the extension of the concept natural number. But this is in effect the smallest or minimal domain containing the natural numbers, and this is what yields the required new number ω, the first after all the natural numbers. (2') thus strengthens principle (*a*). To proceed beyond ω now, e.g. to $\omega + 1$ or $\omega + n$, one need only add finitely many elements to $\{0, 1, \ldots, \omega\}$ ordered after ω, or one could even just make some appropriate order rearrangement of $\{0, 1, \ldots, n, \ldots\}$. To obtain a limit for the $\omega + n$ one might argue similarly that simple permutations of order among the natural numbers will yield $\omega + \omega$. But in any case, one needs (2') again to obtain an ordinal beyond all those possible by rearranging the natural numbers. Mere rearrangements are not enough, as Cantor well knew. (2') *yields* a least domain A including all the finite and countable ordinals. (An ordinal is countable if the set of its predecessors is equinumerous with the set of natural numbers. Cantor called the collection of countable ordinals the second number-class or (II), the first number-class being the set of natural numbers (see §2.2).) A with its elements taken in natural order is well-ordered, and thus (provided it is not Absolute) it has an ordinal number α. But now clearly $A = \{\gamma : \gamma < \alpha\}$ which means that $\alpha \notin A$ and so must be *uncountable*, and moreover it must be the least uncountable ordinal. (Cantor at first denoted this ordinal by Ω; nowadays it is called ω_1, following Cantor's later notation (see §2.2). This method will likewise prove the existence of a first ordinal following any given number-class.

(2′) is clearly a powerful principle for generating ordinal numbers and, of course, *sets*. The important point to notice is that its power, the singling out of *minimal* domains, relies on impredicative sub-domain formation. For if a domain is characterized as the smallest having property ϕ, it will itself have ϕ, and yet be the intersection of all domains having ϕ. Hence, from the point of view of set existence, principle (*a*) is boosted in the strongest possible way. (*a*) itself just says that there is an actually infinite domain embracing all values of a varying quantity. Cantor assumes *in addition* that there exist sub-domains which correspond to impredicative specifications, an assumption which goes most of the way towards acceptance of the existence of *arbitrary* sub-domains, and thus implicitly, when applied to sets, a full power-set operation. For it is the use of impredicative means of singling out subsets of a given set which permits the proof of what is known as Cantor's theorem (see §2.3(*a*)). Cantor's enunciation of principle (*a*) makes clear his belief that mathematics should be grounded on set theory (or at least on collections). The analysis of the existence principles for ordinals begin to show how strong his assumptions of set existence are.

– It should be remarked that in his published work Cantor did not explicitly use this method via (2′) (or even via (2)) to prove (respectively, assert) the existence of his Ω, though it quite clearly is a simple generalization of his method for proving the existence of ω. However, two things must be noted. First, as a matter of later interest, the method is clearly appealed to in the famous letter to Dedekind of 1899 (see van Heijenoort [1967], pp. 115-6, and §2.2). But even as early as 1883 (2′) is at work. Cantor asserts that the two principles of generation 'give us the ability *to break through every barrier* in the formation of the concept of real, whole numbers' (Cantor [1883*b*], p. 197; the italics are Cantor's). This certainly suggests a belief that these methods take the ordinal number sequence *arbitrarily* far. My 'reformulation' of (2) as (2′) just makes explicit the implicit reliance on the existence of sets in matters concerning the ordinal numbers. Once or twice Cantor states that a *third* generating principle is needed to break out of the number-classes, what he calls the 'restricting' or 'stopping' principle. He applies this for instance to collect together the countable ordinals to form (II), which he then proves to be un-countable. But application of Cantor's 'third principle' is actually very close to application of the 'least domain' principle on which (2′) is based.

I turn now to Cantor's attempt to use the ordinals in tackling the problem of infinite cardinal number.

2.2 The scale of number-classes

For Cantor, the problem of powers was not simply to explain how or why 'every well-defined set has a definite power' or even just to explain exactly what the powers or cardinal numbers are, though this of course was a part of the problem. Rather, largely because of the overriding importance of the continuum problem, Cantor faced a *further* difficulty of a more specific kind, namely to show that the powers form a generalized arithmetical scale, that they are genuinely an extension of the finite cardinal numbers. We have seen that it was part of Cantor's conception of the infinite that it should be possible to treat transfinite collections quantitatively (i.e. numerically) and that numbering in the infinite realm should be much the same as numbering in the finite realm. And arguably powers or cardinalities can only be genuine cardinal numbers if they are *ordered*, that is linearly ordered or fully comparable (cf. Hessenberg [1906], p. 549). The special difficulty presented by the ordering problem was of enormous importance, since the search for a solution to this was instrumental in dictating the structural properties of the Cantorian powers.

There is no doubt that comparability of sets was originally built into Cantor's conception of power. Thus, in explaining cardinality orderings in his [1878], Cantor asserts: 'If two manifolds M and N are not of equal power, then either M has the same power as a component [subset] of N or N has the same power as a component of M.' ([1878], p. 119.) But clearly Cantor later came to accept that comparability needs to be argued for, and certainly he realized that arguing for it is not a trivial matter. Moreover, which is to the point here, comparability and the ordering problem clearly exercised him a good deal in the period when he was writing his [1883b]. For instance, the following is a passage from a letter to Dedekind (5 November 1882; Noether and Cavaillès [1937], p. 55) in which Cantor explains the contents of his [1883b] (dated October 1882):

. . . it has pleased Almighty God that I should have the most surprising and unexpected revelations in the theory of sets and of numbers; or rather that I should discover what has fermented in me for years, and what I have searched for for a long time. It concerns not the general definition of a point-continuum which we have spoken of and with respect to which I believe I have made some progress, but something much more general and therefore more important.

You recall that at Harzburg I said to you that I could not prove the following theorem:

'If M' is a part of a manifold M, M'' a part of M' and if M and M'' can be put into complete reversible one-to-one correspondence, that is if M and M'' have the same power, then M' itself has the same power as M and M''.'

(According to Cavaillès [1962], p. 85, the Harzburg meeting referred to was between 7 and 12 September 1882.) The theorem Cantor mentions is clearly necessary for establishing that the powers must be ordered. Indeed, as Cantor later acknowledged, it is equivalent to establishing the antisymmetry property of the cardinality relation between sets, i.e. to showing $A \precsim B$ and $B \precsim A$ implies $A \sim B$, what became known as the Schröder–Bernstein theorem. Proving this would at least show that $\prec$ is a *partial* ordering. However, it is also clear that in his letter, in his [1883b], and certainly in his [1895], Cantor subordinates this problem of partial comparability of sets to the problem of *full* comparability, and moreover subordinates this latter to the problem of setting up a complete *scale* of cardinalities.[2] What Cantor does in his [1883b] is to present a framework within which a solution to this scale problem for numbers can be attempted. This framework, as Cantor goes on to indicate in his letter, is built around the extended ordinal number sequence:

I have now found the source of this theorem and I can demonstrate it rigorously and with the necessary generality, thus filling an essential gap in the theory of sets.
   I arrive at a natural extension, a continuation of the sequence of real, whole numbers [Cantor means the finite ordinals] which leads me successively and with the greatest security to the increasing powers whose precise definition has failed me until today. (Noether and Cavaillès [1937], p. 55)

The solution Cantor refers to (the 'source of this theorem') is in essence the modern *ordinal* theory of cardinal number, the theory adopted in the modern axiomatic theory of sets.

We can understand Cantor's solution somewhat better by first approaching the problem of powers in a quite general way. Having taken cardinal equivalence as the basis of his investigation on size, Cantor recognized that sets are thereby partitioned into equivalence classes. Developing an object theory of cardinal number or power, that is to say a theory in which for every set x there is a corresponding size object *card(x)*, requires the choice of *representatives* for the cardinal equivalence classes. Russell (and one might say Frege also) solved this problem by taken the whole equivalence class to represent the members of the class. But Cantor very early on had pointed out a quite different solution. He remarked ([1879], pp. 141-2) that any set from one of the equivalence classes will act as its representative, since by definition it is similar to the other members of the class in the required sense. For ease of reference, let us call such an approach to the problem an *interior* approach, in contrast to the *whole class* approach of Frege and Russell. It should be noted that to opt for an interior theory of cardinal number is to opt for what might be called a *replacement* theory, that is a theory in which *card(x)* is itself a set (since the equivalence classes are composed of sets), and which represents the set x by acting as a replacement image of it since $card(x) \sim x$. These two features will be highly important in the discussions that follow, particularly in Chapter 3.

---

[2] See, for instance, Cantor [1883b], p. 200-201, and [1895], p. 285. Cantor, it seems, never produced a direct proof of the Schröder–Bernstein theorem, that is a proof without relying on the ordinal theory of powers and the well-ordering assumption. Dedekind, however, did. He sent Cantor a proof in a letter of 29 August 1899 (see Cantor [1932], p. 449). By then Cantor already knew the proofs by Schröder ([1898]) and Bernstein (first published in Borel [1898], pp. 104–6); see his reply to Dedekind, 30 August 1899 (Cantor [1932], pp. 449–50). In this letter again it is the full comparability problem which dominates his thinking.

In opting for an interior theory, the important thing is to choose canonical representatives. One reason for this is that one should know in advance what the number-objects are. But from Cantor's point of view (bearing in mind the comparability problem) there is another crucial heuristic reason for looking for canonical representatives. If one looks at the whole class theory, *card(x)* is rather distanced from the set x in the sense that it is not a direct replacement image of x. Theoretically, this distancing of the number from the set should not make any difference to the development of the theory of number, and specifically to the problem of numerical comparability. In both types of theory, the relation of precedence among numbers reduces to the relation of cardinal comparability between sets. But nevertheless it is not difficult to see that the interior method might be heuristically more suggestive. If we choose inside representatives the comparability problem is reduced to comparability for a restricted class of sets (namely the representative sets). Thus, the 'obvious' move, in order to maximize the possibility of benefits accruing from the interior theory, is to choose canonical representatives with 'nice' properties, i.e. representatives for which one would hope to be able to prove comparability. This, I think, is the most instructive way to look at Cantor's theory of cardinality, for crucially he opted for an interior theory in which the canonical representatives (the number-classes) have extremely nice properties, derived from the nice properties of the ordinal numbers.

Cantor's first step was to take the set of finite numbers as the cardinal representative of the sets with the 'smallest infinite power'. This was quite natural, since denumerability is actually defined as 'being cardinally equivalent to the natural numbers', and it was the isolation of this notion which gave mathematical flesh to the bare bones of the theory of cardinal equivalence. But the problem was, as Cantor noted, that '. . . against this, there has been lacking, as yet, any equally simple and natural definition of the higher powers' ([1883*b*], p. 167). Cantor's idea, given that the transfinite ordinal numbers extend the sequence of natural numbers arbitrarily far, was to use sets of these to represent all the powers of transfinite sets, just as the set of finite ordinal numbers represents the smallest transfinite power. The idea is quite simple. Cantor split the sequence of transfinite ordinals into number classes in the following way:

(I)   (the first number class) is the set of all natural numbers.
(II) $= \{\alpha : \{\beta : \beta < \alpha\} \sim (\text{I})\}$

etc.

That is to say, (II) is the set of ordinals which represent denumerable well-ordered sets, (III) is the set of ordinals which represent well-ordered sets equivalent to (II), and so on.

Cantor explicitly defined only the class (II) in his [1883*b*] (pp. 167 and 197), indicating that (III) etc. can be analogously defined. We should note here that there is something of a difficulty with the generality of the definition.

It is *perfectly clear* that the construction of the number-classes is intended to be quite general, i.e. that for each ordinal number γ there is a corresponding number class (γ). This is explicitly stated, for example, on p. 205 of [1883*b*] (in the passage quoted on p. 42). But while the definition of, say, (II) easily extends to any (γ + 1), Cantor does not explicitly define the limit classes (γ) for limit γ. I shall come back to this later (pp. 70 ff). For the moment, let us just note that Cantor's *programme* was to define number-classes for *all* ordinals, though the details of the full definition took some time to sort out. Certainly the most important theoretical point in this programme is Cantor's claim that the

. . . above mentioned number-classes of the definite-infinite, real, integers are to be seen as the natural representatives, presented in unified form, of the orderly ascending succession of powers of well-defined sets. ([1883*b*], p. 167)

In other words Cantor is saying that *all* powers, and all the cardinal equivalence classes, are represented among the number-classes and these classes form an arithmetical scale. This can be summed up in the two claims:

(1) any set which has a power must be cardinally equivalent to some number-class;

(2) the number-classes form a linearly-ordered collection.

(In his [1887–8] Cantor remarked that they form a *well*-ordered collection; see p. 419, and the various passages quoted on pp. 65–7.) These, together with the assumption

(3) every well-defined set has a power

form the basis of Cantor's initial treatment of cardinality.

(1) is the reason for calling Cantor's theory an *ordinal* theory of powers. It is important at this point to say something about the metaphysical background to this theory before saying more about its precise working. The crucial principle at work here is principle (*b*), Cantor's principle of finitism. According to this, infinite sets should not only be capable of numerical determination, but their numerical treatment should be as similar as possible to that of the finite sets. Cantor's attempt to establish that the ordinal numbers form one homogeneous realm has already been discussed. What his theory of cardinality does is to put this homogeneity to mathematical use. There are two kinds of number associated with finite sets, ordinal numbers (indicating, in effect, the length of an ordering) and cardinal numbers (indicating size). The finite (ordinal) numbers are *counting* numbers. More importantly size here can be determined by counting: the given collection is counted through element by element and the (ordinal) number arrived at is the size of the collection. The reason for this is that the way the elements are counted is quite irrelevant; the *same* ordinal number is arrived at however the elements are enumerated. To put the matter in a more abstract way, counting imposes a special kind of linear ordering (well-ordering) and for a given finite set the *length* of the ordering is the same no matter how the order of

the elements might be changed. The upshot is that in the finite case, ordinality completely and directly determines cardinality: the two kinds of number co-incide. Indeed, one might go so far as to say that cardinality in the finite case is just an expression of the fact that counting leads to unique results. I suggest that the basis of Cantor's theory of power is the thesis, following principle (*b*) and the homogeneity of the ordinals, that the same method can also be applied in the infinite realm. Thus the suggestion is that here also size can be determined by *counting*, that is by counting all elements against a fixed stock of transfinite ordinal numbers.

There is a certain sense in which Cantor took 'counting' in the infinite realm literally. (I shall have much more to say on this in Chapter 3.) Of course infinite sets cannot literally be counted. But the boldness of Cantor's theory is that he proceeded *as if they can*. In other words he deliberately suppresses here a funda-mental intuitive distinction between finite and infinite. This is Cantor stretching to its limits Archimedes's idea in 'The Sand-Reckoner' that no collection is too big to be counted, and it fits in exactly with Cantor's whole project of attempt-ing 'the rational subjugation of the infinite'. (This paraphrases Weyl [1949], p. 36, who has 'unbounded' in place of 'infinite'.) The vital importance of the more or less literal insistence on counting is now this. Two crucial features of counting are that there must be a first element counted (to start the process) and that if any group of elements has been counted and there are more left to be counted, there must be a *first* chosen from the remainder which will then act as the immediate successor of those already counted. The rationale behind this latter condition is the following: it is nonsense to assert that a collection can be counted without allowing that any sub-collection can also be counted; hence if some counting has been completed and some remains to be done, there must be a first element which is taken to begin the counting anew, and this element will be the immediate successor of those already counted. If it is asserted that every set can be counted, these two features alone *guarantee* that it must be possible to put every set into well-ordered form. (It is surely no accident that the two counting conditions *precisely* mirror the key conditions in Cantor's definition of a well-ordering: see p. 52.) Finite sets when counted have a special kind of linear ordering which is certainly discrete and well-founded; the counting assumption for sets in general entails that as much of this special nature as possible must carry over to the infinite realm.

Cantor's assimilation of the infinite to the finite is not bought cheaply. There are two fundamental difficulties (leaving aside for the moment the difficulties over the literal assumption of counting): the dependence on well-orderability in general, and the non-uniqueness of the length of well-orderings possible on a given well-orderable infinite set. Both of these difficulties are summed up in the following passage from Russell's [1903]. Counting, says Russell, '. . . is only applicable to classes which can be well-ordered, which are not known to be all classes; and it only gives the number of the class when this number is finite—a

rare and exceptional case' (Russell [1903], p. 114). Russell took these difficulties as showing that it is wrong to involve counting in a theory of cardinal
number. But it was part of Cantor's achievement to show that the view that
Russell expresses, that counting only gives the size in the *finite* case, is wrong, at
least if one takes well-orderability for granted. This is where Cantor's definition
and use of the number-classes is crucial. The number-classes gather together *all*
ordinals of (or rather, belonging to) the same power. Since these classes are
assumed to represent all the powers it is sufficient to count an infinite set in any
manner; the ordinal resulting from this count can only be in *one* number-class,
and this determines the power. The construction of the number-classes, in effect,
factors out the ordinal numbers by power. The difference between the finite and
transfinite 'counting' numbers in Cantor's theory is now this: the finite numbers
determine size directly, and the transfinite *indirectly*, via the number-classes.

Let us now explain the detailed workings of Cantor's theory in order to be
quite clear about his assumptions. The key to the theory is

(4) given any (well-defined) set there is an ordinal number which enumerates it.

For Cantor this is equivalent, given (3), to (1). Part of Cantor's claim for the
ordinals is that they form a complete sequence of representatives for well-
ordered sets. In other words:

(5) given any well-ordered set there is an ordinal number which enumerates it.

If this is combined with the crucial

(6) any well-defined set can be put into well-ordered form

then (4) follows immediately.

The equivalence (for Cantor) of (1) and (4) is now easy to see. Assume (1);
then any set M is equivalent to some (γ). But (γ) is well-ordered, so M can be
put into well-ordered form. Thus by (5) there is an α which enumerates the well-
ordered M. Thus, (4) follows. (Since the set of predecessors of α must by transitivity of $\sim$ be cardinally equivalent to (γ), α belongs to $(\gamma + 1)$.) Assume (4);
then if M is a set there is an ordinal α which enumerates it. But α belongs to some
minimal number-class $(\gamma + 1)$. Hence, since $M \sim \{\beta : \beta < \alpha\}$ and $\{\beta : \beta < \alpha\} \sim$
$(\gamma), M \sim (\gamma)$. Hence (1) holds.

Thus, although given a set M and an enumerating ordinal α, α does not give
the power directly (i.e. itself), it does give it indirectly because it determines a
unique number-class. By definition, all the numbers which can be used to
enumerate a given M will land in the same number class simply because they
must determine equinumerous segments of the ordinal number sequence.

This virtual reduction of cardinality to ordinality formed the basis of Cantor's
theory of powers. That he saw it as the key to the whole problem of what the
powers are is clearly demonstrated in the following three passages, as well as that
from [1883*b*], p. 167, quoted on p. 62:

One of the most important tasks or problems of set theory, which I believe I have solved in my memoir *Grundlagen einer allgemeinen Mannigfaltigkeitslehre* (Leipzig, 1883) [i.e. Cantor [1883b]] consists in determining the various valencies or powers which can occur both in nature as a whole and which are revealed to our understanding. I believe I have solved this problem through the development of the general number-concept of well-ordered sets, or what is the same thing, through the concept of ordinal number. ([1887–8], pp. 387–8, from 1884.)

In order to take the theory of cardinal numbers . . . into the transfinite with certainty, and to try to give it a more rigorous foundation, one is dependent . . . on the introduction of the transfinite ordinal numbers . . . ([1887–8], p. 420)

Among simply ordered sets *well-ordered sets* deserve a special place; their ordinal types, which we can call 'ordinal numbers', form the natural material for an exact definition of the higher transfinite cardinal numbers or powers . . . ([1897], p. 312)

Clearly the hinge on which Cantor's theory turns is (6). Cantor's attitude to this needs discussing, as does the question of what *precisely* the cardinalities are for him. (For instance, are the number-classes *themselves* the cardinal numbers, or are the cardinals perhaps primitive entities *represented* by the number-classes?) These questions are pursued in some detail in Chapter 3. For the moment I want to concentrate more on the more mathematical, as opposed to the foundational, issues and in particular on the technical problem of getting the number-class scale right (showing that it is genuine number-scale), and, in §2.3 on the mathematical impact of Cantor's theory, especially on the continuum problem.

Cantor's [1883b] by itself does not resolve the scale problem. As I have said, it presents a programmatic framework within which the problem of comparability and the scale problem can be tackled. There is no doubt at all of Cantor's belief that this framework provides the solution, that the number-class theory transforms the notion of power into the stronger notion of cardinal number. For example, in [1887–8] Cantor remarks: 'We will see later that the totality of all *cardinal numbers or powers* (the finite and the supra-finite) also forms a *well-ordered* set if one imagines them ordered according to size.' ([1887–8], p. 419.) (Notice that the reference here to the well-ordered *set* of cardinal numbers offends against Cantor's conception of the Absolute. Cantor later acknowledged explicitly that they cannot form a set. See Chapter 4.) This is then followed on the next page by a passage (quoted above) which refers directly to the characterization of powers in terms of ordinal numbers. In his [1891], Cantor is quite explicit:

I have already shown by other means in my *Grundlagen einer allgemeinen Mannigfaltigkeitslehre* [[1883b]] that the powers have no maximum. There it was even proved that the totality of all powers forms a 'well-ordered set' if we imagine them ordered according to size, so that to every power in nature there is a next greater, and that to each set of increasing powers there is a next greater which follows after them all. ([1891], p. 280)

Moreover, it is no accident that this is followed (in the next paragraph) by the claim that:

The 'powers' represent the unique and necessary generalisation of the finite 'cardinal numbers'; they are nothing other than the actually-infinitely large 'cardinal numbers', and they possess the same reality and definiteness as the finite cardinal numbers; . . . ([1891], p. 280)

the claim to 'numberhood' being precisely what comparability (*a fortiori*, well-orderedness) establishes.

But while the direction of Cantor's thinking is clear, what he claims to have established in [1883*b*] is not in fact established there. Even taking (6) (general well-orderability) for granted, it is not shown that well-ordered sets are comparable, a result which is needed to show that the number-class powers are comparable. Neither is it argued that the ordinal numbers themselves (or sets of them) are well-ordered. More importantly, Cantor certainly did not establish in [1883*b*] that the powers have no maximum, for it is not at all obvious that the number-classes (α) represent definite *increases* in power with increasing α. This result was extremely important for Cantor. For one thing it would seem to tie in with his conviction that the Absolute is the only 'absolute maximum'. If there turned out to be a maximum power, which could be grasped by us, this *could* be taken as a 'quantitative determination' of the Absolute, which is impossible by principle (*c*) (though see §8.3). In addition, the existence of a maximum number-class (in terms of size) would destroy much of the symmetry between ordinals and cardinals and between the finite and the infinite. In the finite, there are just as many cardinal numbers as ordinals, and increasing ordinality necessarily represents increasing cardinality. Cantor certainly wanted this to carry over into the infinite realm, where now, however, an ordinal α 'represents' a cardinal by picking out the cardinality of (α) (see the passage quoted on p. 42). But despite these sentiments and wishes, Cantor does not attempt any demonstration that the number-classes represent an indefinitely increasing scale. Establishing this requires showing that the number-class ($\alpha + 1$) is strictly larger than (α), and that it is the *next* largest power, in other words

(7) $(\alpha) \prec (\alpha + 1)$ and there can be no set M such that $(\alpha) \prec M \prec (\alpha + 1)$.

All that Cantor proves in [1883*b*] is the important special case

(8) $(\mathrm{I}) \prec (\mathrm{II})$, and there can be no set M such that $(\mathrm{I}) \prec M \prec (\mathrm{II})$.

(See Cantor [1883*b*], pp. 197–201, and also [1897], pp. 331–3. I shall come back to this proof in §2.3(*a*).)

But it is clear that (7) is what is aimed at, as is implied by the following passage from [1887–8] which stems from 1884:

This totality of all ordinals of the second number-class constitutes a new *valence*, and indeed the valence *following next* after the power of the first number-class, as I have rigorously shown [1883*b*]. The same path of ideas leads us to higher number-classes and the higher valencies belonging to them. This is a wonderful expanding harmony, the precise working out of which is the theme of transfinite number theory. ([1887–8], p. 390)

Thus one can call (7) the first immediate goal of the Cantorian programme.

Cantor did establish later in [1891] that there can be no maximum power by showing (in effect) that given any set A the set of functions from A to $\{0, 1\}$ has a strictly greater power. (Actually Cantor only proves a special case—where A is [0, 1]—but he notes that the method, the famous diagonal argument, is quite general (see §2.3(*a*)).) But this solves only a part of the problem. The crucial question of whether the (α) represent a workable scale of cardinality remains unanswered, despite Cantor's claim to the contrary in the passage quoted on p. 66. Furthermore, this new proof raises sharply the question of just how the powers of A, $\{0, 1\}^A$ etc., are related to the number-class powers.

The comparability and scale problems still preoccupied Cantor in his next paper [1895]. But again, while faith in the number-class theory is reaffirmed, the crucial result (7) is not argued for. He points out very early on, after introducing the cardinals generally by reference to 'abstraction', that 'at this stage of our chain of reasoning' comparability of cardinals 'is by no means self evident' and 'can scarcely be proved'. He goes on: 'Only later, when we have achieved an overview of the ascending sequence of transfinite cardinal numbers and an insight into their relations can [comparability] be proved.' ([1895], p. 284.) Then later he repeats that the cardinal numbers (and it is quite clear he means cardinals in general) can be well-ordered, and, very much as in [1891], he claims:

To *each transfinite cardinal number* there is a *next greater* which arises out of it by a unified law. And to every unbounded well-ordered set $\{\alpha\}$ of ascending cardinal numbers α there is a *next greater* following them all which arises in a unified way out of that set.

For the rigorous foundation of this, discovered in 1882 and set out in the memoir *Grundlagen einer allgemeinen Mannigfaltigkeitslehre* [[1883*b*]], we make use of the so-called '*ordinal types*' . . . ([1895], pp. 295–6)

(It is clear from the mention of [1883*b*] that Cantor means the *ordinal* theory of powers despite the reference to 'ordinal types', which were not introduced in full generality by Cantor until 1884 (see Grattan-Guinness [1970]). Cantor made this clear later on in [1897], p. 312, in a passage quoted on p. 65.) Again despite this very clear reference to the number-class theory there is no attempt at establishing these crucial claims. One would gather from the remarks following this passage that an explicit treatment of cardinality will follow after the theory of types and ordinals is set out. Indeed, the later second part of the paper ([1897], p. 312) begins by emphasizing the importance of ordinals in the theory of powers; to quote further from a passage already cited:

Among simply ordered sets *well-ordered sets* deserve a special place. Their order-types which we call ordinal numbers constitute the material for an exact definition of the higher transfinite numbers or powers, a definition which thoroughly conforms to that which we gave for the smallest transfinite cardinal number *aleph-null* through the system of all finite numbers ν.

Here at least there is a more systematic treatment of well-ordered sets. For instance, Cantor establishes the two crucial theorems ([1897], §§13 and 14):

(9) any two well-ordered sets are ordinally comparable

and the corresponding

(10) if α, β are two ordinals then either $\alpha < \beta$ or $\alpha = \beta$ or $\beta < \alpha$.

This at least now would guarantee (modulo acceptance of (6)) the comparability of power. But not only does Cantor not draw this conclusion, he does not even go on to define the number-classes or alephs generally, despite the clear evidence in the various passages quoted above that he still held to the number-class theory of power. In fact, alephs in general are not mentioned.

As a piece of notational history, the aleph notation was first introduced to the mathematical public in Cantor's [1895]. $\aleph_0$ is characterized as 'the power of the natural numbers', i.e. of the class (I). Cantor then (at least in one section) makes quite free mention of $\aleph_\alpha$ generally, though no general characterization is given. However, it is implicitly clear that $\aleph_\alpha$ is the power of the class $(\alpha + 1)$. Thus, in [1897], p. 333, $\aleph_1$ is introduced as the power of (II). The aleph notation appears to have been referred to first in a letter to Vivanti of 13 December 1893 (see Meschkowski [1965], pp. 504–8, where the letter is published), though here Cantor has $\aleph_1$, $\aleph_2$, etc. in place of the later $\aleph_0$, $\aleph_1$ etc. The new notation was intended as a simplification of Cantor's earlier and quite impractical $\overset{*}{\omega}$, $\overset{*}{\Omega}$, etc., and $\overline{\omega}$, $\overline{\Omega}$, etc., which are both clumsy and insufficiently general. (One finds this earlier notation, for example, in Cantor's long letter to Gold-scheider of 18 June 1886 (published in Meschkowski [1961], pp. 83–90; see also Dauben [1979], pp. 180–81). In a certain sense the initial ordinal theory of cardinality looks *back* to this earlier notation.) Cantor's choice of the letter aleph to denote the infinite cardinals is explained in a letter to Felix Klein of 30 April 1895:

It has seemed to me for many years indispensable to fix the transfinite powers or cardinal numbers by some symbol, and after much wavering to and fro I have called upon the first letter of the Hebrew alphabet Aleph = $\aleph$. Hopefully the public will soon make friends with it.

We are dealing here with *absolute constants* which have the same special importance as 1, 2, 3, . . . , and which indeed stand under the same concept as these, namely the concept of cardinal number. The usual alphabets seem to me too much used to be fitted for this purpose. On the other hand, I didn't want to invent a new sign, so I chose finally the aleph, which in Hebrew also has the numerical value 1.

$\aleph_1$ is indeed in a certain way a new unity. (Cantor *Nachlass VII*, pp. 142–3)

At this stage $\aleph_1$ still represented the first infinite cardinal. Cantor changes this to $\aleph_0$ in a letter to Klein of 19 July 1895: *Nachlass VII*, pp. 158–9. Cantor was writing all this apropos of his [1895] submitted to *Mathematische Annalen*, of which Klein was the editor.

Cantor's degree of faith in the number-class theory (amply testified to in the various passages quoted above) makes his hesitancy over (7) rather surprising, *particularly* given that (7) is not difficult to prove. Indeed it seems to stem directly from Cantor's earlier strong existence assumptions about ordinals. In terms of alephs what has to be proved is:

(7′) for any α, $\aleph_\alpha < \aleph_{\alpha+1}$ and there is no power strictly between them.

But $\aleph_\alpha$ and $\aleph_{\alpha+1}$ are just the powers of $(\alpha + 1)$ and $(\alpha + 2)$ respectively. By the ordinal existence arguments outlined in §2.1 there must be a first ordinal β *after* all the ordinals in $(\alpha + 2)$ (unless, of course, $(\alpha + 2)$ is Absolute; see below). Suppose now that $\aleph_\alpha = \aleph_{\alpha+1}$, i.e. $(\alpha + 1) \sim (\alpha + 2)$. Then clearly $\{\gamma : \gamma < \beta\} \sim (\alpha + 1)$, which by definition means $\beta \in (\alpha + 2)$. But $\beta \notin (\alpha + 2)$. Hence $\aleph_\alpha < \aleph_{\alpha+1}$. Moreover, it is easy to see that if $C \prec (\alpha + 2)$, then $C \precsim (\alpha + 1)$. For C must be well-orderable; so let γ be an ordinal which enumerates C. It follows from $C \prec (\alpha + 2)$ that $\gamma < \beta$ (β characterized as above). Hence, since β is the *limit* of the ordinals in $(\alpha + 2)$, $\exists \delta \in (\alpha + 2)$, $\gamma \leqslant \delta$. Thus since $\{\xi : \xi < \delta\} \sim (\alpha + 1)$ we must have $C \precsim (\alpha + 1)$. In other words, there can be no C such that $\aleph_\alpha < card(C) < \aleph_{\alpha+1}$, and (7′) and (7) are proved. The elements of this argument are to be found in Hessenberg [1906], pp. 553–4, in proving theorem 38: to every set M of alephs there is a least which is greater than all those in M. The procedure here is rather different from Cantor's of [1883b]. Hessenberg assumes that all sets have a power, and the alephs are characterized as the powers of the well-ordered sets. $\aleph_{\alpha+1}$ is then 'defined' as the well-ordered power first after $\aleph_\alpha$. Thus the need for a proof of (7′) is obviated by the way the alephs are introduced. However, Hessenberg proves theorem 38 to stress the existence of 'limit' powers, i.e. for any set of powers of well-ordered sets there is a *first* power strictly greater than all of them. Hessenberg's procedure is followed by Hausdorff in his [1914], although Hausdorff proves more about the number-classes (see Hausdorff [1914], pp. 122–9, and p. 72 below). In any case, the fact that Cantor himself did not use such a simple argument is puzzling.

However, the situation is not quite so clear cut as it looks at first sight. There were various complications surrounding the number-class theory which, though not destroying Cantor's belief in its correctness or efficacy, might well have contributed to a certain tentativeness in formal presentation. In the first place, there were problems from the beginning with the generality of the definition of the number-classes. Secondly, there were later doubts about the well-ordering assumption (6), and doubts thrown up by the discovery of contradictions.

As the passage from his [1883b] quoted in §1.4 (p. 42) makes clear, Cantor intended that there should be a number-class for each ordinal number. However, the account of the classes there suggests that each class (α) is defined in terms of the *immediately* preceding class. But if λ is a limit-ordinal there is no such predecessor. There is a natural way of resolving this difficulty, namely by defining number-classes instead in terms of *all* the preceding classes. Thus instead of defining (α) as above, one defines

$$(\alpha)' = \{\beta : \{\gamma : \gamma < \beta\} \sim \bigcup_{\xi < \alpha} (\xi)'\}$$

So instead of, say, the third number-class being $\{\beta : \{\gamma : \gamma < \beta\} \sim (II)\}$ it is now defined as $(III)' = \{\beta : \{\gamma : \gamma < \beta\} \sim (I) \cup (II)\}$. This definition keeps the number-classes disjoint as in the original definition, but is much more general. For instance, we now have (if we accept infinite unions) limit classes, e.g.

$$(\omega)' = \{\beta : \{\gamma : \gamma < \beta\} \sim (I) \cup (II) \cup \ldots\}.$$

And certainly the simple proof for (7) now goes through for these $(\alpha)'$.

There are two points to be noted here. The first is that once this step is taken one may as well take the powers as characterized not by the number-classes $(\alpha)'$ themselves, but by the number-classes taken cumulatively. For instance, suppose we define

$$[1] = (I)$$

$$[\alpha + 1] = [\alpha] \cup \{\beta : \{\gamma : \gamma < \beta\} \sim [\alpha]\}$$

i.e.

$$[\alpha + 1] = [\alpha] \cup (\alpha)'$$

and for limit ordinals λ

$$[\lambda] = \bigcup_{\gamma < \lambda} [\gamma]$$

For the number-classes (α) and $(\alpha)'$ are sets of ordinals of a specified power, and the specified power in the definition of $(\alpha + 1)'$ is the power of the cumulative class $[\alpha]$, *not* the immediately preceding class $(\alpha)'$, as it was in the analogous successor case in Cantor's original definition. So, given this, it is more natural to base the characterization of power *from the beginning* on the cumulative classes $[\alpha]$. That is to say, once one moves in this direction is is more natural to characterize $\aleph_\alpha$ as the cardinality of $[\alpha]$ ($[\alpha + 1]$ for $\alpha < \omega$) instead of as the cardinality of $(\alpha)'$ ($(\alpha + 1)'$ or $(\alpha + 1)$ for $\alpha < \omega$).

The second point to note is that this was more or less the procedure Cantor followed in his famous 1899 letter to Dedekind (see van Heijenoort [1967], pp. 115-6). The number-classes as he defines them there are still disjoint, as in the $(\alpha)'$. But it is also clear that what I have called cumulative powers are really what are used. Cantor proceeds recursively as follows. Taking ω_0 (as he now denotes ω) as 'the least transfinite number', $\aleph_0$ is characterized as the power of ω_0. Then he forms the class $Z(\aleph_0)$, denoted also Ω_0, which is the collection of all ordinals of power $\aleph_0$, or as Cantor says the collection of all α with $\omega_0 \leqslant \alpha < \omega_1$, ω_1 being 'the least transfinite number whose cardinal number is not equal to $\aleph_0$'. Then he defines $\aleph_1$ as $card(\omega_1) = card(Z(\aleph_0))$, and states that $\aleph_1$ must be the first power greater than $\aleph_0$. This process is iterated through the ω_n, Ω_n, and $\aleph_n$. But the interesting thing is what happens at the first limit level after all the $\aleph_n$. Here Cantor characterizes ω_{ω_0} as 'the least ordinal not in any of the classes Ω_n' and then $\aleph_{\omega_0} = card(\omega_{\omega_0})$. But, he says, this is also characterizable 'by means of the equation

$$\aleph_{\omega_0} = \sum_{\nu=1,2,3\ldots} \aleph_\nu.'$$

But clearly what this means is that

$$\aleph_{\omega_0} = card\left(\bigcup_{n<\omega} \Omega_n\right) = card(\{\beta : \beta < \omega_{\omega_0}\}),$$

in other words that $\aleph_{\omega_0}$ is characterized as the power of the cumulative class $[\omega_0]$. Moreover, in effect the same is the case already for $\aleph_1, \aleph_2$, etc. For if $\aleph_1$ is the power of ω_1 then $\aleph_1 = card(\{\beta : \beta < \omega_1\}) = card([2])$. So, in fact, Cantor's classes $Z(\aleph_\alpha)$ are just what I called the $(\alpha)'$, and not the (α). Furthermore, Cantor explicitly states that

$$\sum_{i=1,\ldots,n} \aleph_i = \aleph_n$$

(with the claim that it is easy to prove), which is *exactly* what is required in proving that the classes (n) and $(n)'$ are of the same power (and hence identical). For here one has to show that the size of a later class 'dwarfs' that of the earlier, i.e. that

$$card((n)) = card\left(\bigcup_{k\leqslant n} (k)\right).$$

What this suggests is that the switch to the $(\alpha)'$ (or Cantor's $Z(\aleph_\alpha)$), and the basing of the cardinal numbers on them, does not distort Cantor's original intention. This was later confirmed by Hausdorff's work on the number-classes. Hausdorff (in effect following Hessenberg's [1906]) characterizes cardinal numbers (as opposed to powers in general if one chooses *not* to depend on the well-ordering theorem) as the powers of well-ordered sets ([1914], p. 122). He then shows that these numbers must be well-ordered, and that to any set of them there is an immediate successor, a smallest number strictly greater than all the numbers in the set. (His proof is, more or less, the 'simple' proof based on the ordinal existence assumptions sketched above, p. 69.) The alephs are then defined recursively by

$\aleph_0$ = the power of the natural numbers

$\aleph_\alpha$ = the smallest well-ordered power
greater than all $\aleph_\gamma, \gamma < \alpha$.

The number-classes and the initial ordinals are then defined in terms of the alephs (in fact, the inverse of Cantor's proposal). Thus $Z(\aleph_\alpha)$ is the collection of all ordinals with power $\aleph_\alpha$, and ω_α is the first member of $Z(\aleph_\alpha)$. Although arrived at by a different route, these number-classes clearly mirror Cantor's conception. The striking thing is that Hausdorff shows that $Z(\aleph_\alpha)$ not only has the same cardinality as $\{\beta : \beta < \omega_{\alpha+1}\} = [\alpha + 1]$ (or $[\alpha + 2]$ if $\alpha < \omega$), but that they are actually *ordinally* similar. The proof is quite simple (see Hausdorff [1914], p. 126). For

$$\{\beta : \beta < \omega_{\alpha+1}\} = \{\beta : \beta < \omega_\alpha\} \cup Z(\aleph_\alpha).$$

Thus, the corresponding ordinal equation is $\omega_{\alpha+1} = \omega_\alpha + \gamma$. Certainly $\gamma \leqslant \omega_{\alpha+1}$; but suppose $\gamma \neq \omega_{\alpha+1}$, then the cardinality of $\omega_{\alpha+1}$ would be $\aleph_\alpha$, which is impossible because $\omega_{\alpha+1}$ is the first ordinal of power $\aleph_{\alpha+1}$. Hence necessarily $\gamma = \omega_{\alpha+1}$, and so the type of $Z(\aleph_\alpha) = \omega_{\alpha+1}$, the same as the type of $[\alpha + 1]$ (or $[\alpha + 2]$, for $\alpha < \omega$). This strongly suggests that nothing at all is lost (and, because of greater generality, much gained) by concentrating on the cumulative number-classes rather than Cantor's original disjoint classes. (It is worth noting that Hausdorff's proof depends on the result that $\aleph_\alpha + \aleph_\beta = max(\aleph_\alpha, \aleph_\beta)$, and this is just what Cantor asserted in the 1889 letter (see p. 71), at least for finite α, β.)

One feature of Cantor's treatment of the number-classes in the Dedekind letter is the use of the quite general notion of *initial ordinal*. Indeed, the whole weight of the treatment rests on the assumption of the existence of these ordinals, characterized as 'the *least*' transcending a given power. This fits *exactly* into the pattern of strong (impredicative) ordinal existence assumptions described in §2.1 and the assumptions are just what Cantor would require to prove (7) for his $Z(\aleph_\alpha)$. The articulation of this dependence on the initial ordinals is interesting for another reason. According to Kowalewski [1950], Cantor proposed not just defining the alephs *via* the initial ordinals, but actually defining them *as* the initial ordinals. Thus, $\aleph_0$ is ω_0, $\aleph_1$ is ω_1, etc., exactly as in the modern von Neumann theory. (Kowalewski does not say when Cantor proposed this. However, he does say it was Cantor's 'habit' or 'custom' [*Gepflogenheit*] to identify the powers with the initial ordinals.) This is certainly a move in the direction of the cumulative theory of power. For ω_α directly represents (is the type of, and marks the end of) the αth number-class (or if $\alpha < \omega$, the $(\alpha + 1)^{th}$ number-class). Thus, to say that a set A has the power ω_α is to say precisely that it can be put into one–one correspondence with the αth $((\alpha + 1)^{th}$ for $\alpha < \omega)$ *cumulative* number class. But more strikingly, this idea fits rather beautifully with the spirit of Cantor's ordinal conception of cardinality, and is indeed a remarkable simplification of it. Whether one characterizes power by cumulative or non-cumulative number-classes is to a large extent irrelevant (except perhaps for the problem of generality mentioned above). For in both cases to say that a set A has a certain number-class power is to say that it can be *counted* by any one of a whole class of ordinals. Both conceptions then are fundamentally *ordinal* theories of cardinality, and this is the crucial feature of Cantor's idea. The same applies to the initial ordinal theory, only now the cardinal number *itself* is an ordinal just as the finite cardinals are also ordinals.

The hesitation in exploiting the strength of the ordinal assumptions to prove results about cardinals can therefore be partly explained by difficulties over the general characterization of the number-classes. (Though it should be said that we do not know at what stage precisely Cantor moved to the cumulative characterization of power.) But this by itself does not explain why Cantor only dealt with this question (or indicated that he could deal with it) in correspondence, and not in any publication. Zermelo in his [1932] (p. 352, n. 9) also noted that Cantor in his [1895] and [1897], despite the promises and claims catalogued above, restricted himself to investigating the second-number class, and hence $\aleph_1$ only. He conjectured that Cantor's hesitancy is explained by his not being able to prove the 'aleph theorem', i.e.

(11) every power is an aleph

(which is just the aleph version of (1)), *and* by the discovery of the Burali-Forti contradiction (see also Cavaillès [1962], p. 116). Zermelo does not explain Cantor's difficulties over (11). Certainly one main difficulty is the one of generality just discussed, but which we have seen that Cantor is quite capable of dealing with. It is much more likely, however, that Zermelo means difficulties with the crucial (6), the well-ordering assumption. In his [1883b] Cantor stated '. . . it is always possible to arrange every *well-defined* set in the *form* of a *well-ordered* set . . .', and he called this principle a 'fundamental law of thought, rich in consequences and particularly remarkable for its general validity' (p. 169). There is some evidence which might indicate that at some later point he had doubts about it. For in a letter to Mittag-Leffler of 14 November 1884 (quoted at length by Schoenflies [1927], pp. 17–18) he claims to have proved that the continuum could *not* be of the second power, indeed that '. . . it has no power specifiable by a number'. This, of course, would mean that the continuum could not be well-ordered and therefore that (6) is false. Cantor changed his mind about the result the following day (letter to Mittag-Leffler of 15 November 1884 (Schoenflies [1927], p. 18)). But it is possible that the doubt and vacillation that Cantor experienced on these matters led him to believe that well-ordering must actually be proved or argued for, and not just assumed. And the lack of an available proof might well have made Cantor hesitate to publish an explicit full-blooded account of the number-class theory. However, against this one could well argue that if this were the case there was nothing (other than a respect for completeness) which prevented him from publishing a *provisional* account conditional upon the correctness of (6) and to be followed by a later attack on this. But there is no hint of this provisional attitude in any of Cantor's strong statements of belief in the correctness of the full number-class theory.

I shall argue in Chapter 3 that, whatever may have been Cantor's state of intellectual doubt about the well-ordering theorem, there is a sense in which it was *fundamental* to his thinking about sets. In other words, the well-ordering theorem is in some sense a 'law of thought' about Cantorian sets. But the weight of the number-class theory falls not only on (6) but also crucially on the strength of the ordinal existence assumptions. Hesitancy over (11) (or (7)) might then be explained by doubt or confusion about the length of the ordinal number sequence. (After all, to argue that there is a limit number following after a sequence of ordinal numbers, one has to be sure that the sequence is not Absolute.) This supposition is supported by Cantor's own discovery of the Burali-Forti antinomy which he apparently knew as early as 1896. This may have given Cantor pause for thought about his existence assumptions. It is surely significant that Cantor in his famous 1899 letter to Dedekind attempted to tackle *both* the Burali-Forti problem and the aleph theorem (11) in *one argument*. I will delay discussion of this argument until after I have discussed the extent to which well-ordering was an integral part of Cantor's conception of set in Chapter 3. (Cantor stated to Jourdain in a letter of 4 November 1903 that he conveyed his proof of the aleph theorem based on the distinction between *consistent* and *inconsistent* multiplicities (see §4.1) to Hilbert in 1896. This proof, of course, contains an explicit recognition of the Burali-Forti danger. It should be noted, however, that Cantor remarks in this Jourdain letter: 'I have intuitively known the undoubtedly correct proposition that there are no cardinal numbers other than alephs for about 20 years (since the discovery of alephs themselves).' Grattan-Guinness [1971], p. 116.)

2.3 The attack on the continuum problem

(*a*) *The first step : the uncountability of the continuum and the second number-class*

Although he did not prove (7) (at least in print), Cantor devoted considerable attention to proving (8) separately which he succeeded in doing in [1883*b*] with a proof which curiously is *not* generalizable to a full proof of (7). It is not difficult to understand why he concentrated on this special case. His interest in the properties of the class (II) was intimately connected with interest in the continuum problem. This interest was perhaps strengthened by the structure of his proof of (8) and its similarity to his first proof of the non-denumerability of the continuum.

Cantor's first proof of (8), given in his [1883*b*], has what one might *loosely* call a constructive and a non-constructive part. The constructive part shows that given *any* denumerable sequence of denumerable ordinals it is possible to diagonalize and find a denumerable ordinal not in the sequence. I call this part of the proof 'loosely constructive' because it shows that no countable sequence of countable ordinals can contain every countable ordinal, and it does this *without* referring to the impredicatively defined totality (II) of all countable ordinals. (It should also be noted that the proof involves implicit appeal to the axiom of choice, since it depends on the countable union theorem.) What Cantor then does is to adapt the proof to show that if one assumes a one-one function f from the natural numbers (I) onto the set (II) one can derive a contradiction by defining in terms of f an ordinal from (II) which cannot possibly be in the image of f. This part of the proof showing that the power of (II) is greater than that of (I) (the rest of (8) follows easily) is *non*-constructive. The definition of the 'omitted' ordinal is impredicative since, via the involvement of f, it refers to the collection (II) itself. (For Cantor's proof, see [1883b], pp. 197–200. It is repeated, with some modifications, in [1897], pp. 331–3.) Before looking at the impredicative element here, let us look at the similarity between the proof of (8) and Cantor's two proofs that the continuum is not cardinally equivalent to the natural numbers.

In his [1874] Cantor proved

(12) No denumerable sequence of elements of an interval [a, b] can contain *all* elements of [a, b].

The proof of (12) (often overlooked in favour of the [1891] 'diagonal argument' when reference is made to Cantor's proving the non-denumerability of the continuum) was clearly the forerunner of the constructive part of the proof of (8). Cantor's method was the following.

Let $\qquad$ (*) $\qquad a_1, a_2, a_3, \ldots, a_n, \ldots$

be any given sequence of elements of [a, b]. We can assume that $a_1 = a$ and $a_2 = b$ without loss of generality. Let $a^{(1)}$ be the first member of (*) (according to index) lying in (a, b); let $b^{(1)}$ be the first member of (*) lying inside $(a^{(1)}, b)$. Similarly, let $a^{(2)}$ be the first member of (*) lying in $(a^{(1)}, b^{(1)})$, and $b^{(2)}$ the first member lying in $(a^{(2)}, b^{(1)})$. Continue this process as far as possible. If it can only be continued finitely far, then there is a last interval $(a^{(n)}, b^{(n)})$ which (since there is no $(a^{(n+1)}, b^{(n+1)})$) can contain at most *one* member of (*). Thus, we can certainly find an element from $(a^{(n)}, b^{(n)})$, and hence from [a, b], *not* in (*). Suppose, however, that we continue the sequence $(a^{(n)}, b^{(n)})$ indefinitely. Four things must be noted: $a, a^{(1)}, a^{(2)}, \ldots, a^{(n)}, \ldots$ is a bounded increasing sequence with a limit point a' in [a, b]; $b, b^{(1)}, b^{(2)}, \ldots, b^{(n)}, \ldots$ is a bounded decreasing sequence with a limit point b' in [a, b], and since any $a^{(i)}$ is less than all $b^{(k)}$ we must have $a' \leqslant b'$; and lastly for any given n, a_n cannot possibly belong to $(a^{(n)}, b^{(n)})$, simply by the way the nested intervals are constructed. Now, with regard to a', b' two possible cases arise.

(i) $a' < b'$. But in this case no member of (a', b') can possibly be in (*), for if $x \in (a', b')$ then $x \in (a^{(n)}, b^{(n)})$ for all n, and hence $x \neq a_n$ for all n.

(ii) $a' = b'$, which, as Cantor pointed out ([1879], p. 145), is certainly the case if (*) is everywhere dense in $[a, b]$. Here a' cannot be in (*), for again $a' \in (a^{(n)}, b^{(n)})$ for all n, and hence $a' \neq a_n$ for all n.

Either way there are members of $[a, b]$ not in (*). (See Cantor [1874], p. 117. The proof is repeated and modified in [1879b], pp. 143–5. One should note that this proof, as stated, also contains a non-constructive element, namely appeal to the Bolzano-Weierstrass theorem (cf. van Dalen's remarks in Fraenkel, Bar-Hillel and Levy [1973], p. 268).)

The reference to the full completed interval $[a, b]$ is here strictly unnecessary. What is proved, in effect, is that if one starts with a sequence of numbers (*), not necessarily in their natural order though bounded above and below by b and a respectively, then it is possible to find a strictly increasing subsequence a_{n_0}, $a_{n_1}, \ldots$ with $n_0 < n_1 < \ldots$, and a strictly decreasing subsequence $b_{m_0}, b_{m_1}, \ldots$, with $m_0 < m_1 < \ldots$, which necessarily lead us to a number between the given limits a and b but *not* contained in (*). (This presentation in terms of subsequences is more explicit in Cantor's [1879b] version of the proof.) Compare this now with the 'constructive' part of the 1883 proof. Here, Cantor proceeds as follows.

Assume that we are given a sequence of countable ordinals

$$(**) \qquad \alpha_0, \alpha_1, \alpha_2, \ldots$$

not necessarily in natural order. Select from (**) a subsequence increasing in natural order by much the same method as in the 1874 proof. Let $\alpha_{n_0} = \alpha_0$; let α_{n_1} be the first element of (**) greater than α_0 (if there is one); let α_{n_2} be the first element of (**) after α_{n_1} greater than α_{n_1} (if there is one); and so on. If the process peters out with some maximum member of (**), then the desired result follows easily. If it does not, we obtain an infinite increasing sequence

$$(***) \qquad \alpha_{n_0}, \alpha_{n_1}, \ldots, \alpha_{n_k}, \ldots$$

with $n_0 < n_1 < n_2 < \ldots$. Now, says Cantor, if we take (***) and add to it all numbers less than each of the α_{n_k}, we obtain set 'obviously of the *first* power' ([1883b], p. 198). (It is at this point that Cantor depends on the countable union theorem and hence implicitly well-ordering. In his [1878] Cantor claimed that this theorem is 'easily proved', though he did not give the proof. In his [1897] proof of the non-denumerability of (II) he explicitly uses countable union, though here he proves it: [1895], p. 294, and [1897], Theorem J, p. 329.) From this, by the ordinal generating principles, it follows that (***) characterizes a number β which cannot be in (***), yet which must be countable.

Thus it is clear that these two 'constructive' proofs have a good deal in common. (Cantor applied the same method again in his [1884*a*], pp. 215–18, Theorem A, in proving that any perfect set is uncountable.)

The non-constructive part of the proof of (8) is now this. Assume that (II) is given as a sequence $\alpha_1, \ldots, \alpha_n, \ldots$; then by the 'constructive' part of the proof, there is certainly a number β from (II) which is greater than *all* α_n, which is a contradiction. Stated explicitly: if f is a function from (I) one–one onto (II), then $\exists \beta \in (\text{II}) \forall k \in (\text{I}) [\beta > f(k)]$; but this β by construction is either

$$\sup_{k \in (\text{I})} \{f(k)\} \quad \text{or} \quad \sup_{k \in (\text{I})} \{f(k)\} + 1.$$

Thus the characterization of β is impredicative. β is defined in terms of f; f refers to (II) and β is an element of (II). This impredicative method reappears in Cantor's second, and more famous, proof of the non-denumerability of the continuum ([1891]). The general method there is explained not for the case (I) $\not\sim [a, b]$ but rather for the case $L = [0, 1] \not\sim \{0, 1\}^{[0,1]} = M$. I quote the argument in full:

M does *not* however have the *same* power as *L*. For otherwise *M* could be put into unique one–one correspondence to the variables *z* [of *L*], and thus *M* could be thought of in the form of a single-valued function

$$\phi(x, z)$$

of the two variables *x* and *z*, in such a way that through every specification of *z* one obtains an element $f(x) = \phi(x, z)$ of *M* and conversely every element $f(x)$ of *M* could be obtained from $\phi(x, z)$ by specifying a certain *z*. This however leads to a contradiction. For if we understand by $g(x)$ that single-valued function of *x* which takes only values 0 or 1 and which for every value of *x* is different from $\phi(x, x)$, then on the one hand $g(x)$ is an element of *M*, and on the other it cannot be obtained from $\phi(x, z)$ by any specification $z = z_0$, because $\phi(z_0, z_0)$ is different from $g(z_0)$. (Cantor [1891], p. 280)

The impredicativity is clear: $g(x)$ is characterized in terms of $\phi(x, z)$, whose definition refers to *M* of which g is an element. Thus, a very similar trick is used in the proofs of 1883 and 1891. One 'forms' a set according to some specification (here (II) or $\{0, 1\}^{[0,1]}$), and one then uses impredicative methods to boost conclusions about how many members the set has.

Cantor had clearly already added the non-constructive twist to the 1874 proof when he repeated it in [1879*b*]. What is new about the [1891] argument is that it is based on a general principle allowing for the formation given any *A* of a set strictly bigger than *A*. Thus the provability of general cardinal increase lacking up to now in the number-class sequence was established for a different kind of set formation. This introduction of sets of the form $\{0, 1\}^A$ indicates that impredicative and therefore arbitrary subset formation is not far away, adding a further impredicative element to the impredicative ordinal specification that we noted in §2.1. And the willingness to iterate the argument is surely significant too. This shows that the essential ingredients of a power set principle are clearly present in [1891]. (See Chapter 5. However, it is wrong to go quite as far as Wang [1974] (p. 212) and say that Cantor actually has power sets here.) There is often in Cantor a conflict between the strong realist sentiments outlined in Chapter 1 and a 'constructivist' mode of expression, an emphasis on *our* capacity to *perform* a certain operation. (This mode of expression is most noticeable in Cantor's brief discussions of sets, and the conflict is most pronounced in his theory of number. See Chapter 3.) Use of the impredicative methods of set (or ordinal) formation, however, shows that, while some aspects of Cantor's mathematical theory can be recast in a constructivist framework, some parts certainly cannot. Use of impredicative methods is unashamedly realist (see Chapter 6). Moreover it seems that genuine increase in the infinite cardinal size of sets *cannot* be achieved without some impredicative method (see Wang [1954], pp. 562–4). It should be noted, however, as Borel first pointed out, that if one denies that, say, the countable ordinals or the real numbers *can* be all collected together into a set, one can still use the 'constructive' core of the proofs outlined to conclude that no countable sequence of elements of the given kind can include all conceivable elements of that kind. But Cantor *does* assume that such collection is possible. (Borel's views seem to have undergone a transition. In his [1898], Note I, he seems to accept the Cantor proof of the uncountability of $\mathbb{R}$ (pp. 108–9), though there are some hints of a doubt about the treatment of $\mathbb{R}$ as a completed domain, and certainly doubts about forming or defining completed domains of higher powers, for example the domain of real functions. In Note II he seems to be already against the treatment of the second-number class as completed (see Borel [1898], pp. 121 and 123). In his [1900] he seems to have brought the doubt firmly to the arithmetic continuum also. He states clearly: '. . . it is not possible to create the non-denumerable with the denumerable' (p. 143). One would certainly take this to mean that continual 'formation' of Cauchy sequences from the rational numbers (or of new limit numbers from countable ordinals) will never lead outside the realm of the denumerable, and thus never to the classical completed continuum (or, respectively, to the second-number class). In his [1908], he actually suggests that the dichotomy between 'effectively and non-effectively enumerable' is the important one for mathematics, not that between 'denumerable and non-denumerable' which is 'of no practical value' (p. 164). He claims that the 'practical continuum' which mathematicians actually use is denumerable but non-effectively enumerable.)

Before discussing some further implications of Cantor's proof of (8) it is important to note that this proof was by no means a route to a proof of the general statement (7). The core of Cantor's proof (which also rests on the fact that $\aleph_0^2 = \aleph_0$, as the countable union theorem was later stated) is the result that if A is a set of ordinals less than ω_1 of countable type β, then A has a supremum which is less than ω_1. But Hausdorff noticed that this is only generalizable to those ω_α which are *regular*, i.e. not cofinal with any smaller ordinal.

As he pointed out, the sequence $\omega_0, \omega_1, \omega_2, \ldots, \omega_n, \ldots$ of ordinals less than ω_ω provides a counter-example if the regularity condition is dropped. This is the first stage at which the proof method would be disrupted, since ω_ω (the least element of class (ω)) is the first singular ordinal after ω. Hausdorff's generalization under the regularity assumption is stated and proved in [1914] (see pp. 129 and 132). The distinction between singular and regular ordinals is first introduced in [1908], p. 442, where it is recognized that singular initial ordinals ω_α will not be closed under $<\omega_\alpha$ sums of ordinals $<\omega_\alpha$; that the cofinality of ω_ω is ω is pointed out in [1906], p. 125 (see also [1908], p. 442). Cantor's result $\aleph_0^2 = \aleph_0$ *did* generalize without restriction: $\aleph_\alpha^2 = \aleph_\alpha$ was first proved by Hessenberg ([1906], p. 593) and later by Jourdain ([1908]) more smoothly. According to Bernstein ([1905b], p. 150) Cantor had a proof of this which built on his own proof of $\aleph_0^2 = \aleph_0$ given in [1895], pp. 294–5. This is essentially what Jourdain does, in effect using the canonical ordering of *Ord* $\times$ *Ord* to show that it can be injected into *Ord* (see, e.g. Drake [1974], p. 50). Hessenberg proves explicitly that $\aleph_\alpha + \aleph_\beta = \aleph_\alpha \cdot \aleph_\beta = max(\aleph_\alpha, \aleph_\beta)$, thus proving that the arithmetic for multiplication and addition is as simple as possible.

In any case interest was naturally focused, at least initially, on the properties of the class (II), and not on $\aleph_\alpha^2 = \aleph_\alpha$ or (7). The two proofs sketched here show that both the sets (II) and $[a, b]$ (or the reals as a whole) are rich enough in members to transcend the denumerable. The particular result (8) concerning (II) also constitutes the first step towards the scale solution to the continuum problem. Shifting from (I) to (II) *does* increase power, so at least the foundation of an increasing scale is laid. Moreover, assuming the comparability of sets (or at least, the comparability of (II) with the continuum), the continuum hypothesis of 1878 must now be equivalent to:

(13) the continuum is cardinally equivalent to (has the same power as) (II).

So, as Cantor stated it:

I hope soon to be able to answer the question [about the various powers in Euclidean spaces] by giving a rigorous proof that the sought-after power [of the continuum] is none other than that of our *second* number class (II). From this it will follow that all infinite point sets either have the power of the first number-class (I), or the second number-class. (Cantor [1883b], pp. 192–3.)

The continuum problem thus becomes, as Cantor wished, a problem about the position of a given power in a scale. As the second sentence of this passage suggests, Cantor's [1883b] demonstrates that (II) has one of the key properties required for it to be of help in solving the continuum problem as originally stated. For in proving (8) Cantor had shown that

(14) any infinite subset of (II) must either be equivalent to (I) or the whole of (II).

(See Cantor [1883b], p. 200, and also the letter to Dedekind, 5 November 1882,

Noether and Cavaillès [1937], p. 57.) That the continuum also has this property
(*mutatis mutandis*) was precisely the way the continuum hypothesis of 1878 was
formulated.

The fact that (II) has one of the key properties required for it to represent
the power of the continuum no doubt deeply impressed Cantor (see the letter
to Dedekind of 5 November 1882). But there were other aspects of (II) also
revealed in the proof of (8) which perhaps helped to convince Cantor that (II)
was a *likely* candidate for representing the power of the real numbers. The
proof underlines an important structural similarity between (II) and any closed
interval [*a, b*]. There is a striking passage where Cantor touches on this similarity:

. . . ω can in a certain way be considered as the limit which the variable finite
whole number ν approaches, though only in the sense that ω is the *smallest*
transfinite ordinal number, that is, the smallest *fixed* number which is greater
than *all* finite numbers ν. In the same way $\sqrt{2}$ is the limit of certain variable,
increasing, rational numbers; though here it should be added that the difference
between $\sqrt{2}$ and the fractions which approach it becomes arbitrarily small,
whereas $\omega - \nu$ always equals ω. This distinction does not change the fact that
ω is to be recognized as quite as definite and complete as $\sqrt{2}$, nor that it contains within itself traces of the numbers which strive towards it, just as $\sqrt{2}$
contains something of the fractions approaching it.

The transfinite numbers are in a certain sense *new irrationalities*, and in my
view the best method of defining the *finite* irrational numbers is quite similar
to, and I might even say in principle the same as, my method of introducing
transfinite numbers. One can say unconditionally: the transfinite numbers
stand or fall with the finite irrational numbers: they are alike in their innermost
nature [*Wesen*], since both kinds are definitely delimited forms or modifications
of the actual infinite. (Cantor [1887-8], pp. 395-6)

It is difficult to know exactly how far Cantor took the similarity, though there
are several points which should be made here.

Cantor's reference to 'the best method of defining irrational numbers' is to
his [1872] where reals are defined *via* Cauchy sequences of rationals. The final
remark in the passage points to one clear similarity in the method of definition,
as well as recalling Cantor's claim explained in §2.1 that the transfinite numbers
are just as real as the finite numbers. Both (II) and the real numbers are actually-
infinite domains 'created out of', or more accurately, defined in terms of, a given
actually-infinite set (I) or the rational numbers respectively. In both cases, the
elements of the new collection are defined by considering rearrangements of
(elements of) the base set: for the real numbers, arbitrary Cauchy sequences of
rationals; for the countable ordinals, arbitrary well-ordered arrangements of the
elements of (I). Each arrangement is an infinite sequence (a 'definitely delimited
form of the actual infinite'), yet it determines or defines a definite number.
Cantor's 'stand or fall' in the above passage is doubtless a further reference to his
claim (§1.2) that use of the actual-infinite is *essential* in defining the reals. But
there is another important similarity which Cantor touches on in the first part
of this passage, and this has to do with the notion of limit.

Cantor explicitly mentions that both $\sqrt{2}$ and ω can be considered as limits of the sequences of elements used to define them. The behaviour of the real numbers (or any interval $[a, b]$) with respect to limits is one of the crucial features in the first proof of non-denumerability. What enables the proof to go through is the fact that any increasing or decreasing sequence of elements of $[a, b]$ defines a limit element also within $[a, b]$. Rather strikingly, the set (II) has a very similar limit property: any (countable) increasing sequence of elements of (II) defines an element of (II). It seems quite natural to suggest that it was recognition of this which led Cantor to formally apply the proof method for the uncountability of $[a, b]$ to (II), thus explaining why the 1874 and 1883 proofs are so similar. Put more dramatically, recognition that (II) has a similar limit property leads to the recognition that it is a candidate for representing the power of the continuum. It is impossible to say whether Cantor further concluded that this makes (II) a *very plausible* candidate for representing the power of the continuum. What we can say, though, is that if so then he was severely misled, for it seems that this particular similarity between the reals and (II) is counterbalanced, indeed outweighed, by certain striking dissimilarities. For example, the limit property just mentioned can be precisely formulated for (II) by treating $(\omega_1, <)$ as a topological space (using the open-interval topology). But then *unlike* the real line (with the same topology) $(\omega_1, <)$ will be highly discrete ($\{\alpha\}$ is an open set just in case α is a successor ordinal) and highly disconnected (there is a disconnection between any α and its successor $\alpha + 1$). Cantor's analysis had shown that (II) (or $(\omega_1, <)$) is in some sense rich in points. But despite this, topologically speaking one is forced to say that whereas the real line is densely populated, $(\omega_1, <)$ is *sparsely* populated. This 'sparseness' does not mean that (II) cannot represent the power of the real numbers: indeed we know it is consistent to assume it can, and the terms 'sparse', 'dense' are being used here in a topological sense, and need not have anything to do with cardinality. But it does mean that one should be wary of seeing confirmation of the continuum hypothesis in the topological richness of (II) as Cantor possibly did. This, though, is a lesson of hindsight, and one should remember, as film director Billy Wilder once remarked, that 'hindsight is always twenty-twenty' (Cohen and Cohen (eds), *Penguin Dictionary of Modern Quotations*, revised edition, Penguin, Harmondsworth, 1980, p. 356).

(b) What did Cantor achieve?

Having created a scale, or a framework for a scale, Cantor had shown that the continuum hypothesis can be phrased in scale-terms. The next step was the development of an *arithmetic* for powers, and with it the exhibition of the number-classes as an *arithmetical*-scale. Much of this came in his [1887-8] and [1895], after Cantor had started using the abstraction treatment of power. (Actually, the relevant part of [1887-8] stems from 1884 p. 411, n. 1.) In the former we find a definition of the addition of two powers based on the union of two disjoint sets (p. 414). Cantor says that this definition holds also for 'several' powers, but does not expressly say whether this covers the infinite case. But the way he also defines multiplication of two powers here implies that he intended his definition to be quite general. For multiplication is based on a notion of a *product M·N* of two sets. *M·N* is the set 'which arises out of *N* when one puts *in place of each* single element of *N* any set which is equivalent to *M*' (p. 414). But this must mean that one takes the *union* (or general addition) of the new set. It should be noted here that the step to *M·N* from *M* and *N* involves properly not only duplicating enough (disjoint) copies of *M* but also a replacement argument on *N*. In [1887-8], though, there is no treatment of cardinal exponentiation. This came explicitly only in [1895], pp. 287-8. Given two sets *M* and *N* Cantor first calls a function from *N* to *M* a 'covering' of *N* by *M* [*Belegungsfunktion*] (see also Zermelo [1932], p. 352, n. 3). The power of *M* raised to the power of *N* is then based on the set of all 'covering functions', i.e. the set of all functions from *N* to *M*. (In his [1895] Cantor also changes the definition of the product set *M·N*. This is now defined as the 'connection set' [*Verbindungsmenge*] of *M* and *N*. This is (in effect) the set of all ordered pairs of elements of *M* and *N* (see [1895], p. 286).)

The characterization of exponentiation was a natural generalization from Cantor's [1891] paper, where the argument that there are more real numbers than natural numbers is presented in terms of the 'covering' set of $\{m, n\}$ by (I); the argument is then generalized by using the covering of $\{0, 1\}$ by arbitrary *A*. Through this and the explicit definition of [1895], it becomes clear that the power of the continuum is an exponential power. As Cantor notes in [1895], p. 288, through the binary representation of real numbers, the power of the continuum must be 'represented by the formula $2^{\aleph_0}$, among others'. (Cantor was already using the binary representation in 1885; see letter to Vivanti, 3 December 1885, Cantor *Nachlass VI*, p. 36.)

With this representation of the exponentiation of powers and the cardinality of the continuum, the continuum hypothesis of 1883 (see (13), §2.2(*a*)) is transformed once again. For what Cantor's (13) shows is that, in the aleph notation, the power of the second-number class is just $\aleph_1$. Thus the continuum hypothesis, arithmetically stated, is simply $2^{\aleph_0} = \aleph_1$. And with this formulation the first aim of the 1883 scale approach to cardinality was achieved (though, curiously, Cantor nowhere in his [1895] actually states $2^{\aleph_0} = \aleph_1$). Also the conjectures of [1883*b*], p. 207, n. 10, that the set of all *continuous* real functions and of *arbitrary* real functions have the powers of number-classes (II) and (III) respectively now became that these totalities have cardinality $\aleph_1$ and $\aleph_2$ respectively.

But an arithmetical statement of the problem was far from being an arithmetical *solution*. It was shown by Cantor's successors that the operations of addition and multiplication with the aleph numbers are quite simple. That is to say, $\aleph_\alpha + \aleph_\beta = \aleph_\alpha \cdot \aleph_\beta = max\{\aleph_\alpha, \aleph_\beta\}$ or, put more picturesquely, each new stage in the hierarchy of alephs dwarfs all previous stages. It is very probable that these equalities were known to Cantor by 1899 (see §2.2, p. 71 and §2.3(a), p. 79, n. 1). They were apparently first publicly conjectured by Whitehead [1902], and partial results were obtained by Jourdain [1904a]. Full proofs were given by Jourdain [1904b], Hessenberg [1906], and Jourdain [1908]. But the continuum hypothesis poses a problem in exponentiation or *infinite* multiplication. Indeed $2^{\aleph_0}$ is the simplest non-trivial infinite multiplication we can consider. And neither Cantor nor his successors could find any way of determining what the result of this multiplication is. Infinite exponentiation was and remains a mystery.

There is, of course, *some* information about the exponentials $2^{\aleph_\alpha}$. For example, Cantor's theorem of 1891 shows that $\aleph_\alpha < 2^{\aleph_\alpha}$. And it is easy to show that if $\aleph_\alpha < \aleph_\beta$ then $2^{\aleph_\alpha} \leqslant 2^{\aleph_\beta}$, which is a (very weak!) relative bound result on $\aleph_\alpha$. There is also 'König's theorem', more strictly called the Zermelo–König Theorem, which shows that $2^{\aleph_0}$, more generally $2^{\aleph_\alpha}$, cannot assume certain values in the aleph scale. The Zermelo–König theorem can be stated as follows (we are assuming here that all the cardinals are alephs):

(15) Let $\{a_i : i \in I\}$ and $\{b_i : i \in I\}$ be two families of sets such that $card(a_i) < card(b_i)$; then

$$\sum_{i \in I} card(a_i) < \prod_{i \in I} card(b_i).$$

From this it follows immediately that the cofinality of $2^{\aleph_\alpha}$ must be greater than $\aleph_\alpha$. (The *cofinality* of α, denoted $cf(\alpha)$, is the least ordinal ξ such that there is a ξ-sequence β_δ (i.e. $\delta < \xi$) of ordinals less than α such that

$$\alpha = \lim_{\delta < \xi} \beta_\delta.$$

If $cf(\alpha) = \alpha$ then α is called regular; if $cf(\alpha) < \alpha$ then α is *singular*. In the modern axiomatic theory the alephs are already by definition ordinals. If in a pre-axiomatic framework one does not insist on this then this definition of co-finality can be carried over to the alephs by specifying $cf(\aleph_\alpha) = cf(\omega_\alpha)$.) For assume that $\lambda = cf(2^{\aleph_\alpha}) \leqslant \aleph_\alpha$. Then there will certainly be a sequence of sets $\langle a_\delta : \delta < \lambda \rangle$ such that each $card(a_\delta) < 2^{\aleph_\alpha}$ and

$$\sum_{\delta < \lambda} card(a_\delta) = 2^{\aleph_\alpha}.$$

Applying (15) we get

$$\sum_{\delta < \lambda} card(a_\delta) < 2^{\aleph_\alpha} \, card(\lambda).$$

But the right-hand product is just $2^{\aleph_\alpha}$ since $\lambda \leqslant \aleph_\alpha$. Thus $2^{\aleph_\alpha} < 2^{\aleph_\alpha}$, which is a contradiction. Hence

(16) $\aleph_\alpha < cf(2^{\aleph_\alpha})$.

From this it now follows, for example, that $2^{\aleph_0} \neq \aleph_\omega$. For, because $\aleph_\omega$ is singular and of cofinality ω, and $\aleph_0$ is also of cofinality ω, $2^{\aleph_0} = \aleph_\omega$ would contradict (16). Thus, we know *some* values that $2^{\aleph_0}$ *cannot* take, $\aleph_\omega$, $\aleph_{\omega_\omega}, \dots$ etc.

Theorem (15) was first proved by Zermelo in his [1908*b*], pp. 212–13, though it is stated there directly in terms of sets and not cardinalities. (Zermelo was attempting to do without cardinal (or ordinal) numbers: see Chapter 7.) The idea of the theorem and the proof (for which see, e.g., Levy [1979], p. 107) stems however from König. König proved (15) for the case where $card(a_i) \geqslant \aleph_0$ and $I = \omega$ and a_n is a family of strictly increasing cardinality ($card(a_n) < card(a_{n+1})$). His theorem originally formed part of a proof that $2^{\aleph_0}$ cannot possibly be *any* aleph (see König [1904]). König proved (under the conditions stated)

$$\sum_{n \in \omega} card(a_n) < \prod_{n \in \omega} card(a_n) \leqslant \left(\sum_{n \in \omega} card(a_n) \right)^{\aleph_0}.$$

Assume now that $2^{\aleph_0} = \aleph_\beta$ for some β. König then applied a theorem from Bernstein's doctoral thesis which stated $\aleph_\alpha^{\aleph_0} = \aleph_\alpha \cdot 2^{\aleph_0}$. From this one obtains

$$\aleph_{\beta+\omega}^{\aleph_0} = \aleph_{\beta+\omega} 2^{\aleph_0} = \aleph_{\beta+\omega} \aleph_\beta = \aleph_{\beta+\omega}.$$

But from König's theorem it follows that $\aleph_{\beta+\omega} < \aleph_{\beta+\omega}^{\aleph_0}$ which gives a contradiction. König concluded from this that $\forall \beta [2^{\aleph_0} \neq \aleph_\beta]$, which refutes the aleph theorem. However, Zermelo quickly pointed out (apparently within a day of König's lecture at the Heidelberg International Congress) that Bernstein's proof of his equality works only when α is a successor ordinal. This destroys König's application of the equality. As a result, what we are left with from König's theorem is the negative information that $2^{\aleph_0}$ cannot be any cardinal of cofinality ω. (In the printed version of his lecture, König [1904] concluded definitely only that $2^{\aleph_0}$ cannot be equal to any $\aleph_{\mu+\omega}$, and provisionally that *if* Bernstein's formula is correct it cannot be any aleph. For the story of the conference sitting where König dropped his bomb, see, for example, Fraenkel [1930], p. 216, [1932*b*], p. 473, or Dauben [1979], pp. 247–50.)

Interestingly, at much the same time as Zermelo was criticizing Bernstein's theorem, Hausdorff established a relation between alephs somewhat like Bernstein's, namely the Hausdorff recursion formula:

$$\forall \alpha, \beta(\aleph_{\alpha+1}^{\aleph_\beta} = \aleph_{\alpha+1} \aleph_\alpha^{\aleph_\beta}).$$

He also stated that his proof would not work for limit ordinals in place of $\alpha + 1$, and thus considered that Bernstein's theorem was unproven. He adds that 'its correctness seems all the more problematical' *because* König had shown with its help that the continuum would not have an aleph as its cardinal and that 'there would be cardinal numbers greater than all alephs' (Hausdorff [1904], p. 571). This latter remark, coming as it does before Zermelo's proof of the well-ordering theorem, suggests that the Cantor–Jourdain theorem and proof were already well known. (Cantor and Jourdain both proved the aleph theorem by first proving that any cardinal must be an aleph or greater than all alephs. They then used the theory of absolute collections to deny that there is a cardinal greater than all alephs. (See §§4.1 and 4.2.))

The information from König's theorem has remained the best information about the value of $2^{\aleph_0}$. Gödel summed up the situation as follows in 1947:

But, although Cantor's set theory now has had a development of more than seventy years and the problem evidently is of great importance for it, nothing has been proved so far about the question what the power of the continuum is . . . Not even an upper bound, however large, can be assigned. . . . Nor is the quality of the cardinal number of the continuum known any better than its quantity. It is undecided whether this number is regular or singular, accessible or inaccessible, and (except for König's negative result) what its character of cofinality is. (Gödel [1964], p. 260)

At the end of the paper where Cantor first proposed the continuum hypothesis ([1878], p. 132), he adds a strange and enigmatic remark. That the continuum is cardinally larger than the natural numbers shows that point-sets fall into at least two classes according to power. Cantor then goes on:

Through a process of induction [my italics], which we will not go further into here, I came to the theorem [wird der Satz nahe gebracht] that the number of these . . . classes of linear manifolds is finite and indeed that it equals two.

So far as I am aware, there is no further reference to this 'induction process' in Cantor's published work. However, in a letter drafted (and presumably sent) to Vivanti of 6 November 1886 Cantor does mention it again. What he says there somewhat clarifies the meaning of this reference to induction, and seems already to hint at the hopelessness of proving the continuum hypothesis by arithmetic means. The remark is apropos of Tannery's paper [1884], where Tannery claims to prove the continuum hypothesis. As far as I can understand it, Tannery's argument amounts to this. If we start from the exponential n^m (n and m are natural numbers) and let n increase indefinitely the limit is $\aleph_0$; however, if we let m increase indefinitely, keeping n constant, the limit will be the power of the continuum. I presume that Tannery's idea is that if we compare these two limit processes we obtain a 'jump' from $\aleph_0$ to $2^{\aleph_0}$, and this suggests that there is no power in between. Cantor is rightly very critical of this line of argument, as the following passages show. Here he uses gothic letters to stand for the numbers which were later denoted by alephs; thus $\mathfrak{o}_1$ became $\aleph_0$, $\mathfrak{o}_2$ became $\aleph_1$, and so on.

He [Tannery] believed he had given a proof for the theorem first stated by me 9 years ago that only two [equivalence] classes [by power] appear among linear point-sets, or what amounts to the same thing, that the power of the linear continuum is just the *second*. However, he is certainly in error. The facts which he cites in support of this theorem were all known to me then, as anyone can see, and form only a part of that induction of which I say [reference to text] that it led me to that theorem. I was convinced at that time that this induction is *incomplete* and I still have this conviction today. (Cantor, *Nachlass VI*, p. 87)

In other words, Cantor seems to be saying that these facts *suggest* the theorem, or the conjecture, but do not prove it. Thus it seems that by 'came to the theorem' [*der Satz wird nahe gebracht*] in the [1878] passage he means 'came to the idea of' and not 'established'. Indeed, this is exactly what Cantor says further on in the letter:

If we understand by n the power of the linear continuum, then the theorem to be proved is

$$n = o_2.$$

The facts on which Herr T. believes he can base the theorem are only these:

$$|\nu| + o_1 = o_1, \quad o_1 + o_1 = o_1,$$
$$o_1 \cdot |\nu| = o_1, \quad o_1^{|2|} = o_1, \quad o_1^{|\nu|} = o_1;$$
$$o_1^{o_1} = n; \quad |2|^{o_1} = n; \quad |3|^{o_1} = n,$$
$$\ldots\ldots, |\nu|^{o_1} = n$$

($|\nu|$ is the finite cardinal number or power corresponding to the finite ordinal number ν).

These facts suggest the conjecture that n should be the power o_2 following *next* after o_1; but they are a long way from furnishing a rigorous proof for it, ... (Cantor, *Nachlass VI*, pp. 87–8)

In determining the numerical value of the power of the continuum we have come very little further since.

About the behaviour of the exponential $\aleph_\alpha^{\aleph_\beta}$ in general we *do* know a little more than Cantor and his immediate successors. (For an exposition of the information here, see Levy [1979], pp. 186-9, or Jech [1978], pp. 48-52.) And there is certainly one result due to Silver which yields more *relative* information about the behaviour of $2^{\aleph_\alpha}$. In particular, Silver proved (in *ZF + AC*) that if $\aleph_\alpha$ is a singular cardinal of *uncountable* cofinality such that $2^{\aleph_\gamma} = \aleph_{\gamma+1}$ for all (or a great many) $\gamma < \alpha$, then it must be that $2^{\aleph_\alpha} = \aleph_{\alpha+1}$. (For a proof of Silver's theorem, see Levy [1979], pp. 181-5, or Felgner [1979a], pp. 191-3. Both contain reviews of other relevant results.) Even such relative information, however, is not available (at least in *ZF + AC*) about singular cardinals (like $\aleph_\omega$) of *countable* cofinality. And the situation is even worse with regard to regular cardinals.

In one sense, then, Cantor's arithmetical project is, as Gödel says, a 'pronounced failure' ([1964], p. 260), for transfinite arithmetic gives only partial and meagre answers to the one central question. But one cannot let this statement stand without a serious attempt to appraise the consequences and the importance of the theory Cantor created.

The attempt to achieve a scale of comparable numbers was achieved through the scale of the number-classes. This in turn (if the scale is to be complete) depends on the well-ordering theorem. This dependence looked like, and was widely taken to be, a serious difficulty for Cantor, both *before* and *after* Zermelo's proof. For example, Russell remarked in *The Principles*:

Cantor, it is true, regards it as a law of thought that every definite aggregate can be well-ordered; but I see no ground for this opinion . . . although Cantor professes that he has a proof that of two cardinals one must be the greater, I cannot persuade myself that he does more than prove that there is a series, whose terms are cardinals of which any one is greater or less than any other. That all cardinals are in this series I see no reason to think. (Russell [1903], p. 364)

And *after* Zermelo's proof, he noted:

It is also necessary to say that Zermelo's theorem, namely that if the axiom [of choice] is true then any class can be well-ordered, gives grounds for believing that the axiom is false; for it is hardly believable that every class can be well-ordered. (Russell [1911], p. 172)

But in retrospect it appears that Cantor's intuition in basing cardinality upon well-ordering was perfectly correct. For Hartogs [1915] showed that any system of infinite cardinal numbers (where by 'numbers' we understand objects which are comparable, that is that the sets they represent are cardinally comparable) must depend on the well-ordering theorem. Hartogs showed that not only does the well-ordering assumption imply comparability of sets but it is actually *equivalent* to it. This fundamental dependence on well-ordering was further confirmed in subsequent work. Suppose one tries, as Russell did, to develop cardinality without reference to well-ordering. (Russell's numbers, with the exception of 0, are all 'contradictory' or 'inconsistent' sets (see §4.1 and especially §4.2) and so there are difficulties with his definition of number. However, in axiomatic set theory of the Zermelo–Fraenkel kind they can be mimicked by the so-called 'Scott trick', provided that the axiom of foundation is present, that is to say, provided that the universe is the von Neumann cumulative hierarchy with sets divided into ordinally indexed ranks. Instead of taking $card(x) = \{y : y \sim x\}$ as Russell did, one takes $card(x) = \{y : y$ is of least rank and $y \sim x\}$. For more details, see Fraenkel, Bar-Hillel, and Levy [1973], pp. 97–8.) Then sooner or later one has to prove the theorem

(*)   If A and B are pairwise disjoint and $A \sim B$, and there is a one–one mapping f from A onto B such that $f(a) \sim a$, then $\mathbf{U}A \sim \mathbf{U}B$

(*) is clearly vital in a theory of cardinality, for it establishes that in calculating a sum of cardinals the total is independent of which representative sets are chosen. But for infinite A (*) cannot be proved without the axiom of choice, which, of course, is equivalent to the well-ordering assumption. *Given* for each a in A a mapping $\psi(a)$ from a one–one onto $f(a)$, one can immediately define a mapping of $\bigcup A$ one–one onto $\bigcup B$. But the hypothesis $a \sim f(a)$ only tells us that for each a the set of one–one mappings from a onto $f(a)$ is non-empty. One therefore has to select a ψ from this set. If A is infinite one has to make infinitely many simultaneous choices. And for this one needs the axiom of choice. Levy gives a dramatic example of what can go wrong here without choice or well-ordering: the cardinal sum of denumerably many de-numerable sets can consistently be taken to be either $\aleph_0$ or $2^{\aleph_0}$! (See Fraenkel, Bar-Hillel, and Levy [1973], p. 75, n.1, where this whole subject is discussed.)

It seems then that a theory of infinite cardinality is indissolubly bound up with the assumption of well-orderability, whether or not one takes the position that measures of size *must* be comparable. (For further remarks on this, see §3.5.) Cantor's original tying of cardinality to well-ordering thus seems in-stinctively right. And if one has *eventually* to appeal to the axiom of choice or the well-ordering theorem why not make the best use of it from the beginning and build, as Cantor did, an ordinal theory of powers?

Let us relate the situation here to the axiomatic demands of a theory of cardinality in modern set theory. There are two minimum conditions on cardin-ality:

(*i*) the operation $card(x)$ is defined for all sets x
(*ii*) $x \sim y$ iff $card(x) = card(y)$.

As the remarks on the summing of cardinalities show, one already needs the well-ordering or choice assumption to develop even this theory. But in a theory of sets it is important to add the reductionist condition that cardinal number objects actually be *sets*, i.e.

(*iii*) for every x, $card(x)$ is a set,

a condition which Cantor himself adopted (see Chapter 3). This condition, as Levy's work on the definability of cardinal number showed, commits one to the axiom of foundation, that is to the thesis that the universe is arranged in an ordinally indexed hierarchy. This pushes the theory of cardinality further into the arms of ordinality. (See Levy [1969]. Levy showed that there cannot be a *set* operator $card(x)$ (i.e. satisfying (*iii*)) satisfying the essential (*ii*) *unless* the axiom of foundation is present. More precisely he showed that if we add a new term $card(x)$ to ZF (without foundation or choice) and a new axiom for $card(x)$ tantamount to (*ii*), then there will be theorems not involving $card(x)$ provable in the new system which are not provable in ZF alone.) But Cantor added a fourth condition to these, namely,

(*iv*) for every x, $card(x) \sim x$

(see Chapter 3). (*iii*) and (*iv*) together stipulate that the theory will be a representative theory of cardinality, that cardinal numbers will represent the cardinal equivalence classes from *inside* (see §2.2). What Cantor or Cantor's programme showed was that using the well-ordering assumption and the theory of ordinal numbers the further demands (*iii*) and (*iv*) can be met, joining the theory of cardinality completely to the theory of ordinality with a beautifully streamlined theory of canonical cardinal representatives. The only known way of satisfying (*iii*) and (*iv*) in axiomatic systems is von Neumann's method of initial ordinal numbers, a method very close to Cantor's. (The major difference is that von Neumann shows clearly how to effect the reductionist step (*iii*), something which Cantor left vague (see Chapter 3).)

If one accepts conditions (*i*)-(*iv*) then Cantor's theory, it seems, is *inevitable*. If one only accepts (*i*)-(*ii*), or (*i*)-(*iii*), then some of the detail of Cantor's theory (specifically the ordinal representation of cardinals) is dispensable. But since even here well-ordering must enter at some point (even the ordinals and replacement if one accepts (*iii*)), again *why not* exploit it to the full and use it from the beginning to obtain (*iv*) and the full Cantor theory as well? This was in effect the course which the development of axiomatic set theory eventually followed, since Cantor's theory has been adopted not just in general outline but in specific detail (via von Neumann). Moreover, there is a clear sense in which the desire to take over and incorporate the spirit of Cantor's notion of cardinality dictated the choice of the key set-theoretic axioms. (This will become much clearer in the later chapters on Zermelo and von Neumann.) So while one might say that Cantor's theory failed to solve its central problem (the continuum problem), it is a very important 'failure': modern mathematics is based on it!

It should be noted that, while the demands of Cantor's theory of cardinality played an enormous and profound part in dictating the structure of modern axiomatic set theory, analysis of this axiomatic theory in its turn showed just how far from solving the continuum problem Cantor's theory was. Cohen's independence result showed that the continuum hypothesis is undecidable in the standard axiomatic framework. Not just this, but within the bounds of König's theorem (p. 83), $2^{\aleph_0}$ can take practically any value. Easton's extension of Cohen's result shows that for regular cardinals the exponential operation $2^{\aleph_\alpha}$ is as undetermined as it possibly could be. (Cohen's original proof is in Cohen [1963-4] or his [1966]. For a rather different exposition, see Bell [1977]. Easton's result is in his [1970]; for a discussion of the consequences of this theorem, see Felgner [1979a], pp. 182-3, or Levy [1979], pp. 181-2.)

There is another, quite distinct, sense in which Cantor's theory of ordinals was of fundamental importance, namely its productive effect on neighbouring branches of mathematics, particularly real and complex analysis. Two particularly important classical analytic theorems were first proved using Cantor's transfinite ordinal numbers, Mittag-Leffler's theorem and the Heine-Borel theorem. The fact that these new numbers could lead to profound and far-reaching results in mainstream mathematics was rightly recognized by mathematicians as a strong sign of their mathematical importance, and doubtless played no small part in their acceptance by the mathematical community. (The original proof methods for the two theorems mentioned above, and an indication of the reactions of mathematicians to them, are outlined in Hallett [1979a], §§2 and 4.) There were other important applications of the transfinite ordinals later, not least their role in the classification of elements in hierarchies, for example the *Baire hierarchy* of real functions. This latter hierarchy played a key part in Lebesgue's formulation of his notion of integral. (See Hallett [1979a], §5.)

There is more to be said on the acceptance of the ordinal numbers by mathematicians particularly in view of various attempts to dispense with them in 'mainstream' theories. And this discussion will tie up with Cantor's own work on the continuum problem and its modern extensions.

The conception of the transfinite ordinal numbers originally arose out of the theory of derived sets. The extension of this theory did not, as Cantor had hoped, lead to a solution to the continuum problem. But it did lead to results and definitions of fundamental importance. One of these results was the Mittag-Leffler theorem mentioned above, which was based on a sophisticated analysis of sets of singularities of complex functions made possible by Cantor's analysis of derivation. Another result of key importance was the Cantor-Bendixson theorem, which was the best result Cantor *himself* could attain in attempting to solve the continuum problem.

The development of a fully fledged and precise theory of transfinite ordinals in 1883 allowed Cantor to answer many of the questions raised by the earlier work on derived sets. The achievement of such precision was indeed one of the purposes behind the analysis of the ordinal indices themselves outside the context of derivation (see the Introduction). After 1883 it became clear, for instance, how many ordinals are required for a satisfactory and general treatment of derived point sets. Cantor could prove now that the ordinals of (I) and (II) are sufficient, since the derivation operation $P^{(\alpha)}$ always becomes stationary at some denumerable α, i.e. $\exists \alpha \in (II) P^{(\alpha)} = P^{(\alpha+1)}$ (see Cantor [1884a], p. 218). One part of this result was the theorem stated by Cantor as follows:

(17) If P is a point-set [n-dimensional] whose first derivative $P^{(1)}$ has the power of the first [number-] class then there is a whole number α belonging to the *first* or *second* number class for which $P^{(\alpha)}$ identically vanishes, and of such numbers there is a smallest. ([1883b], p. 171)

But the most important part of the analysis was the Cantor-Bendixson theorem. Cantor already knew ([1883a], p. 159, Theorem II) that if $P^{(1)}$ is countable, then P must be countable. The Cantor-Bendixson theorem was the result of Cantor's attempt to discover what happens when $P^{(1)}$ is uncountable.

Cantor's first statement of the theorem came in [1883*b*] :

If $P^{(1)}$ has the power of the second number-class, then it can always and in only one way be divided into two sets R and S such that

$$P^{(1)} \equiv R + S \text{ [i.e. } P^{(1)} = R \cup S]$$

where R and S have extremely different properties:

R is so constituted that by repetition of the derivation process it can be reduced to annihilation, so that there is always a first number γ of the number classes (I) or (II) for which

$$R^{(\gamma)} \equiv 0;$$

such point-sets I call reducible.

S on the contrary is so constituted that with these point-sets the derivation process brings no alteration because

$$S \equiv S^{(1)}$$

and consequently

$$S \equiv S^{(\gamma)}.$$

Such sets S I call *perfect* point-sets. We can therefore say: if $P^{(1)}$ is of the power of the second number-class, then $P^{(1)}$ splits into a definite *reducible* and a definite *perfect* point-set. (Cantor [1883*b*], p. 193)

Cantor gives no proof of this theorem. If he *had*, it would have constituted a proof of the continuum hypothesis, since he showed shortly after that any non-empty perfect set must have the power of the continuum.[4] Thus, $P^{(1)}$, which by assumption has the power of the second number-class, must have also the power of the continuum, making these two powers equal. It is not really clear why Cantor assumed that $P^{(1)}$ has the power of the second class here. It might simply have been a slip, since in his later statements of the theorem, which all assume only that $P^{(1)}$ is non-denumerable, he does not mention this alteration. Yet he does mention other alterations, namely those caused by Bendixson's counterexample (in his [1883], pp. 416–18) to the assertion that the R Cantor mentions in the theorem is always a reducible set. Cantor's mistake here is rather intriguing. He writes on this in his [1884*a*] as follows:

From the condition that R is at most of the first power, and at the same time a component of $P^{(1)}$, I believed that it should follow that R is the derivative of a certain component of P, and concluded quite correctly from this, that there must be an α such that $R^{(\alpha)} \equiv 0$. However, it has now been shown that in general R need not be the derivative of another set.

This *important* remark was first made by Herr Ivar Bendixson in a communication to me in May 1883. (Cantor [1884*a*], p. 224)

[4] This is first proved in Cantor [1884*a*] pp. 237–44, and [1884*b*], pp. 252–5. The result must have been known, or at least conjectured, by Cantor in 1882, for in his [1883*b*] he bases his definition of a continuum on perfect sets (perfect, polygonally connected). In his [1883*c*] he proves that any perfect set is uncountable (also [1884*a*], pp. 215–18).

It may be that Cantor emphasizes Bendixson's remark as 'important' for the following reason. Let R be any subset of the line. Then trivially there must be a set P, such that $R \subseteq P^{(1)}$. Now, *if* for *any* set R contained in $P^{(1)}$ there must be a $Q \subseteq P$ such that $R = Q^{(1)}$, then the continuum hypothesis would have been proved. For the Cantor-Bendixson results effectively solve the continuum problem (positively) for sets which can be represented as first derived sets. Did Cantor then believe that any set can be represented as a first derived set? Or, at least, did he believe that proving this was the best way to extend the Cantor-Bendixson theorem to a proof of the continuum hypothesis? Certainly, though, Bendixson's result seems to show that this route was now effectively blocked.

The upshot was the following reformulation of Cantor's [1883b] theorem (see Cantor [1884a], pp. 221-2):

(18) If P is such that $P^{(1)}$ is uncountable, then $P^{(1)}$ can be split into two components R and S such that S is perfect and R is either finite or of the first power.

Then from Bendixson's work it could be added that there is always a countable ordinal α such that $R \cap R^{(\alpha)} = \emptyset$ (see Cantor [1884a], p. 224).

What follows from the Cantor-Bendixson theorem, as already remarked, is a partial result about the continuum hypothesis. For it follows from (17) and (18) (and the fact that if $P^{(1)}$ is denumerable so is P) that any infinite set P which is contained in its first derived set must have either the power of the natural numbers or of the whole continuum. In his [1884a] Cantor calls the sets P for which $P^{(1)} \subseteq P$ closed [*abgeschlossen*] sets. *Thus, the continuum hypothesis is true for closed sets.*

This was Cantor's best result on the continuum problem, but, as we shall see, by no means the last word on the 'direct attack'. In 1883-4 Cantor was still optimistic about proving the continuum hypothesis fully. At the end of his [1884a] he notes (p. 244): 'That this remarkable theorem [on closed sets] holds also for non-closed linear point-sets, and likewise for all n-dimensional point-sets, will be proved in later sections.' Cantor apparently intended to continue the sequence of papers, and he promises the result again on p. 257 of his [1884b]. On 26 August 1884 he wrote to Mittag-Leffler on one possible way of applying the result on closed sets (see Schoenflies [1927], pp. 16-17): 'Thus, you see that everything reduces to defining a closed set having the second power. When I have sorted it out, I will send you the details.' Unfortunately, neither Cantor nor anyone else ever sorted it out.

Let us look more closely now at the role of the ordinals. Cantor's theories of cardinality and transfinite ordinal numbers were both intimately involved in the birth of point-set topology, a discipline which at the hands of Fréchet and Hausdorff later blossomed into general topology. For one thing Cantor's investigation showed that cardinality maps are too coarse to distinguish between structures like $\mathbb{R}$ and $\mathbb{R}^3$ which clearly have quite different properties. And this was certainly one spur to the study of similarity under continuous maps. (See Cantor [1879a] or Johnson [1979].) For another, the investigation of set derivation led to the isolation of various key classes of sets, in particular sets *dense-in-themselves* (i.e. those P such that $P \subseteq P^{(1)}$) and, as we have seen, *closed* sets, the latter being a notion of inestimable importance in the study of the structure of the continuum as well as in more general topological settings. The importance of these later developments as well as the key theorems already mentioned undoubtedly reflected credit on the whole net of Cantorian ideas, and thus on the theory of transfinite numbers which, for Cantor at least, formed an indissoluble part of that net.

But despite this success—indeed, acknowledged success—the transfinite numbers have been continually dogged by the view that they are out of place in mainstream mathematics, particularly in the study of the continuum and its subsets, that their success there is somehow accidental, and that they are consequently dispensable. This view was given support initially by the fact that crowning 'classical' results like the Heine–Borel and Cantor–Bendixson theorems, although first proved using transfinite ordinals, do not require the ordinals for their statement. Not unnaturally, then, there were attempts to remove the ordinals altogether, to keep the transfinite ordinals quite separate from the study of the continuum and point-set theory. And there were early victories here, too. Borel, who had philosophical objections to the acceptance of completed higher infinities was the first abolitionist, and he quickly found a proof of the Heine–Borel theorem which did *not* use the ordinals, as did Lebesgue (see Hallett [1979b], pp. 23-5, for details). Lindelöf [1905] was the first to give a proof of the key Cantor–Bendixson theorem without relying on transfinite numbers. Cantor's proof *had* used them, via his theorem D of [1884a], p. 221, which shows that if $P^{(1)}$ is non-denumerable then there will be points belonging to *all* subsequent derivatives $P^{(\alpha)}$. The set of all these is $P^{(\omega_1)}$ which is perfect. (Cantor also easily showed that there must be an α in (II) such that $P^{(\alpha)} = P^{(\omega_1)}$; see [1884a], pp. 223-4.) But Lindelöf took a very clear abolitionist or separationist line, as his [1905] shows. He remarks:

. . . . the most important part of the [Cantor–Bendixson] theorem concerns the division of a closed set into a perfect set and a denumerable set. But this seems quite estranged from the notion of the transfinite and transfinite numbers, and it is therefore desirable to find a proof where this does not play a part. (Lindelöf [1905], p. 183)

Similar sentiments are expressed by another pioneer, W. H. Young (see Young and Young [1906], p. 284). The proofs given by Lindelöf and Borel supported these sentiments, and indeed helped to confirm the earlier impression of irrelevance.

Two fairly obvious points arise with regard to 'abolition' or 'separation'. First, *post hoc* removal does not at all detract from the heuristic importance of the ordinals as agents of mathematical discovery. As I argued in [1979*a*] the original proof of the Heine–Borel theorem establishes a connection between two key properties (bounded infinite sets having upper bounds and compactness or, alternatively, bounded increasing and decreasing sequences having limits and compactness) and the transfinite numbers clearly helped in finding this connection. Once it was known, it was natural to want to simplify the argument, as Borel and Lebesgue did, one way being to see if the 'stepwise' (or recursive) approach characteristic of the involvement of transfinite ordinals can be replaced by a 'one stroke' argument (see Hallett [1979*a*], pp. 23-5). This *was* the case with Heine–Borel, and it was the case also with Lindelöf's proof of the Cantor–Bendixson theorem. Secondly and relatedly, one might claim that heuristic fecundity in the past is a good reason for not 'banning' the transfinite numbers, and that it indicates the value of a more holistic approach than the abolitionists were prepared to admit. Does the success of the transfinite numbers perhaps indicate that they are not so 'foreign' to point-set theory and reasoning about point sets as Lindelöf and Borel supposed? Both these points were hinted at by Hausdorff in comments on the Cantor-Bendixson theorem and Lindelöf's proof:

With Cantor the theory of ordinal numbers played an essential role in the theory of point-sets. Indeed, ordinal number theory was originally constructed precisely for this purpose. E. Lindelöf recognised that the concept of condensation point [*Verdichtungspunkt*] (which already appears in Cantor) allows one to dispense with the ordinal numbers, and our presentation also follows this tendency. For the kernel A_k [of a set A] which in Cantor appears as the end term in a well-ordered series of sets is here defined with one blow as the largest dense-in-itself subset of A. Nevertheless the ordinal numbers maintain their position in the theory of point-sets for both historical and technical reasons. They make possible a finer analysis of the structure of given sets by differentiating between different orders of accumulation points [*Häufungspunken*]. This analysis has performed valuable service in function theory (as is shown by the theorems proved by Mittag–Leffler concerning the existence of functions with prescribed singularities) even though it comes not a step nearer solving the problem of powers for point-sets than the other more summary treatment of the uncountable. (Hausdorff [1914], p. 275)

Put more forcefully, logical minimality, while interesting in itself, is not everything in mathematics and rigid pursuit of it may serve to hide deeper connections. Subsequent developments have borne out the points made here, and have indeed shown that the ordinals are not just heuristically useful in point-set theory, but in some sense are *indispensable*: it has proved impossible to exclude the ordinals from the theory of even relatively 'simple' subsets of the continuum. Zermelo, no doubt emboldend by the success of Borel, Lindelöf, and others, elevated the thesis of 'dispensability of transfinite numbers for Euclidean continua' to the 'principle of the dispensability of transfinite numbers for the whole of set theory'. He was followed in this by Fraenkel and Kuratowski who continued to look for general methods of avoiding ordinals and ordinal reasoning, for example, by looking for substitutes for definition by transfinite recursion. This bold form of abolitionism will concern us in §7.2, where the background and some of the effects are discussed. Now von Neumann's original motivation for adding the axiom of replacement to the Zermelo axioms (for it was he who did so, not Fraenkel) was precisely to *restore* the ordinal numbers which Zermelo and Fraenkel had regarded as unnecessary. But it has turned out that the system *with* the axiom (ZF) has a consequence about Borel subsets of the continuum which is unprovable in the system without the axiom (Z). Thus the ability to prove the existence of ordinals from $\omega + \omega$ upwards allows us also to prove more about Borel subsets, so acceptability of these ordinals is directly connected to what we can establish about rather 'simple' subsets of the continuum. Thus, two domains which Lindelöf, Borel, and others had thought were and should be quite separate, are in fact connected. Not just this, though: recent research on descriptive set theory (the theory of relatively simple subsets of the continuum, which even the tough-minded French mathematicians regarded as falling well within the boundaries of acceptable set theory) has revealed that to answer even rather elementary questions we have to appeal to *very* strong assumptions about the existence of ordinals, assumptions like the existence of a measurable cardinal. And these are not just questions obviously concerning the transfinite (transfinite power, for instance) but 'mainstream' questions too, Lebesgue measurability for instance.

Thus, the theory of ordinals is closely, though perhaps rather subtly and mysteriously, interwoven with the theory of point sets. Abolition or separation is no simple or obvious matter.

The phenomenon of interweaving we witness here is in fact one aspect of the merging of Cantor's two originally distinct lines of attack on the continuum problem, and this is part of the story of §2.3 (*c*). Before we discuss that, and partly by way of introduction, it is worth looking more closely at Lindelöf's new proof of the Cantor–Bendixson theorem.

Let us begin by explaining Hausdorff's remarks on his own and Lindelöf's proof method. What Cantor's iteration procedure through the $P^{(\alpha)}$ does is gradually to strip away limit or accumulation points [*Häufungspunkte*] from P (assuming P closed) until one is left with a *perfect* core S. If one now removed points from S one would then start creating sets Y such that $Y \subseteq Y^{(1)}$, i.e. sets of the kind that Cantor called 'dense-in-themselves' ([1884*a*], p. 228), but which are not also *closed*. S itself is, of course, *both*. This might lead one to conjecture that S is the *largest* dense-in-itself subset of P, what Hausdorff in his book calls the kernel A_k of a set A. Indeed, this is the case. (S is also the set of condensation points of P. A point p is called a condensation point of P if every neighbourhood of p contains uncountably many points of P. This notion goes back to Cantor [1885*b*], p. 268.) Thus, the idea behind the Lindelöf–Hausdorff proof is to *start* by defining S as the largest dense-in-itself subset of $P^{(1)}$ (or, alternatively, the set of condensation points of $P^{(1)}$), put

$R = P^{(1)} - S$, and then proceed to show, assuming $P^{(1)}$ is uncountable, that S is non-empty and perfect and R is countable. A key part in Hausforff's presentation is played by the Young–Hausdorff theorem

(*)   If A is non-empty and dense-in-itself, then $A^{(1)}$ has cardinality $2^{\aleph_0}$

from which follows Cantor's result that every non-empty perfect set has power $2^{\aleph_0}$. The proof is beautiful (see Hausdorff [1914], p. 319, or Felgner [1979a], pp. 168-9), and important for the extensions of the Cantor–Bendixson theorem discussed in §2.3 (c). Let A be dense-in-itself and non-empty, and a_1 and a_2 any two elements of A, with V_1 and V_2 closed *disjoint* spheres with centres a_1, a_2 respectively. But a_1 and a_2 must be accumulation points of A; thus the interiors U_1 and U_2 of V_1 and V_2 contain infinitely many points of A. Choose two of these different from a_1 in V_1, and denote them a_{11} and a_{12}; similarly choose a_{21} and a_{22} in V_2. Now let V_{11} and V_{12} be closed disjoint spheres about a_{11} and a_{12} respectively, chosen such that the radii of V_{11} and V_{12} are less than half the radius of V_1. Choose V_{21} and V_{22} in the same way, and then V_{111}, V_{112}, V_{121}, V_{122}, V_{211}, V_{212}, V_{221}, V_{222} with radii less than quarter of the radius of V_1 and V_2 etc. The picture (for the first three stages) is therefore as follows:

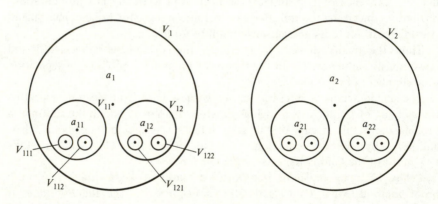

This inductive process can be assumed taken to the limit, yielding a multiplicity of sequences V_{x_1}, $V_{x_1 x_2}$, $V_{x_1 x_2 x_3}$, ... (where $x_n = 1$ or 2). Each such sequence is a nested sequence of closed neighbourhoods whose radii tend to zero. Thus the intersection of each such sequence contains just one point, which we can denote $a_{x_1 x_2 x_3}$ ..... It is easy to see that points corresponding to different sequences are different, and that for any sequence $x_1, x_2, \ldots, x_n, \ldots$ ($x_n = 1$ or 2) there must be a corresponding point in $A^{(1)}$ defined in the process above. Thus, if the points of $A^{(1)}$ so defined are denoted A^*, then we have shown that there is a one–one onto map between A^* and the set of sequences $\{1, 2\}^\omega$. Thus A^* has cardinality $2^{\aleph_0}$, and so also must $A^{(1)}$. By examining the above argument, it is also easy to see that since the points a_1 and a_2 were arbitrarily chosen, *any* point of A is a *condensation* point [*Verdichtungspunkt*] of A, i.e. $A \subseteq A_\gamma$, where A_γ is the set of all condensation points of A.

Theorem (*) is also proved by Young in Young and Young [1906], pp. 64-5, though Hausdorff's presentation, followed here, is incomparably clearer. The proof is, in effect, a modification of Cantor's proof ([1884a], pp. 216-8) that every perfect set is uncountable. A simple extension (see below, where we assume A is closed) gives a very nice proof that any perfect set has the power of the continuum, a much nicer proof than Cantor's original proof of [1884b].

Cantor's partial result on the continuum hypothesis now follows immediately. Let A be closed and uncountable. Being uncountable A must have a non-empty dense-in-itself subset B ($= A \cap A_\gamma$). (This is first established in Cantor [1885b], p. 268. See also Young and Young [1906], p. 53, Hausdorff [1914], p. 268, and Lindelöf [1905], p. 184.) Thus by (*) $B^{(1)}$ has power $2^{\aleph_0}$. But $B^{(1)} \subseteq A^{(1)}$ and since A is closed $A^{(1)} \subseteq A$. Thus A has power $2^{\aleph_0}$.

The full partition result of the Cantor–Bendixson theorem can also now be easily obtained. For any set A, A_γ is a perfect set. But since each condensation point is an accumulation point, we have $A_\gamma \subseteq A^{(1)} \subseteq A$. Thus A is split into $S = A_\gamma$ and $R = A - A_\gamma$. But $x \in A - A_\gamma$ is not a condensation point of A, and thus cannot be a condensation point of R. Thus $R \cap R_\gamma = \emptyset$. But Hausdorff showed that this is a sufficient condition for R to be at most countable. (See Hausdorff [1914], p. 268, and also Lindelöf [1905], p. 188. Young and Young [1906], p. 53, also establish that if $R_\gamma = \emptyset$ then R is countable, which is good enough here.) Thus

(**) If A is closed and uncountable then it can be split into two sets S and R which are perfect and at most countable respectively.

And this is just the content of (18) obtained by a different route.

(Lindelöf's proof is similar. He proves for any uncountable set P, P_γ (staying with Hausdorff's notation) must be non-empty and perfect (Lindelöf [1905], pp. 184, 186) and that $R = P - (P \cap P_\gamma)$ must be countable (p. 185). If P is closed, then $P_\gamma \subseteq P^{(1)} \subseteq P$, and so $R = P - P_\gamma$. Thus, given that any non-empty perfect set has power $2^{\aleph_0}$, (18), and the theorem that the continuum hypothesis holds for closed sets, follow.)

The approach here uses the set A_γ 'in one blow' instead of Cantor's 'successive' derivations. We could also use Hausdorff's kernel A_k. For we know that A_γ is dense-in-itself, and thus for a closed set A, $A_\gamma \subseteq A_k$, this being the largest dense-in-itself set in A. But since A_k is dense-in-itself we can use the proof of (*), and this proof shows that any neighbourhood of any point of A_k contains uncountably many points of $A_k^{(1)}$. But since $A_k^{(1)} \subseteq A^{(1)} \subseteq A$ any point of A_k is a condensation point of A, that is belongs to A_γ. Thus $A_k = A_\gamma$, so we might just as well have started with A_k.

Hausdorff's presentation is interesting because he proves the results stated here not just for Euclidean point-sets but for more general topological spaces. He builds up to them by gradually strengthening the conditions on the space. Particularly important here is the so-called *second axiom of countability* in the proof that if $A \cap A_\gamma = \emptyset$ then A is countable. This is used twice above, once to show that $A_\gamma \neq \emptyset$ and once to show that R is countable. The axiom states in effect that the topology of the space has a countable neighbourhood base, and given this the theorem stated follows easily from the countable sum theorem (the countable union of countable sets is countable). Lindelöf [1905], p. 188, proves that every finite-dimensional Euclidean space has the property stated by the axiom, and this property is then applied, as Hausdorff later used it, to show that if $A \cap A_\gamma = \emptyset$, then A is countable. Hausdorff thus turns Lindelöf's *theorem* into an *axiom*.

(c) The continuum hypothesis and later developments

Although it was the best result Cantor himself could achieve, the Cantor-Bendixson theorem was not the last word on the continuum problem. Both Cantor's direct attack, culminating in this theorem, and the indirect numerical scale approach were subsequently and extensively pursued. And while our knowledge of the truth or falsity of the continuum hypothesis (CH) has not been essentially advanced, partial results have been obtained and, more importantly here, the methods used raise questions which deserve a preliminary discussion since they touch on matters relating to the nature and substance of the Cantorian enterprise.

One can crudely describe the idea behind Cantor's scale approach as follows: to provide (or show the existence of) enough ordinals to guarantee that all sets, and particularly the continuum, can be counted and will therefore find a place in a numerical scale. Whatever difficulties Cantor himself encountered in carrying through this project, it was fully carried through in the axiomatization of Zermelo supplemented by the work of von Neumann (see Chapters 7 and 8). Providing that the axiom of replacement is present, there are more than enough ordinals to count the continuum, and indeed enough to count all sets, and there is, as Cantor wanted, a complete numerical scale of infinite size. (If replacement is left out, one can only prove the existence of ordinals up to, but not including $\omega + \omega$. Thus, there are by no means enough ordinals to count $P(\omega)$, the powerset of the von Neumann ordinal ω, or the continuum.) While von Neumann showed how to transplant the Cantorian programme into an axiomatic framework, the metamathematical investigation of this framework through Gödel and Cohen showed that Cantor's full goal cannot be achieved in this framework: CH is independent of the standard axiom systems. The systems give us a beautiful numerical scale, but they cannot tell us where in this scale the continuum appears. There are apparently enough ordinals, but not enough about the continuum is decided.

In a lecture given in 1946 Gödel put forward the suggestion that despite appearances there may *not* be enough ordinals, and that the lack of 'decisiveness' in the standard axiom system could be rectified by adding *more* ordinals. (See Gödel [1946], [1947], and the discussions in Kreisel [1976], p. 115, or Kreisel [1980], pp. 56-9.) The suggestion goes back to Gödel's earlier experiences ([1931]) with statements which in some sense should be decided by an adequate (complete) axiom system but which are not. Gödel showed that for any formal system sufficiently strong to contain recursive arithmetic there will be sentences in the language of the system which can be neither proved nor refuted by the axioms. Nevertheless, Gödel remarks that if one extends the system in a natural way the undecided sentences become decided. Since he bases his proof in [1931] on a system of finite types stemming from *Principia Mathematica*, he remarks that the undecided Gödel sentence can be decided when we add a new type level ω. And he adds: 'An analogous situation prevails for the axiom system for set theory.' (Gödel [1931], p. 610, n. 48*a*.) More generally he remarks:

. . . the true reason for the incompleteness inherent in all formal systems of mathematics is that the formation of ever higher types can be continued into the transfinite . . . , while in any formal system at most denumerably many of them are available. (Gödel [1931], p. 610, n. 48*a*)

Of course, in his [1931] Gödel seems to be talking about piecewise extensions of the formal systems. However, in his [1946] he suggested using an idea which is central to his hierarchy of constructible sets in the famous consistency proof for the Generalized Continuum Hypothesis (GCH), i.e. taking ordinals as given, not constructed, and adding a type level for any ordinal whose existence is admitted. In the consistency proof for GCH these are just the ordinals whose existence can be proved by the use of the standard *ZF* axioms. But what Gödel suggests in [1946] is that type levels be added for ordinals which are *beyond* any whose existence can be proved by the *ZF* axioms. Thus we can now add potentially very many new types in one fell swoop.

How does this connect with the incompleteness result of [1931]? This is not hard to see. The 'missing' sentences exploited by the incompleteness theorems are metamathematical in content, in effect those which state 'The system concerned is syntactically consistent.' To extend the system in the right way (that is, if we believe in consistency!) one could add such a statement as an axiom, or correspondingly the statement 'The system has a model.' In terms of the *ZF* axiom system of set theory it is natural following this line to add the statement 'There exists a set M which is a model for the *ZF* axioms.' However, the Zermelo-von Neumann theory of the ordinal-indexed cumulative hierarchy of sets makes it possible to frame this statement about set-models as a statement about *ordinals*. Namely:

(A): There exists an ordinal α large enough such that the cumulative level R_α is a model for the usual *ZF* axioms.

In any system ZF plus a statement of the form (A), the Gödel sentence of ZF is positively decided. This actually ties up with the direction in which Zermelo wanted to take set theory, as Gödel implicitly says in his [1947], p. 520 ([1964], p. 264). Zermelo in his [1930] suggests that there is not a single domain of set theory, but rather a succession of domains or universes where each universe appears in a subsequent extension as a genuine set. To make this picture precise he uses exactly the idea of a level hierarchy (the Zermelo–von Neumann hierarchy) marked by ordinal numbers. The first such level which acts as a genuine 'universe' (or model) for the ZF axioms will involve an ordinal whose existence cannot be established by the ZF axioms (cf. Zermelo [1930], pp. 44–5); indeed if the ZF axioms are consistent this is a necessary limitation, as Gödel's [1931] showed. Thus exactly as with our statements (A) the existence of these universes depends just on the postulated existence of the new non-ZF ordinals.

But Gödel's observation of 1946 adds a crucial new element to Zermelo's, i.e. as one shifts upwards through the hierarchy of new axiom systems with its corresponding hierarchy of new ordinals (and new models) more and more statements undecided in the systems lower down become decided. However, Gödel adds something else, something which goes beyond the kind of sentences dealt with in [1931]. As we mentioned, the undecided sentences of the incompleteness theorems are metamathematical in nature. But Gödel suggested that we should apply the same methods to *any* statements which are undecided by the basic axioms, and thus to statements which are originally and fundamentally set-theoretic in nature:

In set theory, e.g., the successive extensions [of the formal system, and therefore of the concept of demonstrability] can most conveniently be represented by stronger and stronger axioms of infinity. . . . Such a concept of demonstrability might have the required closure property, i.e. the following could be true: Any proof for a set-theoretic theorem in the next higher system above set theory . . . is replaceable by a proof from such an axiom of infinity. It is not impossible that for such a concept of demonstrability some completeness theorem would hold which would say that every proposition expressible in set theory is decidable from the present axioms plus some true assertion about the largeness of the universe of all sets. (Gödel [1946], p. 85)

Thus we have the clear suggestion: questions may remain undecided in ZFC because there are *not* enough ordinals. And this then becomes a clear heuristic programme: try to answer questions undecidable on the basis of the standard axioms by adding new large ordinals. To come back to the CH now, it is clear from his [1947] that Gödel was already assuming that this is ZF-undecidable (cf. [1947], p. 519, [1964], p. 263; see also Wang [1981], pp. 657–8, and n.9) even though at that time only its relative consistency had been clearly established. So Gödel was assuming that there is already a prime candidate for a problem to be tackled by use of axioms of infinity. In short, although the continuum can be counted in *ZFC* the continuum problem may still be undecidable because there are not enough *ZF* ordinals.

This suggestion gave new, indeed sharp, interest to the 'exotic' properties of ordinals which Mahlo, Hausdorff, Tarski, Kuratowski, and Ulam had previously explored, and of course inspired the search for new exotic properties. Set theory appears again, as Zermelo had described it in [1930], as open ended. And adding layers of ordinals at the 'top end' increases our knowledge of what lies near the bottom, as Gödel remarked in [1947] now explicitly tying up the remarks of 1946 to the continuum problem:

That these axioms [of infinity] have consequences far outside the domain of very great transfinite numbers, which are their immediate object, can be proved: each of them (as far as they are known) can, under the assumption of consistency, be shown to increase the number of decidable propositions even in the field of Diophantine equations.

He goes on:

As for the continuum problem, there is little hope of solving it by means of those axioms of infinity which can be set up on the basis of principles known today (the above-mentioned proof for the undisprovability of the continuum hypothesis, e.g. [i.e. the proof via the inner model of constructible sets], goes through for all of them without any change). But probably there exist others based on hitherto unknown principles; . . . (Gödel [1947], p. 520; cf. [1964], pp. 264–5)

This, then, is a direct invitation to investigate suitable 'largeness' properties, and thus to find suitable axioms of infinity. And this programme is a direct extension of Cantor's attack on the continuum problem, namely to solve this problem through the erection of a suitable numerical scale, suitable now meaning large enough, and certainly larger than we previously thought necessary.

Together with the metamathematical examination of various axiom systems, pursuit of the large cardinals programme has formed the core of modern research in pure set theory. It is impossible to trace its development here; not only would this require a book on its own, but it would take us too far away from our present subject. (For an excellent short review of results on the continuum hypothesis, including large cardinals, see Felgner [1979a] and also Martin [1976]. For an extensive review of large cardinal axioms, see Kanamori and Magidor [1978].) Nevertheless, two questions deserve to be raised and briefly tackled. Did Gödel's programme lead to any substantial increase in our knowledge of the continuum? How do the proposed axioms of infinity fit with the Cantorian view of the set-theoretic universe?

First we should ask: what reason do we have for thinking that a new large cardinal principle will decide the continuum hypothesis, or even add to our knowledge of what goes on at this relatively 'humble' level of the cumulative hierarchy? Gödel gives one clear reason: namely, as he states in the passage quoted above, we know that addition of axioms of infinity provides more information even about the domain of natural numbers. Why not also about real numbers and subsets of real numbers? (Why should coded metamathematical statements be privileged statements about numbers?) Secondly, the case of the axiom of replacement already shows the effect that extending the hierarchy can have. This axiom was conceived by Fraenkel originally as a kind of axiom of infinity since it was designed to add sets of power $\geqslant \aleph_\omega$ (see §8.2). And indeed it allows the addition of *many* more layers to the Zermelo–von Neumann hierarchy than can be proved to exist by the Zermelo axioms alone.[5] In the present context, replacement has two key effects. First, it adds very many apparently quite simple sets, for example the countable ordinal $\omega + \omega$. Of course, these sets are not 'low down' in the sense of the cumulative hierarchy, but they are very low down in terms of cardinality and thus in terms of *Cantorian* simplicity. ('Simple' sets can be quite complicated in respect of their cumulative structure.) This already has some impact on CH. For the numerical version of CH is false in the Zermelo system without replacement: $P(\omega)$ is larger than any (Zermelo) set of countable ordinals. Of course, this tells us really nothing about CH itself, only that the Zermelo system lacks the means to formulate the numerical hypothesis adequately (i.e. to provide an uncountable set of countable ordinals). But nevertheless, adding replacement (higher levels of the hierarchy) rectifies this and so is perhaps of some passing heuristic relevance to CH. Perhaps adding even more levels, then, *could* affect our knowledge of the continuum and its relation to the second number-class, maybe by adding more functions, thus allowing us to find one taking $\aleph_1$ onto $P(\omega)$, or adding enough real numbers to be able to show that there can be no such function. Not only this, but adding the new layers by replacement actually does allow us to establish much more about subsets of the continuum, as is shown by Martin's theorem that all Borel games are determined (see §8.3), which is proved in *ZFC* but cannot be proved if replacement is taken out. Moreover it is a statement not about any of the 'exotic' objects added by replacement but solely about Borel subsets of the continuum. Thus adding new layers *does* affect our knowledge of the continuum. Indeed, Kreisel even goes so far as to call the axiom of replacement and Martin's theorem '. . . easily the *most convincing contribution to Gödel's programme so far*.' (Krieisel [1980], p. 60; the italics are in the original.) He adds that 'even if not for Gödel, for mathematical practice the assumption of C_{ω_1} [the first level of the set hierarchy with uncountable index] is an "axiom of infinity" '.

Thirdly, and perhaps a little less to the point, there is a kind of analogy to Gödel's programme in number theory, where in order to increase our knowledge of natural numbers we shift to analytic number theory, which is an example of adding higher (infinite) levels in order to learn more about what goes on in the finite levels. (See Kreisel [1980], pp. 67–8.)

[5] Levy's reformulation of *ZFC* in his [1960] using the reflection principle gives clear sense to the claim that replacement is an axiom of infinity. See pp. 117–18 below.

In any case, the conjecture that large cardinal axioms would affect our knowledge of the continuum and the behaviour of its subsets proved correct. For example, Scott [1961] showed that the existence of a measurable cardinal contradicts the axiom of constructibility ($V = L$). (For definitions of measurable and the other exotic cardinals mentioned in what follows see Drake [1974] or Kanamori and Magidor [1978].) Gödel predicted in [1947] that CH is false, and had remarked that, say, the existence of inaccessible and Mahlo cardinals if consistent remain consistent when $V = L$ (and thus GCH) is added. Scott's result was thus the first indication that a large cardinal axiom could break the traditional pattern. And indeed Rowbottom later showed that the axiom of a measurable cardinal implies the existence of a non-constructible real number, thus that it does have effects 'low down' in the cumulative hierarchy. (Rowbottom [1971] —the work stems from 1964—or Drake [1974].) It also gives information about projective subsets of the continuum (see below).

But despite this promise, and despite a wealth of ingenuity in inventing axioms of infinity, CH remains undecided. Certainly if Gödel is right in believing that CH is false, none of the major large cardinal axioms so far discovered will help establish this, since they are all relatively consistent even with GCH. (This holds for inaccessible, Mahlo, Π_n^m indescribable, weak compact, Ramsey, measurable, compact and supercompact cardinals, to name but many. See Jech [1978] for details.) Thus none of these principles can *disprove* the CH. Suppose Gödel is wrong. Do these principles help in establishing or undermining CH? This question is a little subtler, and ties in with the extension of Cantor's direct attack on the continuum problem, to which we turn now.

The direct approach via derived sets led to the desired results only for closed sets. Cantor himself then tried a new decomposition idea using the notions of *coherence* and *adherence* of an arbitrary set of real numbers. The *adherence* of P, denoted Pa, is just the set of isolated points of P, while the *coherence*, denoted by Pc, is just the set of limit points which actually belong to P, i.e. $Pc = P \cap P^{(1)}$. Clearly this decomposition was designed to surmount the difficulty that with arbitrary point-sets there is no simple connection between P and $P^{(1)}$. Cantor, following the previous pattern, iterated his new constructions through the countable ordinal numbers. This did lead to an important new theorem, namely that for any uncountable P successive iteration will lead finally to a 'fixed point', namely a coherence Pc^α which is uncountable and dense-in-itself. (See Cantor [1885b], p. 268. The theory of ahderence and coherence was also presented in his unpublished paper of 1884.) But this was as far as it went. No doubt Cantor attempted to show that every uncountable dense-in-itself set is of the power of the continuum, but as far as we know did not produce a proof.

Despite the lack of success of this new method of decomposition (at least so far as CH was concerned), this was by no means the end of the direct appoach. Young and Hausdorff (using in particular the notion of condensation point introduced by Cantor) returned to the Cantor-Bendixson theorem and attempted to extend it. (The set Pc^α mentioned above is the set of all condensation points which belong to P.) It was Young who first opened the way, not least by establishing a proof method which can be used for all the classical extensions of the Cantor-Bendixson theorem (the Young-Hausdorff theorem proved in §2.2(*b*)). Young and Young commented:

As already remarked the theory of open [i.e. non-closed] sets is still so far imperfect that we cannot say whether or no potencies [powers] other than the finite integers, a and c [i.e. $\aleph_0$ and $2^{\aleph_0}$], exist on the straight line. The theory of density in itself, however, as developed by Cantor, . . . , enables us to make the same assertion with respect to a very general type of open sets, that we were able to make with respect to closed sets, namely that potencies other than those mentioned do not exist. (Young and Young [1906], p. 63)

(Young does not use 'open' in the modern topological sense as meaning 'complement of a closed set'; open here just means Cantor's non-closed.) They then proceed to prove the theorem:

An ordinary inner limiting set which has a component dense-in-itself has the potency c [i.e. $2^{\aleph_0}$]; otherwise it is countable. (Young and Young [1906], p. 64, theorem 31)

Since Young proved that any uncountable set A has an uncountable dense-in-itself subset, the theorem amounts to showing that in the given case the power must be $2^{\aleph_0}$. (See Young and Young [1906], pp. 54-5. The relevant subset is presumably the set $A \cap A_\gamma$; see above, p. 97, or Hausdorff [1914], p. 269. The Youngs' account is rather ambiguous since it could be read as saying that always $A_\gamma \subseteq A$, which is false. As remarked, the theorem stems from Cantor [1885*b*].) Young's definition of an 'ordinary inner limiting set' is not crystal clear. They are either intersections of countably many intervals, or countable unions of such intersections. It seems from the proof as if the first is meant, though the cardinality result clearly goes through for the latter too. Using the Young-Hausdorff theorem proved in §2.2(*b*), theorem (*), the proof is almost immediate. Let $A = \bigcap_{n \in \omega} I_n$ and let B be a non-empty dense-in-itself subset of A, and a_1 and a_2 be points of B as at the start of the proof of (*). The spheres (or intervals) $V_{x_1}, V_{x_1 x_2}, \ldots$ etc. ($x_n = 1$ or 2) are chosen in the same way, except that since a_1 and a_2 are in *all* I_n we can clearly demand that V_1 and V_2 are contained in I_1, while $V_{11}, V_{12}, V_{21},$ and V_{22} are contained in I_2, and so on. But then clearly the intersection of any of the descending sequences of V intervals is always contained in $\bigcap_{n \in \omega} I_n = A$. Thus all the points $a_{x_1 x_2 x_3} \cdots$ are in A, so A must be of power $2^{\aleph_0}$.

Young's proof, although divided up rather differently, is very similar to Hausdorff's. Hausdorff's formulations, however, are overall incomparably clearer. In any case, Hausdorff was able to get much more out of this proof.

In his [1914] he introduced the first classification according to complexity of Borel sets. The closed sets are called F sets, the domains [*Gebiete*] G sets. (Domains are what we now call open sets, i.e. X is a domain if all its points are interior points, i.e. for Euclidean spaces, if to each $x \in X$ there is an open interval (sphere) I such that $x \in I \subset X$.) Countable intersections of G sets are denoted G_δ, countable unions of F sets F_σ, and vice versa, yielding G_δ, $G_{\delta\sigma}$, $G_{\delta\sigma\delta}$, ... and F_σ, $F_{\sigma\delta}$, $F_{\sigma\delta\sigma}$, ... As Hausdorff points out the F sets are all complements of G sets, the F_σ of G_δ, the $F_{\sigma\delta}$ of $G_{\delta\sigma}$, and so on. (See Hausdorff [1914], pp. 304-5.) The collection of all these sets forms the realm of Borel sets. The class of Borel sets can also be defined as the smallest class containing the open sets and closed under complementation and countable union. Alternatively we can define them as a hierarchy by recursion over the countable ordinals. Thus

$\Sigma_1^0 = $ the class of open sets

$x \in \mathbf{\Pi}_\alpha^0$ iff x is the complement of some $y \in \Sigma_\alpha^0$

$x \in \Sigma_\alpha^0$ iff there are sets y_i ($i \in \omega$) such that $y_i \in \mathbf{\Pi}_{\beta_i}^0$, where $\forall i \in \omega$ $[\beta_i < \alpha]$, and $x = \bigcup y_i$.

The Borel sets are then the sets in $\bigcup_{\alpha \in \omega_1} \Sigma_\alpha^0$. This matches Hausdorff's classification by

$$\Sigma_1^0 = G, \qquad \mathbf{\Pi}_1^0 = F$$
$$\Sigma_2^0 = F_\sigma, \qquad \mathbf{\Pi}_2^0 = G_\delta$$
$$\Sigma_3^0 = G_{\delta\sigma}, \qquad \mathbf{\Pi}_3^0 = F_{\sigma\delta} \text{ etc.}$$

Hausdorff used this classification to begin a more systematic extension of the results on the power of point-sets. Young's theorem concerns a subset of G_δ. Hausdorff showed first that exactly the same proof works for arbitrary G_δ sets ([1914], pp. 319-20). Thus

(19) An infinite G_δ ($\mathbf{\Pi}_2^0$) subset of the real numbers has either power $\aleph_0$ or power $2^{\aleph_0}$.

The same clearly holds for F_σ (Σ_2^0), $G_{\delta\sigma}$ (Σ_3^0) and $F_{\sigma\delta}$ ($\mathbf{\Pi}_3^0$). But Hausdorff took the extension further, first to $G_{\delta\sigma\delta}$ (and thus to $G_{\delta\sigma\delta\sigma}$ and the corresponding complements and unions). The proof again makes essential use of the Young–Hausdorff method, but this time it is complicated by the fact that the given uncountable set is not a direct intersection of open sets but the intersection of sets which are the unions of intersections of open sets. The proof actually operates with these latter open sets, which is essentially why the Young–Hausdorff method can be applied. The trick, though, is to arrange for them to be unpacked. (See Hausdorff [1914], pp. 465-6.)

In his [1916] Hausdorff extended the method and the result to *all* Borel sets, a result which he had conjectured already in [1914], p. 466. Again the complexity of an arbitrary Borel set requires attention before the Young-Hausdorff theorem can be used. What Hausdorff does first is to arrange the Borel sets in a hierarchy B_α, indexed by the countable ordinals, and which uses countable intersections and unions but not complementation. (This is enough; for example, every closed set is already G_δ (see Hausdorff [1914], p. 306).) Again if A is some Borel set the purpose is to reach and operate with the domains which appear at 'base level'. But since each stage is formed by using intersections or unions of sets from earlier stages, then any descending path from the given set A down through sets involved earlier is indexed by a descending sequence of ordinal numbers, and thus must arrive at the *first* stage after a finite number of steps. But the first stage is simply the collection of all open sets (G sets). Thus, open sets are finitely accessible by descent through the complexity of A construction. Having shown this, Hausdorff could then apply the technique of the Young–Hausdorff theorem, assuming A uncountable, to show that A must have a subset of power $2^{\aleph_0}$. Thus

(20)   Any infinite Borel set has either power $\aleph_0$ or power $2^{\aleph_0}$.

(This was also proved by Alexandroff [1916] via the perfect subset property.)

It should be noted that in pushing this result through Hausdorff brings in again the ordinal numbers which Lindelöf and Young had thrown out. Actually here they can be dispensed with again once the idea of projective sets are introduced. For the Young-Hausdorff proof can be repeated for G_δ sets of the *plane*, and crucially for projections onto the line of these; since every Borel subset of the line is also representable as the projection of a G_δ subset of the plane, (20) will follow. (For details, see Felgner [1979a], pp. 172-3.)

The Cantor–Bendixson theorem, and the core of its revised proof, thus goes surprisingly far. What emerges from this is thus a modified, weaker version of Cantor's direct approach, namely: 'Attempt to establish the continuum hypothesis by proving the desired property first for relatively "simple" sets, and then extend this to more and more complicated ones', a programme which begins with Young and Young [1906]. Hausdorff's classification of the Borel sets gives some substance to the vague terms 'simple' and 'more complicated', since it characterizes these subsets in terms of complexity of definition. And the subsequent introduction of the projective hierarchy extends this further. This gave the modified programme solid ground to work on. The spirit of the programme has carried over to what is now called 'descriptive set theory'. As Martin writes:

In descriptive set theory one restricts one's attention to *simple* sets of real numbers; sets of simple topological structure or sets which are definable in a more simple way. [One of the advantages of such restriction is]:
Many questions which seem unanswerable for arbitrary sets (the continuum hypothesis) can be answered for sufficiently simple sets. (Martin [1977], p. 784)

The programme has some plausibility. Concentration on the simple cases may lead to methods which are more generally applicable, as the Cantor–Bendixson theorem already shows. Moreover, even if the work does not get beyond simpler sets, the results for these might constitute *evidence* for or against the continuum hypothesis. However, once we bring up the subject of evidence we run into trouble. Already in his [1914] Hausdorff warned against reading too much into the restricted results on the power of subsets of real numbers:

If we knew for all sets of a Euclidean space what we know for the closed or G_δ sets, namely that they are either finite, countable or of the power $\aleph$ [of the continuum] then $\aleph$ would be the next power after $\aleph_0$, and the continuum problem as posed by Cantor's hypothesis $\aleph = \aleph_1$ would be decided. However, in order to see how far we still are from this goal, it is sufficient to recall that the system of sets F or G_δ forms only a vanishingly small part of the system of all point sets. (Hausdorff [1914], p. 321)

On p. 305 Hausdorff points out that the classes F and G_δ have only the power of the continuum, whereas the class of all point sets must have power $2^\aleph = 2^{2^{\aleph_0}}$. On p. 271 Hausdorff also remarks that the property of being closed or of being a domain imposes an 'enormous restriction on the generality of a set'. He points out on the other hand that the class of dense-in-itself sets has the power $2^{2^{\aleph_0}}$, the same power as that of all subsets. Even after extending the Cantor–Bendixson result to *all* Borel sets, Hausdorff remained cautious. Concluding his proof he remarks

Thus the question of power is clarified for a very inclusive category of sets. Nevertheless, one can scarcely see this as a genuine step towards the solution of the continuum problem, since the Borel sets are still very specialised, and form only a vanishingly small subsystem (having power $\aleph$ of the continuum) in the system of all [sub]sets (which has power $2^\aleph > \aleph$). (Hausdorff [1916], p. 437)

With his term 'specialized' here Hausdorff touches on the main difficulty of the 'piecemeal' approach. Namely, there is a danger that the 'simple' sets we deal with (or know how to define) are just *too* simple. What matters is not so much their infrequency or relative scarceness in the domain of all subsets of real numbers (and this is what Hausdorff's emphasis on cardinalities stresses) but rather that they have rather special properties. Indeed, this is exactly what we find, for as far as the results on the continuum hypothesis go hitherto, all the uncountable simple sets satisfying it have a rather special property: they contain a non-empty perfect subset, such a perfect set, of course, having power $2^{\aleph_0}$. Young and Hausdorff do not actually prove this (except of course in the case of the *closed* sets) and indeed do not seem too interested in this property. But their proof method can be simply adapted to show exactly this. Thus, as far as the available proofs were concerned, the fact that the continuum hypothesis holds for Borel sets goes hand in hand with the stronger result, that uncountable Borel sets possess the perfect subset property. (Alexandroff [1916] proves (20) by this route.) And unfortunately possession of a perfect subset is a rather special property of uncountable sets. There are only $2^{\aleph_0}$ perfect sets, and crucially as Bernstein proved in [1908]:

(21) If the continuum can be well ordered then there is a decomposition of the reals into two disjoint uncountable sets A, B such that neither has a perfect subset.

(For a proof, see Moschkovakis [1980], p. 81, or Felgner [1979a], p. 175.) This theorem indicates that it is rather hopeless to pursue the continuum hypothesis by methods which will also prove that any uncountable set has a perfect subset. More importantly, it suggests that the Borel sets are structurally too special to act as a guide to the continuum hypothesis *overall.*

This suggestion was confirmed in the attempt to extend the Hausdorff results into the next hierarchy of subsets of real numbers, the so-called *projective hierarchy*. As a counter-example to a claim of Lebesgue, Souslin showed in 1916 that projections onto the line of Borel subsets of the plane are not necessarily Borel subsets of the line. (See Kuratowski [1980], p. 68-9, and Souslin [1917], pp. 90-1.) Souslin called such projective sets *analytic*. Moreover, Souslin was able to prove two highly important theorems:

(22) Every uncountable analytic set has the power $2^{\aleph_0}$ (has a perfect subset)

(announced in Lusin [1917], p. 94) and

(23) The Borel sets are precisely those sets which can be represented both as analytic sets and as complements of analytic sets.

(See Souslin [1917], pp. 90-1.) This represents another considerable extension of the 'piecemeal' approach to the continuum problem. But the previous pattern is repeated since *again* the result goes hand in hand with the perfect subset property. And indeed it is at this point that the programme begins to run out of its original steam.

In his [1925], Lusin noted that Souslin's analytic sets have many nice properties, including the perfect subset property. He adds:

There remains here only one important gap: we do not know if every uncountable complement of an analytic set has the power of the continuum.

He goes on:

My efforts towards settling this question have led to an unwelcome [*inattendu*] result: there is a family of effective sets admitting a mapping onto the continuum such that we do not know *and will never know* if any uncountable set of this family has the power of the continuum, or is of the third category, nor even if it is measurable. (Lusin [1925], p. 1572)

This family he proceeds to define, and it is what we now know as the family of projective sets. The linear projective sets are defined as follows:

$\Sigma_1^1 = $ the analytic sets

$= \{A : \exists B \ [B \text{ is a Borel subset of } \mathbb{R} \times \mathbb{R} \text{ and } A \text{ is the projection of } B] \}$

$\Pi_1^1 = \{A : \exists B \in \Sigma_1^1 \ [A = \mathbb{R} - B] \}$

$\Sigma_2^1 = \{A : \exists B \text{ which is } \Pi_1^1 \text{ in } \mathbb{R} \times \mathbb{R} \text{ and } A \text{ is the projection of } B\}$

$\Pi_2^1 = \{A : \exists B \in \Sigma_2^1 \ [A = \mathbb{R} - B] \}$

. . . .

and so on. Lusin points out that he and Souslin were not originally interested in the projection operation itself, to which Souslin's work of [1917] had drawn attention. But Lebesgue had insisted that because of its essential and intuitively geometric nature the projection operation was a natural way of extending the notion of 'definable set' (Lusin [1925], p. 1574). Lusin's pessimism is thus interesting: he doubts the provability of the continuum hypothesis even for this rather clear class of sets. As Kreisel points out (and as became clearer with Gödel's work on constructible sets) the CH 'lacks stability': 'which sets and which maps do we want to know about?' ([1980], p. 65). Lusin is thus saying that the continuum problem is hopeless even for this kind of 'definable set'. The pessimism was moreover justified. Souslin's (23) shows that the projective hierarchy is related to the Hausdorff hierarchy, since the Borel sets form a part of its first level, and also (22) shows that just as the Hausdorff–Borel hierarchy begins with sets which satisfy the continuum hypothesis, so does this projective hierarchy. But the analogy ends there: for Souslin's result appears to be the best that can be achieved on the basis of the standard axioms for set theory. Souslin himself established the further partial result:

(24) Every Π_1^1 set is the union of $\aleph_1$ Borel sets

and Sierpinski showed (Lusin and Sierpinski [1925], Sierpinski [1926]) how to extend this to

(25) Every Σ_2^1 set is the union of $\aleph_1$ Borel sets

from which, since $\Pi_1^1 \subseteq \Sigma_2^1$, follows

(26) Every uncountable Σ_2^1 set has power $\aleph_1$ or power $2^{\aleph_0}$.

But under the usual axiomatic assumptions this result cannot be strengthened. Given the perfect subset idea it is natural to ask whether uncountable Π_1^1 or Σ_2^1 sets have perfect subsets. But this cannot be established even for Π_1^1 sets: Gödel announced in [1938], p. 556, that in *ZFC*

(27) The axiom of constructibility implies the existence of an uncountable Π_1^1 set of reals with no perfect subset.

(For a proof see Jech [1978], p. 529.) What this shows is that it is consistent with *ZFC* to assume *both* the correctness of CH *and* that not even every uncountable projective (thus relatively simple!) subset of the reals has a perfect subset.

Thus as early as 1938 it was shown that pursuing the perfect subset property cannot lead to a proof of CH even for sets 'low down'. Yet Gödel's result also seems to confirm that as far as our knowledge is expressed in proofs hitherto, CH *is* connected with the perfect subset property. This is frustrating: it seems that to prove CH in the projective hierarchy we are condemned to pursue a method which we know, given Bernstein's (21) and the axiom of choice, will not work overall. Bernstein's theorem with its application of the axiom of choice, contains several ironies. First, the proof involves a form of diagonal argument not dissimilar to that by which Cantor proved that the real numbers are uncountable, the proof which first sharply posed the question of the diversity of infinite size. Moreover, we have seen that the well-ordering theorem and hence the axiom of choice is crucial to Cantor's *scale* approach to this question. Yet Bernstein's result shows that *precisely* this assumption renders hopeless what turned out to be the central method (search for perfect subsets) of Cantor's other (and more successful) direct approach. The axiom giveth and the axiom taketh away. (Solovay [1970] showed that if *ZFC* plus the existence of an inaccessible cardinal is consistent, so is *ZF* plus the axiom of dependent choices plus 'every uncountable subset of reals contains a perfect subset'. Thus (given the first strong consistency assumption) without AC the perfect subset property *could* be universal.)

These partial results and the 'speciality' of sets up to $\mathbf{\Pi}_1^1$ brings us back to the difficult question of 'evidence' for the continuum hypothesis. The *possibility* that the Borel and analytic sets may be taken as a test of CH was clearly formulated by Martin:

We might test [CH] by looking at sets of reals which are, in some sense, *simple*. If such simple sets do not provide counterexamples we can try to see whether there are reasons to suspect that the simple sets considered are in a relevant way different from arbitrary sets of reals. (Martin [1976], p. 87)

But Martin quickly concludes:

Can we regard these facts about $\mathbf{\Sigma}_1^1$ and $\mathbf{\Sigma}_2^1$ sets as evidence for CH? I do not think we can. For the results about the cardinalities of $\mathbf{\Sigma}_i^1$ sets, $i = 1, 2$, are corollaries to stronger results. Every $\mathbf{\Sigma}_i^1$ set, $i = 1, 2$, is countable or has a perfect subset (assuming MC for $i = 2$). Now by a simple application of the axiom of choice, there exists an uncountable set with no perfect subset. Thus, while our simple sets have the cardinalities required by CH, this is so because they have an *atypical* property, the perfect subset property. (Martin [1976], p. 88)

Thus, the establishment of these consequences of CH cannot be taken to *support* it, says Martin. (MC here is the assumption that there is an uncountable measurable cardinal (see below).)

The question of evidence was first raised by Gödel in a famous passage from his [1947] ([1964]), where he suggested that there is evidence *against* CH. He listed a number of consequences of CH concerning the existence of 'strange' subsets of the real line. The existence of such sets, says Gödel, is implausible, 'paradoxical', and would 'seem to indicate that Cantor's conjecture will turn out to be wrong' (Gödel [1947], p. 523; [1964], pp. 266-7). He adds:

One may say that many of the results of point-set theory obtained without using the continuum hypothesis are also highly unexpected and implausible. But, true as that may be, still the situation is different there, insofar as in those instances (such as, e.g., Peano's curves) the appearance to the contrary can in general be explained by a lack of agreement between our intuitive geometrical concepts and the set-theoretical ones occuring in the theorems. (Gödel [1947], p. 524; [1964], p. 267)

Against this Martin remarks:

While Gödel's intuitions should never be taken lightly, it is very hard to see that the situation *is* different from that of Peano curves, and it is even harder for some of us to see why the examples Gödel cites are implausible at all. (Martin [1976], p. 87)

(For similar remarks, see Martin and Solovay [1970], p. 177; cf. also Kreisel [1980], p. 65.)

It seems to me that Martin is right, and for reasons which, combined with the situation so far on simple sets, show the problem of evidence to be a hopeless one. First, the problem with Peano curves is not so much a 'lack of agreement' between geometrical intuitions and set-theoretical concepts, but rather that the set-theoretical reconstruction of the continuum goes far beyond what geometrical intuition demands. We do presumably have some geometrical 'intuitions' about, say, continuity. But these were not far reaching or precise enough to allow the development of a clear analytic theory of the continuum. Indeed it was for this very reason that they were replaced by a set-theoretic *theory* of the continuum. One might say, too, that this theory is *faithful* to the geometric intuitions in the sense that it allows the derivation of the intuitively correct consequences. But it brought with it too such things as Weierstrassian continuous nowhere-differentiable functions or Peano curves. These are strange because they were unforeseen; that is, we could not conceive of them before just on the basis of this intuition. (Even now we cannot *picture* what they look like, and the suggestion of conflict came perhaps through the mistaken assumption that what we can picture is all there is. Anyway, with the arrival of the set-theoretic continuum this assumption is firmly dropped.) What now of the case of the CH? According to Gödel, if we add CH to the set-theoretic theory of the continuum we arrive at some consequences which were not just unforeseen but for Gödel implausible. Now in order to reach this conclusion Gödel must appeal to some picture of the continuum which conflicts with these consequences. This cannot be that presented by codified set-theory, so it must be some other informal 'intuitive' picture, and moreover one which goes much further in heuristic strength than geometric intuitions do. Now the set-theoretic theory of the continuum is based on no strong intuitions of how the continuum should be *set theoretically*, at least none such as would yield any detailed information about the structure of the continuum. What then *does* Gödel appeal to? It could be some intuitive concept of set. One could argue that this was only first relatively clarified by an axiomatization, *and* that the conflict between this and the consequences of CH just does not arise. (Indeed, this is the whole problem!) But let us suppose, against this, that there were/are strong and clear intuitions of sets, based in part, perhaps, on Cantor's conception of numbers and how they should relate to the collections they number (see §2.2(*a*) and Chapter 3). Let us even suppose (as Gödel presumably did (see §6.2)) that this notion of set is clear enough to yield the full cumulative hierarchy. (Kreisel, for example, takes this view; see among others, his [1973].) Even now we still need to fill out the picture by recourse to axioms (why otherwise do we need them?) and, to repeat, these simply do not yield the right amount of information to counter the consequences Gödel points to.

Martin's mention of MC in one of the passages quoted above brings us back to large cardinal axioms. The use of these in descriptive set theory marks in a way the merging of Cantor's two lines of attack on the continuum problem, for here we use properties of the numerical scale in order to further the piecemeal programme. Indeed, this is the latest twist in the continuing story of the cyclical abolition and subsequent rehabilitation of the ordinals in point-set theory. And surprisingly, but in striking further confirmation of Gödel's prediction, large cardinal axioms do affect our knowledge of subsets of the continuum, even as far 'down' as the projective hierarchy. Again it is the axiom MC which provides this confirmation, for we have:

(28) MC implies that every uncountable Σ_2^1 set of reals has a perfect subset

and also

(29) MC implies that every Σ_3^1 set is the union of $\aleph_2$ Borel sets. Thus, every uncountable Σ_3^1 set has power $\aleph_1$, $\aleph_2$, or $2^{\aleph_0}$.

((28) is due to Solovay [1969] and (29) to Martin [?]; see Martin [1977], pp. 806-7.) Thus, the measurable cardinals assumption extends the classical results (22), (25), and (26), shifting them up one stage both in terms of levels in the projective hierarchy and in terms of the cardinals mentioned in (26).

But while these results are remarkable and impressive, they do not do much to clear up the confusion over the continuum hypothesis. Indeed, they seem to confirm the impossibility of finding 'evidence' to decide CH. For example, (28) matches the classical pattern in again tying uncountability to the perfect subset property. Thus Σ_2^1 sets are also still 'special', and it still seems that we are stuck with 'specialness'. (29) is somewhat different, since it introduces a new cardinality $\aleph_2$. This *might* suggest that as we move up the projective hierarchy more and more cardinalities become 'available' for subsets, and that this would perhaps be provably revealed if we could find an appropriate large cardinal assumption beyond measurability. Thus, the suggestion would be that $2^{\aleph_0}$ is some $\aleph_\alpha$ with $\alpha > \omega$. The problem with this is that this possibility is already opened up classicially by (26), and yet adding the measurables assumption closes it again ((28)). Thus, in terms of its suggestive strength about CH, MC is quite ambiguous; it seems to underline the impotence of set theory with respect to the continuum problem. And (28) and (29) are the *best* results the assumption of measurability can provide, for Silver [1971] showed that it is consistent with MC that there is a $\mathbf{\Pi}_2^1$ set with no perfect subset. There are no other large cardinal candidates which provide any other information about the cardinalities of projective sets. (Descriptive set theory does use other new principles which are not large cardinal axioms, for example the axiom of projective determinacy: 'all projective games are determined'. See Martin [1977].) Moreover, with these radical new principles one has always an additional problem: suppose it *does* yield interesting information about the continuum—is it nevertheless, a plausible addition to the axiom system? Are such principles 'intrinsic to set theory' in the same way, say, that the axiom of choice is? (See §3.5.)

With this we come back to the question of the legitimacy of altering the Cantorian scale, of extending the scale of numbers beyond those proved to exist by *ZFC* axioms. What I want to examine briefly here is how far large cardinal principles may be seen to fit the philosophical and heuristic doctrines underlying Cantor's theory. Recall first Cantor's position with respect to existence: if a concept 'coheres' (is consistent) with an existing, accepted body of concepts (or a theory), then there must be objects falling under it; the objects in questions must exist. This was precisely how Cantor himself justified the introduction of transfinite numbers to begin with. The problem of the *consistency* of large cardinal axioms relative to the basic *ZFC* axioms is rather hopeless. But coherence, being less precise, is perhaps more helpful: if one can plausibly argue that a concept like 'uncountable inaccessible cardinal' is coherent, then from the Cantorian point of view such objects must exist. And in this case, it would be legitimate and quite within the spirit of the Cantorian programme to add a new axiom to this effect. But how does one approach coherence? One striking thing is that large cardinal principles are remarkably well behaved in that they seem to fit nicely into a linear hierarchy (see Kanamori and Magidor [1978], pp. 265-6). For example, every strongly compact cardinal is measurable, every uncountable measurable is weakly compact, every weakly compact cardinal is Mahlo, and every Mahlo cardinal is inaccessible. For a range of cardinals stemming from such a diversity of concepts this regularity is remarkable, and to this extent they form a 'natural' system of extensions of the *ZFC* cardinals. In addition one might look at the consequences of the axioms. In this respect the axiom of uncountable measurable cardinals is most interesting. It surprisingly 'predicted' that there are non-constructible sets, and, in so far as the axiom of constructibility has *not* been adopted as a genuine permanent addition to *ZF,* one might say that this 'prediction' gives MC some confirmatory weight. Moreover, although MC does not yield any solid 'evidence' for or against CH, it does confirm patterns already suggested by the *ZFC* axioms. For example, *ZFC* shows that every uncountable Σ_1^1 set has a perfect subset, and also that every Borel game is determined. The axiom of measurable cardinals extends both results, the first to Σ_2^1 sets, as we have seen, and the second to analytic sets. (See Jech [1978], p. 561.) Moreover, the same pattern outlined above for Σ_1^1 and Σ_2^1 and the perfect subset property is repeated for other key properties like Lebesgue measurability and the Baire property (see Martin [1977], pp. 798 and 806). If these results are not exactly confirming evidence for the axiom of measurable cardinals, they certainly show that its consequences 'cohere' with or match those of *ZFC*.

But from the Cantorian point of view the most interesting approach to coherence comes from principles (*b*) and (*c*) of Chapter 1—respectively that infinite sets, and therefore infinite numbers, should have the same basic properties as finite numbers, and that the Absolute can never be mathematically determined. Various authors have attempted to give philosophical (or heuristic) arguments in favour of large cardinal axioms, for example Wang [1974], [1977], Reinhardt [1974], Kanamori and Magidor [1978], and Kreisel [1980]. (Wang [1974] also attributes the principles he states to Gödel.) I want to look a little more closely at two arguments often used, what Kanamori and Magidor [1978], p. 104, call *generalization* and *reflection* respectively, because these *do* have obvious connections with the Cantorian principles (*a*) and (*b*).

Generalization expresses the idea that there should be cardinals which stand in a similar relation to the numbers below them as ω does to the finite numbers. In some sense, ω is the first 'traumatic' or 'catastrophic' stage in the development of numbers, that is, radically different from all the numbers below. The idea expressed by generalization is that ω should not be the only such catastrophic point. There is a similar suggestion in Wang's *principle of uniformity*:

Uniformity of the universe of sets (analogous to the uniformity of nature): the universe of sets does not change its character substantially as one goes over from smaller to larger sets or cardinals, i.e. the same or analogous states of affairs reappear again and again (perhaps in more complicated versions). (Wang [1974], pp. 189–90)

Such suggestions are somewhat reminiscent of the Cantorian 'finitism' which I discussed in §1.3, for this too is a principle of uniformity. But whether this similarity can be stretched so far as to say that particular properties of ω must be repeated further up the ordinal sequence is difficult to say. This partly depends on *which* properties one considers. One might for example claim that it is part of the nature of the relationship between ω and the finite numbers that ω cannot be 'reached' by finite arithmetic operations alone. It is exactly for this reason that we require a new principle to yield ω (see §2.1); axiomatically expressed, we need an axiom of infinity. But why should the finite numbers be 'privileged' among all numbers in having an 'inaccessible' number sitting above them? Why not a new number given by a new axiom, which is inaccessible to all the ordinals generated by cardinal exponentiation and addition, the two basic arithmetic operations of the *ZF* axioms? And then why not unlimited repetition, or a principle which formalizes unlimited repetition?

On the face of it the definition of inaccessibility looks rather *ad hoc* for it is fashioned purely in terms of what we know the *ZFC* axioms are capable of producing. (α is (strongly) inaccessible if (*a*) $\beta < \alpha \rightarrow 2^\beta < \alpha$; (*b*) if for any $\gamma < \alpha$ we have $\alpha_\beta < \alpha, \forall \beta < \gamma$, then $\bigcup \{\alpha_\beta : \beta < \gamma\} < \alpha$.) The property of measurability on the other hand is fashioned quite apart from our understanding of *ZFC*. But actually this 'artificiality' of inaccessibles is no worse than is already the case with ω. For this is defined as the set of natural numbers, or the least set containing all natural numbers, and by natural number one intuitively means something (finitely) generated by addition of 1 from 0. Thus ω is the least number *not* generated by this process. Looked at in this way, the definition of inaccessibility is no more *ad hoc* than what lies behind the definition of ω (see also the remarks on reflection below). Since ω is also measurable (if we demand $< \kappa$ additivity instead of countable additivity) as well as inaccessible, it seems that the axiom of uncountable measurable cardinals can also be justified by appeal to generalization. Here one might argue that the measurability of ω is a reasonable guide to the difference in extent between the finite and ω. Thus, by definition points of ω (that is, natural numbers) are each assigned value 0—they are, so to speak, of insignificant extent—while ω itself has value 1—it is of significant extent. Moreover, by finite additivity, no finite collection of points, or no finite collection of sets finitely generated from a finite collections of points, will yield a significant measure. (Maybe one could put it thus: one already needs ω insignificant pieces to obtain anything significant.) Since this failure to match the significant extent of ω is a plausible way of 'characterizing' finiteness, one might now legitimately apply generalization, or Cantor's principle of finitism. Why should the finite be privileged? Why are not other initial segments of the ordinals 'dwarfed' in a similar way?

Certain large cardinals have been generated by the generalization to infinitary languages of a property possessed by first-order predicate logic (compactness), thus yielding the so-called compact cardinals (ordinarily, strongly and super). One can plausibly justify the axiom of choice as simply the generalization to the infinite of the ordinary principle of existential specification. And this does seem to fit with Cantorian finitism, for it is a direct extension to *all* sets of something agreed to hold for finite sets (see §3.5). But use of the compactness property to obtain large cardinals is rather different. Here the argument is not directly about numbers, but about the existence of an infinitary language partially mimicking first-order logic; it thus depends on the legitimacy of using infinitary languages at all, and this must come (if at all) from logical rather than set-theoretical considerations. More interesting, perhaps, is the fact that measurability can be cast as a form of restricted compactness (see Kanamori and Magidor [1978], p. 114), and that strong compactness can be rephrased as a form of generalized ultrafilter theorem (Kanamori and Magidor [1978], p. 114). (This is not *too* surprising given the involvement of the standard ultrafilter theorem in the model-theoretic proof of the compactness theorem; see Bell and Machover [1977], pp. 180-2.) But again, what good Cantorian reasons are there for us to accept the need for generalizations of Boolean algebras which satisfy a generalization of the ultrafilter theorem?

Briefly stated there is some connection between Cantorian finitism and the idea of generalizing cardinal properties. But the difficulty, which we already encountered in examining the doctrine with Cantor, is to say *which* properties of the finite are basic, and thus which should be extended. There is no doubt that the property of numerability considered in §1.3 and this chapter can be taken as a basic property of the finite. And since this was the subject of Cantor's main application of 'generalization', to argue that a cardinal generalization is legitimately Cantorian one should, properly speaking, argue that what one is generalizing is also a basic property. But here our ideas may be just *too* vague.

A much better, clearer, and more clearly Cantorian approach to large cardinality comes through what Kanamori and Magidor call 'reflection'. As Wang [1974], p. 189 states it: 'Reflection principle: the universe of sets is structurally undefinable.' (See also Wang [1977], p. 318, or Kreisel [1980], p. 58.) Now this *is* very close the Cantorian doctrine of the Absolute, that 'the Absolute cannot be mathematically determined', and is certainly close if we allow that Cantor's original idea of the Absolute approximates to the modern notion of a universe of set theory. (Reinhardt [1974], pp. 190-2 is one author who explicitly recognizes the connection between Cantor's doctrines about the universe and reflection principles.) The similarity is strengthened as soon as we look more closely at what lies behind the reflection principle. For this says, more or less exactly, that any attempt to characterize the universe of sets will fail, and one will instead have characterized only a set. Thus, if we should reasonably expect that the universe possesses a property, then one can also reasonably expect that there exists sets which have it. And with regard to this property, the universe is *reflected* in these. (In such arguments 'universe' is sometimes replaced by 'class of all ordinals'. Such replacement is quite natural given the tie up between ordinals and the universe through the cumulative hierarchy. Moreover, this also fits with Cantor's view that the ordinals are a natural 'expression' or 'representative' of the Absolute.) Thus, to give one application, if we take the *ZF* axioms, the class of *ZF* ordinals is determined by the operations of addition and exponentiation. But it cannot be that such an intuitively simple characterization yields the class of all ordinals (the universe); on the contrary it must point only to a set, an ordinal, the first non-*ZF* ordinal (or its corresponding stage). Thus we can legitimately add an axiom to *ZF* which permits such an ordinal.

The resemblance between the reflection idea and Cantor's doctrine of the Absolute is close not only because of the similarity of the terms 'structurally undefinable' and 'mathematically indeterminable', but also because Cantor himself had effectively used such a reflection argument before. Cantor states that before him the first number-class was taken as an 'appropriate symbol of the Absolute'. But

. . . precisely because I regarded that infinity as a tangible or comprehensible idea, it appeared as an utterly vanishing nothing in comparison with the absolutely infinite sequence of numbers. (Cantor, [1883b], p. 205, n. 2)

(The whole passage is cited on p. 42.) In other words, there must be numbers beyond this 'first attempt' to capture the Absolute, or rather this attempt succeeds only in determining the limit of the finite numbers, a new number ω. One might ask why Cantor did not repeat this trick, and why he allowed that the collection of all Cantorian ordinals *is* an appropriate symbol of the Absolute. But there is a clear answer. The first number-class is determined just by the operation of 'addition of one'. Thus, speaking Cantorianly such a simple characterization (a 'tangible idea') cannot possibly capture the mystery of the Absolute (of God), and therefore this operation must be transcended. But Cantor gave no such clear characterization of *all* ordinals in terms of arithmetic operations. Thus, there were no clear operations specified which could be simply transcended. Against this the *ZF* axioms do give a rather clear and simple statement of what the ordinals are. Thus, one can plausibly say—following Cantor—that such a clear statement cannot possibly capture the Absolute: the *ZF* ordinals determine not *all* ordinals but only a set, a new ordinal. Thus, certainly Cantor can allow that the class of *all* ordinals is indeed an appropriate symbol of the Absolute, because he does not attempt to make precise what all ordinals are. Indeed, according to this view we shall never know what all ordinals are. The universe is perpetually open ended.

As it was first used the idea behind what Wang calls the 'reflection principle' synthesizes rather beautifully Cantor's doctrine of the Absolute both with Gödel's 'axioms of infinity' programme and with Zermelo's idea of ever expanding universes (Zermelo [1930]). This synthesis is first made clear through Levy's work on axioms of infinity in his [1960]. Gödel's basic idea (see above) is the addition of set models to the system. Starting just from the Zermelo axioms Z minus the axiom of infinity, Levy formulated in his [1960] the so-called Levy reflection principle (or schema) according to which for any given formula ϕ we can find a *set u* which is already a model of the Zermelo axioms and which satisfies ϕ just as well as the universe can. For this ϕ, *u reflects* the universe. Levy showed that adding this principle to Z yields exactly the system *ZF*. Not just this, but he showed that if we do exactly the same for *ZF* we obtain a theory which contains many inaccessible cardinals. Since Levy's work showed that the reflection principle for Z is equivalent in Z to the axioms of replacement and infinity, it suggests that these two working together (in *ZF*) constitute a broader axiom of infinity which is then *naturally* generalized to a strong axiom of inaccessible infinity. (Levy states this explicitly in [1960], p. 223.) All this comes from applying Gödel's extension programme quite literally, and, based on an analysis of how *ZF* extends *Z*, it shows how the programme can yield natural extensions of *ZF*. It should be clear how the Levy reflection principle fits beautifully with the Cantorian thesis of the 'undeterminability of the Absolute (the universe)', for it follows from it that if *s* is any sentence of the (first-order) language of set theory, there is already a set in which it is true if it is true at all. Thus we cannot hope in this language to say anything which is true *only* of the universe.

We have here the beginnings of a clearly Cantorian coherence argument for large cardinal principles. Since the work of Levy and Bernays (see Bernays [1961]) attempts have been made to extend these principles by first extending the language of set theory and admitting that even then only some *set* will be characterized by various properties and not the universe itself. Such a project is by no means easy. (See for example Wang [1977], pp. 318–26.) But the clear connection between reflection and Cantor's doctrine of the Absolute suggests that this is a good path for a Cantorian theory to follow.

What all this means is that one might find perfectly good Cantorian extensions of set theory. But it seems that Cantorian set theory will still not be able to answer the continuum problem, the problem which inspired its development.

3

CANTOR'S THEORY OF NUMBER

In this chapter, I want to pay off some of the explanatory debts accumulated in Chapter 2. I stated there that Cantor's mathematical theory of cardinal number is as an *interior* theory with the number-classes as the interior representatives of power. But it is interesting to ask: what exactly are the numbers for Cantor? It might be argued that although he developed the number-class structure it does not follow that he takes the various number-classes to be *themselves* the powers or cardinal numbers of the infinite sets. As we have seen, there were difficulties with the number-class theory, not least over well-ordering. But although these did not force Cantor to give up the theory (§2.2 (*a*) shows clearly that he fully realized its potential and remained committed), equally the use of number-classes as representatives of power does not force one to assume that they themselves are the numbers. Other possibilities are available. For example, it is quite possible to read Cantor as supporting the view that cardinal numbers are actually primitive non-set objects. This is perhaps suggested by the later characterization of $\aleph_0$ and $\aleph_1$ as 'the cardinal *numbers* belonging to' the first and second number-classes respectively. ('We call the cardinal number belonging to it [the totality of all finite cardinal numbers] "Aleph-zero", written $\aleph_0$.' Cantor [1895], p. 293.) Here, while the number-classes would *represent* all the infinite numbers, they are not actually the numbers themselves. Another possibility is that Cantor adopted the whole-class notion of cardinal number adopted by Frege and, later, Russell. This is a frequent interpretation of Cantor, not least because, beginning in 1884 (Grattan-Guinness [1970], p. 86), and then in correspondence and in his [1887-8], [1895], and [1897], Cantor proposed an *abstractionist* account of number which is *apparently* quite independent of well-ordering, the ordinals, and the number-classes, *and* which has some resemblance to the Frege–Russell whole-class definition of number. Again, if this interpretation is right, the number-classes cannot be themselves the cardinal numbers.

No doubt Cantor himself was never completely clear on these matters. However, I want to argue here that the natural Cantorian position is indeed that the number-classes themselves (or better the cumulative classes or the initial ordinals—it makes little difference which) are the cardinal numbers. One simple point should be made immediately. The thesis that the number-classes are *not* the numbers is theoretically and ontologically wasteful. It is crystal clear from Cantor's own statements that the central pillar of his theory of cardinality is the representation of all infinite sizes through the number-classes.

To claim in addition that these representatives are not actually the numbers, that some other entities have this privilege, is to offend against a natural simplicity requirement, namely 'do not invoke what you do not need'. Since the number-classes do the hard theoretical work of numbers, it is prodigal to bring in other entities to stand as merely titular cardinal numbers and to claim that the number-classes are merely surrogates useful for simplifying the mathematics of the theory. Thus, when Cantor says, referring to the number-classes, that 'the ordinal numbers form the natural material for an exact *definition* [my italics] of the higher infinite cardinal numbers or powers' ([1897], p. 312) one is inclined to take him literally. At the very least one can say that identification of the cardinal numbers with the number-classes (and the construal of these cumulatively) is an obvious step in any later simplification of the theory. Such steps are certainly reflected in Cantor's 1899 reworking of the theory, and in his identification of the cardinal numbers with initial ordinals. Such simplification is also particularly natural in any attempt (like von Neumann's) to remodel the theory axiomatically. For here reduction of complexity, particularly ontological complexity, is a crucial prior step in the search for simple axioms.

Thus, whatever Cantor's own precise position, this is already enough to show that the natural Cantorian position is that according to which the number-classes *are* the cardinal numbers. Thus, in this sense (and in others, as we shall see (Chapter 8)) von Neumann's theory of number is naturally Cantorian (see the remarks in §2.2(*a*)). To strengthen the point slightly, we saw in §2.2 that while Cantor might not have declared openly for the initial ordinal theory, in practice by 1899 this was what he was using.

Nevertheless, one should not ignore Cantor's abstractionist account of number. However, I argue that this account does not fit with either the Frege–Russell or the primitive entity construals of number. Indeed, it takes us right away from both, back to the key thesis of [1883*b*] that numbers are themselves sets which *directly* represent the sets they number. Moreover, while Cantor's abstractionist account has severe difficulties, a serious attempt to overcome these reveals again an *ordinal* concept of cardinality very close in spirit to the number-class construal.

3.1. Cantor's abstractionism, set reduction, and Frege–Russell

Untangling Cantor's views about the status and structure of numbers is no easy task. What *is* clear is that from the mid-1880s on Cantor stressed again and again that numbers are 'abstractions' from the sets they number. Before attempting to fathom Cantor's explanation of abstraction, it is worthwhile considering *why* he should have taken an abstractionist position. There are two plausible possibilities. One is that having failed to find a proof of the well-ordering theorem Cantor wanted to present in his published work a notion of cardinal number which does not openly depend for its explanation on the assumption of well-ordering. The abstraction idea as presented avoids this, and moreover is not only general in this sense, but can be applied *mutatis mutandis* to ordinal numbers and order-types. The second possibility leads on from this. We saw that in the [1883b] presentation the argument for the existence of ordinals is tied directly to the existence of well-ordered sets. Since, according to Cantor, mathematics must deal with infinite sets, it must then also admit infinite numbers: such at least was the trend of the argument. However, we also saw that its weakest point was the connection between the set (admitted to exist) and its number. Cantor called an ordinal number a 'representation in our inner intuition' of the given set. Not only is this as it stands quite unclear, but it also suggests appeal to subjective capacities, and such a subjective element might be regarded as unfortunate in a realist conception of mathematics. It is thus at least a possibility that Cantor's abstractionist account, like Frege's, was designed partly to shift away from this vague 1883 theory towards a clear existence proof. After all, Cantor made clear statements *opposing* the intrusion of psychologistic elements into the theory of number. For example in his review of Frege [1884] Cantor writes approvingly:

It should also be recognized that the author has adopted the correct position in so far as he has demanded that spatial as well as temporal intuition and likewise all psychological elements must be kept out of the concepts and basic propositions of arithmetic. Only in this way can rigorous logical purity be achieved, and thereby the justification for the application of arithmetic to imaginable objects of cognition. (Cantor [1885a], p. 440)

What, with hindsight, makes these aims particularly plausible is recollection of the Frege and Russell definitions of number. These definitions are often called abstractionist, and Russell certainly saw himself as coherently explaining the notion of abstraction (Russell [1903], p. 305 and pp. 114–15). Neither definition is connected with well-ordering or the ordinals (another of Russell's aims was to avoid such a connection, and Frege, too, wished to avoid appeal to orderings); both completely dispense with subjectivist elements (this was one of Frege's express aims) by giving a clear set-theoretical and thus reductionist existence argument. Nevertheless, while Cantor's abstractionist account is explicitly reductionist, it is not clear that he avoids either well-ordering and the ordinals *or* subjectivist elements. Indeed, reading Cantor's explanation of abstraction closely it seems much more like an attempt to *explain* the 'representation of a number in our inner intuition' by explaining numerical structure rather than an attempt to avoid reference to such 'representations' altogether. At least, so I shall suggest here.

According to the Frege–Russell definition of number, the cardinal number associated with a set or a concept is the set of all sets or concepts equinumerous to it.[1] The association with Cantor's notion of cardinal is based on an alleged

---

[1] Russell gives an analogous definition of the order-type of an ordered set. Cantor's conception of order-types will be discussed later. 'Russell' here refers to the Russell of

similarity between the above definition and Cantor's famous description of the cardinal number of a given set M as '. . . the general concept which arises from the set M when we abstract both from the nature of its various elements m and from the order in which they are given.' (Cantor, [1895], p. 282.) The appeal to such an abstractionist concept goes back to 1884. For example, the following passage from [1887-8] goes back to 1884 (according to Cantor's footnote here):

Let M be a given set, thought of as a thing itself, and consisting of definite well-differentiated concrete things or abstract concepts which are called the elements of the set. If we abstract not only from the nature of the elements, but also from the order in which they are given, then there arises in us a definite general concept (*universale, unum versus alia* [universal, one as opposed to others] with the meaning: *unum aptum inesse multis* [one thing shared by many]) which I call the *power* of M or the *cardinal number* belonging to M. (Cantor [1887-8], p. 411)

(This statement, judging by Cantor's footnote, almost certainly goes back to the unpublished paper in Grattan-Guinness [1970], which also contains an abstractionist account. Other abstractionist formulations differing from one another only slightly can be found in [1887-8], pp. 379, 387 (which Cantor says stems from 1884, n.1), and in [1885a], p. 441. The same formulation as the second cited here appears in letters to Illigens (Cantor, *Nachlass VI*, p. 59) and Goldscheider (*Nachlass VI*, p. 69; also Meschkowski [1961], p. 84).) This number so described will be, in Cantor's words, 'the general concept under which fall all and only those sets which are equivalent to the given set' ([1885a], p. 441), which can be interpreted as a statement of the fundamental condition (*ii*) on cardinal number considered in §2.3(*b*), namely:

(*ii*) for any $x, y, card(x) = card(y)$ iff $x \sim y$

(Cantor, of course, gives an analogous description of the order-type of an ordered set M where 'we abstract only from the nature of the elements' retaining their order; see e.g. Cantor [1895], p. 297, and §3.4.)

As it stands, this description of cardinal number is already quite different from the Frege-Russell treatment. First, on the logical level, as Russell himself points out in the *Principles*, Cantor's statements cannot stand as *definitions* of number; so-called 'definition by abstraction' is not a definition. What grounds do we have for allowing, without demonstration, that given a set M there *is* a 'general concept' under which fall all and only those sets equivalent to M; and even if we assume that such 'concepts' exist, what grounds do we have for assuming that there is only *one* such concept? As Russell notes, Cantor's description

the *The Principles of Mathematics* and not the Russell of *Principia*. Moreover by joining Russell and Frege I am deliberately overlooking crucial differences between them, for example their respective concentration on sets (or classes) and concepts.

. . . is merely a phrase indicating what is to be spoken of, not a true definition. It presupposes that every collection has some such property as that indicated— a property, that is to say, independent of the nature of its terms and of their order; depending we might feel tempted to add, only upon their number. (Russell [1903], pp. 304–5)

Judging from a remark of Cantor's in reply to a letter from Benno Kerry it seems that Kerry had already raised the possibility that there may be several quite different general concepts which all sets of the same power share. (Cantor to Kerry, 21 March 1887, Cantor *Nachlass VI*, p. 106.) Cantor does not answer the point. A somewhat similar criticism was put by Borel:

According to Cantor, the notion of power is arrived at by starting from the idea of the set and performing a *double* abstraction: abstraction from the *order* and from the *nature* of the objects. But it does not seem to me that this remark suffices to allow reasoning with powers independently of any substratum. (Borel [1898], p. 103, n.1)

Russell's approach, naturally, does *not* suffer from this defect. Using the principles of comprehension and extensionality, Russell proves, given x, the existence of a unique set X_x which contains all and only sets y such that $y \sim x$. X_x can then be taken to be *card*(x).

Nevertheless, one might argue that Cantor is not far away from the Frege-Russell definition. Cantor's discussions of number are (to say the least) vague; one is therefore forced to *interpret* him. And certainly his frequent references to cardinal number as 'the general concept . . . etc.' *do* suggest an affinity with Frege whose definition he knew. (Frege's definition appeared in his [1884] which Cantor reviewed ([1885a]).) It is not altogether surprising, therefore, that both Frege [1892], p. 164, and Russell [1903], p. 305, reflecting on Cantor's comments on cardinal number saw their own definitions as sharp delineations of something Cantor had been fumbling towards. Indeed the affinity with Frege is underlined in some remarks Cantor made in the unpublished paper of 1884 (Grattan-Guinness [1970]) where he comes very close to embracing the Frege definition of number. For, he states that '. . . the [equivalence] class of a set M is nothing other than the extension . . . of the general concept belonging to the set M which I have called *the power of the set M*' (Grattan-Guinness [1970], p. 86).

But, natural or not, it is a mistake to see Cantor's remarks as anything like a nod in the direction of the Frege–Russell theory. For one thing it will emerge later that Cantor's conception of the Absolute rules out the 'whole class' approach to number, as Cantor himself was probably aware. But for another this association ignores Cantor's own account of abstraction. And this leads us in a quite different direction.

Immediately after Cantor's 1895 description of cardinal number (p. 122) there is the following attempt to explain the process of abstraction which gives rise to cardinal number:

We denote the cardinal number or power of [a set] M, the result of this two-fold act of abstraction, by $\overline{\overline{M}}$. Since each individual element m if we disregard its nature becomes a 'one', the cardinal number $\overline{\overline{M}}$ is itself a definite set composed of nothing but ones which exists in our mind as the intellectual image or projection of the given set M. (Cantor [1895], pp. 282–3)[2]

And Cantor insists that not only is $\overline{\overline{M}}$ a *set,* but, moreover, a set *equivalent* to the base set M(i.e. $\overline{\overline{M}} \sim M$) (Cantor [1895], p. 284). Thus note that Cantor is adding to the fundamental condition (*ii*) above, the two further conditions (*iii*) and (*iv*) familiar from §2.3(*b*):

(*iii*) $card(x)$ is a set, for any set x.

(*iv*) $card(x) \sim x$, for any set x.

(For the analogous conditions imposed on order-types, see §3.4.)

To our sophistcated eyes, while these conditions on number seem quite natural, the theory which gives rise to them seems rather strange. Nevertheless Cantor was to a large extent relying on a traditional explanation. For example in Euclid we find: 'A *number* is a multitude composed of units' where 'An *unit* is that by virtue of which each of the things that exist is called one.' (Heath [1925], Vol. 2, p. 277.) And according to Heath, this idea is much older than Euclid, going back to Pythagoras. Certainly it was expounded with great regularity until Frege, Cantor, and Dedekind began to shape what we might call the 'modern' theory of number. If we compare Euclid's statements with Cantor's, the *monas*, which Heath translates as *unit*, corresponds to Cantor's one [*Eins*] and the *multitude* of units corresponds to Cantor's 'set of ones'. Cantor saw himself as following directly in a line from Euclid. In a letter to Peano of 21 September 1895 he wrote:

I conceive of numbers [*Anzahlen*] as 'forms' or 'species' (general concepts) of sets. In essentials this is the conception of the ancient geometry of Plato, Aristotle, Euclid, etc. except that with me it is clearer and sharper and the Ancients had no idea of the transfinite. (Cantor *Nachlass VII*, p. 186)

However, Cantor insisted ([1887–8], pp. 380–1) that his conception differed from Euclid's precisely in replacing 'multitude' by 'set'. For Cantor stressed that numbers are not only composed of units, but must themselves have unity:

The cardinal numbers and the ordinal types are *simple* conceptual pictures: each of them is a *true unity* (*monas*), because in it a multiplicity and manifold of ones [*Einsen*] is joined together. (Cantor [1887–8], p. 380)

---

[2] These 'ones' seem to appear first in section 8 of [1887–8], p. 423, which Cantor tells us dates from 1884. But here they are only directly referred to in the characterization of order-types. However, n.1., p. 413 (same section) mentions 'the "ones" of a cardinal number'. 'Ones' are referred to freely in connection with all numbers in the introduction to [1887–8].

Stressing that the collection of 'ones' is itself a set should certainly be enough to guarantee for Cantor that it has the requisite unity to be a single object or a logical individual. Indeed, that numbers are sets seems forced on Cantor by interpreting them already as collections. Certainly numbers must be 'objects of mathematical study' or 'mathematical determination'; and so according to Cantor's theory of infinity if the collection of 'ones' arising from a set (possibly infinite) is to be itself an *object* then it too *must* be a set, since it must fall in the realm of object-collections or object multiplicities, that is the combined finite, transfinite realm, or the realm of finite and infinite sets. (Cantor sometimes suggests that numbers have *more* unity than ordinary sets; see e.g. [1887-8], pp. 380-1. I come back to this in §3.3.)

It is at this point that Cantor's set-theoretic reductionism is at its strongest. In the unpublished paper of 1884 Cantor describes his theory of order-types, a generalization of the notion of ordinal number, as follows:

It forms a large and important part of *pure set theory*, thus also of *pure mathematics*, since this latter according to my conception is nothing other than *pure set theory*. (Grattan-Guinness [1970], p. 84)

This is a very strong reductionist statement. As we have seen in §2.1 the [1883*b*] account of (ordinal) numbers does not live up to this strong claim. I pointed out that there is certainly a reductionist tendency in [1883*b*]. Ordinals are taken to exist, it seems, simply because (well-ordered) sets exist and because we have access to or can form representations of them. There is no explicit claim there that these numbers must actually be sets: Cantor merely suggests that there is some intuitive 'causal' or at any rate necessary link between numbers and sets forcing us to accept the former given the latter. This suggestion of a linkage is reiterated in the various [1887-8] statements on number. For example:

If we look at Euclid's definition of finite cardinal number, then it must be recognized first that, just as we do, *according to its true origin*, he relates numbers to the *set*, . . . (Cantor [1887-8], p. 380)

And then explicitly bringing in infinite sets, he says

Once the actual-infinite in the form of actually infinite *sets* had in this way [i.e. through a proper analysis of potential infinity] asserted its citizenship in mathematics, then the development of the actually infinite number-concept became inevitable, through appropriate, natural abstractions, just as the finite number-concept, the material of arithmetic hitherto, had been achieved through abstraction from finite sets. (Cantor [1887-8], p. 411)

Again, Cantor says in a comment on Gutberlet's thesis [1878] that while there are actually infinite sets, there are no actually-infinite numbers:

. . . infinite number and set are indissolubly linked; if one gives one of them up, then one no longer has any right to the other. (Cantor [1887-8], p. 394)

The difference now is that the link is further explained via the coexistence of multiplicities of ones, and these, as we have seen, must for Cantor be multiplicity-objects, therefore *sets*. Thus with his account of abstraction Cantor has moved somewhat closer to fulfilling the reductionist 'manifesto' of 1884, as close as he was to get. Whether Cantor's attempt can be described as successful must depend ultimately on what view one takes of the strange 'ones', and this we shall come to later. But it at least seems clear from the discussion so far that Cantor was *not* assuming that numbers are a special kind or form of entity outside the realm of sets. So one possible rival to the thesis that the number-classes are the infinite cardinal numbers is ruled out.

We can also see clearly that the Frege–Russell construal of Cantor's account of abstraction is ruled out as well, despite the fact that Cantor's statements of 1884 might seem to be close to the Fregean account. The quasi-'Euclidean' (or set representative) explanation of abstraction in [1887-8] (and later) contradicts such a construal. Cantor stresses that $card(x) \sim x$ and this is impossible in the 'whole-class' theory. The Frege–Russell numbers (0 excepted) are vastly bigger than the sets they number.

One might add here another argument, relating to Cantor's theory of the Absolute, to explain why he would not accept the Frege–Russell definition of number. The size of the collections chosen by Frege–Russell as numbers in effect puts them beyond the realm of Cantor's transfinite. Certainly they cannot be genuine mathematical objects according to Cantor's *later* theory of consistent and inconsistent (or absolute) collections (see §4.1). For the Frege–Russell collections (0 excepted) are all of the same cardinality as the whole universe of existents (what Cantor in 1899 called 'the totality of everything thinkable'), and thus on Cantor's theory are much too extensive to be sets. It is not clear how precisely Cantor could or would have put this argument before he transformed the rather vague theory of *the* Absolute into a more precise theory of absolute collections. It would certainly have been clear to Cantor, though, that the universe of all existents is a collection which 'reflected' or 'involved' the Absolute. He had clearly recognized in [1883b] that even the ordinal number sequence reflects the Absolute, and the same would hold *a fortiori* for the universe itself. (This is partly because intuitively speaking there can be nothing larger, but also because according to Cantor's conception it mirrors the capacity of God's intellect (see §1.4).) This *might* suggest recognition that anything equally extensive would fall outside the transfinite, and thus outside the province of mathematics.

There is *some* indication in Cantor's review of Frege [1884] that he was aware of this difficulty. While he approved of Frege's critical analysis of earlier theories of number, he was himself critical of Frege's positive contribution. He calls it 'an unhappy idea'

. . . to take 'the extension of a concept' as the foundation of the number-concept. He [Frege] overlooks the fact that in general the 'extension of a concept' is something quantitatively completely undetermined. Only in certain cases is the 'extension of a concept' quantitatively determined, then it certainly has, if it is finite, a definite natural number, and if infinite, a definite power. For such quantitative determination of the 'extension of a concept' the concepts 'number' and 'power' must previously be already given from somewhere else, and it is a reversal of the proper order when one undertakes to base the latter concepts on the concept 'extension of a concept'. (Cantor [1885a], p. 440)

Cantor then goes on to explain (by appeal to abstraction) what the notion of power or cardinal number of a *set* is. These remarks can be interpreted in the light of his theory of infinity as follows. Only some 'extensions of concepts' are sets, that is to say, fall within the range of mathematical study and the domain of mathematical objects; furthermore only (and just) sets can have numbers belonging to them. From this together with the assumption that numbers are sets (which admittedly Cantor does not bring out in [1885a]) it follows that only a collection that can *have* a number (that is to say, something which is an object, a set) can possibly *be* a number. Thus, Cantor seems to say, given an arbitrary extension, before one can show that it is a possible number, one has to show that it is a *set*, which is tantamount to showing that it *has* a number. This, according to Cantor, makes operation with arbitrary 'extensions of concepts' problematic, not least in any attempt to *define* number. Cantor's approach, it seems, does not suffer exactly from this problem. His theory of number deals only with *sets,* and he assumes ('axiomatically' or by definition) that 'every well-defined set has a definite power' ([1883b], p. 167). For this reason his treatment of number is best viewed, not as an attempt to define number, but as an attempt to explain, given a set, what the number belonging to it is like (see §3.2). In his [1885a] he explains it only vaguely (and poorly, as he might have learned from Frege) as a certain kind of concept. His more elaborate abstractionist account, stemming from around this time, at least attempts to make it clearer by explaining it as a certain kind of set. In terms of clarity the result is still arguably much less successful than Frege's. But my purpose here is only to try to understand why Cantor rejected Frege's account.

   If Cantor does operate just with sets, he does not at this time give any indication of how one decides in general which among all extensions *are* the sets. But suspicion of operating with arbitrary extensions long before the appearance of contradictions is certainly interesting (see §4.1). It suggests that the set concept is delimited, and this in turn (if only very hazily) suggests an 'axiomatic' approach to sets. For if sets are the essence of mathematics, and not *any* collection can be a set, then for the sake of precision one must eventually say what the sets are. At least one must attempt to specify (by some rules, or guidelines, or 'axioms') some of the collections which are sets. Interesting in this connection is the close bond Cantor points to here between sets and numbers. He seems almost to suggest that to show whether a collection is a set one has to show that it has a number. 'Having a cardinal number' becomes then almost a criterion for sethood. In the nearest he came to an 'axiomatic' approach to set theory, the 1899 letters to Dedekind, Cantor moves a little further in this direction. Here, in effect, sets are distinguished as those collections which can be (ordinally) numbered. This brings with it its own special difficulties, as we shall see.

3.2 Difficulties with the strange theory of 'ones'

So far, I have said a little about what Cantor's theory is *not*. It is time now to look a little more closely at what lies in and behind his explanation of abstraction.

The reductionist intention in Cantor is clear, at least by [1895]. But the argument for the *existence* of the number sets is still not so clear. In Russell [1903], for example, the argument for the existence of *card*(*M*) is simply the claim that given a set *M* another *set* $\{N : N \sim M\}$ exists, based on implicit appeal to the principle of comprehension. Whatever one might think of this, particularly because of the later difficulties with comprehension, the argument is at least relatively straightforward, and certainly not tainted with psychologism. However, despite the plausibility of the supposition that Cantor was trying to eschew psychologism, the same cannot be said of *his* conception. Cantor's existence argument is of a totally different kind. He claims that given *M*, $\bar{\bar{M}}$ exists in our mind as 'the intellectual image or projection of *M*' once we have abstracted from the order and nature of the elements '*with the help of our active capacity of thought* [*Denkvermögens*]' ([1895], p. 282). And in his presentation in [1887-8] (stemming from 1883-4) there is an explicit reference to 'reflection': the cardinal number or power of *M* is that general concept which

. . . one obtains when one abstracts [from the nature and order of the elements of *M*] and reflects only on what is common to all sets which are equivalent to *M*. (Cantor [1887-8], p. 387)

(See also the unpublished paper of 1884, Grattan-Guinness [1970], p. 86.) And again

The act of abstraction with respect to nature and order . . . effects or rather awakens in my intellect the concept 'five' [for example]. (Cantor [1887-8], p. 418, n.1)

Or, again in [1887-8] Cantor states:

Are not *a set* and *the cardinal number belonging to it* quite different things? Does not the *first* stand to us as an object, whereas the latter is an abstract image *in our* intellect?[3] (Cantor [1887-8], p. 416)

Certainly then the later abstractionist account of number seems, as I have already suggested, to be an attempt, not to eschew the earlier notion of number

---

[3] The italics are Cantor's. The same passage is important in a different context; see §3.3, p. 135.

as a 'representation in our inner intuition', *but to explain it*. And the theory of 'ones' is the main element in this attempt.

The object theory of cardinal number as such demands only that for each set there is a unique object governed by condition (*ii*) above. The representational object theory goes much further. The key idea here is that what is grasped when we grasp a cardinal or ordinal number is a 'generalization' of those aspects of the given set we are focusing on; that is, either the diversity of its members or their diversity and arrangement (order). This is exactly what Cantor means when he refers to something '*common* to all sets equivalent to *M*'. He does not just mean something common to all these sets in the sense of belonging uniquely to them, which is all (*ii*) demands. Rather, he means 'common' in the sense of sharing the same form or structural properties as these sets, or so his explanation proposes. When measuring the size or diversity of a collection, a particular member is important *only* by being there; that is, it is important only in so far as its presence contributes to the size. Now in this respect every element of *M* (or even every element of any set equivalent to *M*) plays exactly the same role because it contributes exactly the same amount to the overall size, and thus to the cardinal number. Thus, it is natural to regard every element *m* of a set *M* as being 'transformed' into a token representative, what Cantor calls here a 'one', which for convenience we can denote *E* (for *Eins*). And moreover, this *E* should be the same for each *m* in *M*, and indeed for each *m* in *any M*. Or at least, if there are *several* units, they should be virtually indistinguishable. Apart from this 'elementwise' transformation, it is *assumed* according to Cantor's account that nothing else changes: each element contributes a 'one' or a 'unit' to the size and to the number, and what we are left with after the transformation is a set of 'units'. In a sense, then, this should be enough to explain what is *common* to all sets equivalent to *M*. $\bar{\bar{M}}$ as a set of representatives of the elements of *M* (and its equivalents) represents the number of *M* (and each of its equivalents).

There is at first sight some plausibility in this explanation. However, on closer inspection there are manifold difficulties, not least with the contention that if $\bar{\bar{M}}$ represents *M* cardinally, and $N \sim M$, then $\bar{\bar{M}}$ also represents *N*. But there are two central difficulties which tower above the others. One is Cantor's explanation of the transformation of *M* into $\bar{\bar{M}}$ as 'abstraction', and the other is the crucial insistence that what is arrived at is a *set*. Both difficulties were discussed at length by Frege in his [1884]. There he devotes considerable space to various versions of the theory that numbers are composed of units, paying close attention particuarly to the views of Locke and Leibniz among the earlier writers and to Thomae, Lipschitz, Jevons, and Schröder among his contemporaries. Cantor is *not* criticized since he expounded the 'ones' theory only *after* the publication of Frege's book. Since Frege delivered a damning attack on 'unit' or 'one' theories and since Cantor in his review mentions with approval Frege's criticism of earlier views on number, it is puzzling why he later adopted or adhered to the 'ones' theory without at least an attempt to say how the difficulties raised by Frege can be surmounted. Certainly in considering Cantor's theory it is essential to read Frege [1884] (§§28–44) on similar accounts.

The first major difficulty concerns abstraction itself. The key to the 'ones' theory is that each element m of a set is 'transformed into' or 'gives rise to' a unit or 'one', E. Suppose we attempt to explain this elementwise transformation by the intellectual removal or abstraction of properties from m. It seems that this is just what Cantor means when he says that 'out of each single element m, if we ignore its nature, a "one" arises' (Cantor [1895], p. 283). But this implies that when we start from any element, and 'remove' properties, we end up with something (a property?) which should be common to all things— presumably their oneness or unity. But as Frege says it seems impossible to achieve this through a process of thought or intellectual reflection:

For suppose that we do, as Thomae demands, 'abstract from the peculiarities of the individual members of a set of things', or 'disregard, in considering separate things, those characteristics which serve to distinguish them'. In that event we are not left, as Lipschitz maintains, with 'the concept of the Number of the things considered'; what we get is rather a general concept under which the things in question fall. The things themselves do not in the process lose any of their special characteristics. For example, if I, in considering a white cat and a black cat, disregard the properties which serve to distinguish them, then I get presumably the concept "cat". Even if I proceed to bring them both under this concept and call them, I suppose, units, the white one still remains white just the same, and the black black. I may not think about their colours, or I may propose to make no inference from their difference in this respect, but for all that the cats do not become colourless and they remain precisely as before. (Frege [1884], p. 45)

Thus intellectual 'transformation' of one object (m) into another (E) is impossible.

Having discussed this difficulty Frege now considers another which is pertinent to Cantor's insistence that numbers are sets. As Frege says: 'We cannot succeed in making different things identical simply by dint of operations with concepts.' He goes on:

But even if we did, we should then no longer have things in the plural, but only one thing; for as Descartes says [Principles of Philosophy; I, §60], the number (or better, the plurality) in things arises from their diversity. (Frege [1884], p. 46)

Interpreted in the light of Cantor's insistence that numbers be sets, Frege's criticism amounts to this. If we could reduce every element to a unit, it should be to the same unit, say E, but then according to the principle of extensionality for sets the derived set $\bar{\bar{M}}$ would always just consist of one element; thus $\bar{\bar{M}}$ would always be the same number $\{E\}$ for every M. $\bar{\bar{M}}$ cannot represent M, since it always collapses. From this we get a hopeless tangle. We have said that we want units to be the same to represent the same numerical contribution, and yet we must demand that they be distinguishable if we do not want to fall foul of extensionality. As Frege remarks, it seems we are pulling in opposite directions: we are forced '. . . to ascribe to units two contradictory qualities identity and distinguishability' (Frege [1884]). (For similar criticisms by Frege of this type of theory see his [1891-2], [1903], and [1906].)

So it seems, then, that the 'ones' theory of number collapses into incoherence, not least because of the demand that the result of abstraction is an extensional *set*. With some justice then Frege later commented ([1892], p. 164), after seeing Cantor's theory of 'ones', that Cantor uses '. . . those unhappy ones which are different even though there is nothing to differentiate them'. And that Cantor '. . . clearly has not the slightest suspicion of the presence of this difficulty which I have dealt with expressly in sections 34 to 54 of my *Grundlagen* [[1884]].'

But is Cantor's theory *so* incoherent in this particular respect? In the first place, Cantor only once (in [1895]) refers to abstracting from the nature of *each* individual element. And this in any case seems to run counter to his other views about number. Focusing on the *individual* transformation $m \mapsto E$ implies that we allow our mind to settle separately on the elements and to run through and successively assemble a set of 'ones'. But Cantor would certainly be against this. For example, in reference to Leibniz's comments on number, Cantor remarks:

The addition of ones, however, can never serve for a definition of a number, since here the specification of the main thing, namely *how often* the ones must be added, cannot be achieved without using the number itself. *This proves that the number is to be explained only as an organic unity of ones achieved by a single act of abstraction.* From this it follows in addition how fundamentally false it is to wish to make the concept of number dependent on the concept of time or on so-called temporal intuition. (Cantor [1887-8], p. 381, n.1)

Here the key sentence is the second, which I have italicized, with its stress on a single act of abstraction. In this respect, it seems as if Cantor has certainly shifted from his position in [1883*b*]. For there he says:

The series (I) of positive, real whole numbers 1, 2, 3, . . . , ν, . . . has its genesis in the repeated setting down and union of underlying unities looked upon as equal. The number ν is the expression both for a definite finite number of such successive setting downs, and for the union of the unities to a whole. (Cantor [1883*b*], p. 195)

The involvement of succession, and thus perhaps of temporal duration, is perfectly clear, and in clear contrast with the later statement, which rejects temporality and therefore genuine succession in forming a number. Thus the later emphasis that 'abstracting from the nature of the elements' (as Cantor usually puts it, not 'from each single element') is a *single* act seems not to refer to a process of reflection on individuals or to a gradual stripping away of properties, but rather to an *immediate* switch to a new collection, the collection of 'ones'. This seems much more like the use of a classically conceived functional operation than of mental reflection, much more like the immediate replacement of elements of a set by new marking element(s) than abstracting in the reflective sense that Frege criticizes. As Frege says, if we start with m and 'abstract', we get nowhere—m remains m! Thus it is much simpler to assume that m is 'replaced', or better 'marked', by *another* element E. Then there is no pretence to the removal or annihilation or change of m or M; it is only assumed that M is functionally related (one to one, no pun intended) to a different set $\bar{\bar{M}}$.

The natural idea of replacing each element of a set by a new element which is meant to *mark* it not only avoids the difficulties of 'abstraction' but is extremely natural. Indeed it lies behind the Euclidean conception of number. The most primitive idea of assessing the size of a collection of objects is to count through them replacing each object by a tally mark, say a raised finger, or a stone, or even a formal sign, for example a vertical stroke |. Thus, if one has to count four apples, then after the count one is left with four raised fingers or four stones or the row of strokes ||||; the particular nature of the elements being counted has now conveniently been ignored, though they have not been destroyed, or altered, or 'abstracted away'. It is, of course, only a short step from this to treating the group of strokes *as* the number of apples, or whatever, in the given collection. Then, of course, one might assign numbers or number symbols rather than just strokes to elements. So the result of counting the four apples would now look like this: (|) (||) (|||) (||||). The number last cited gives the size of the set. The inconvenience of this system for large numbers explains the use of special symbols to replace lengthy strings of strokes. These new symbols would represent the introduction of a new higher-level unit.

There seems no doubt that this tally mark conception is actually the origin of our notion of number. For example, it is directly reflected in the ancient Babylonian (*c.* 2000 B.C.) and Egyptian (3500-1700 B.C.) symbolism for natural numbers, for this is a tally mark symbolism of exactly the simple kind mentioned above (see Kline [1972], pp. 5, 16). And one finds much the same with the Greeks. According to Kline: 'The Pythagoreans usually depicted numbers as dots in sand or as pebbles.' (Kline [1972], p. 29.) Of course, to follow this procedure rigidly is to make number either irredeemably formalistic (this, it seems, is a mistake which Kronecker later made (see Frege [1903], p. 155, n.1., or Cantor [1887-8], p. 382)) or materialistic. The realist tradition however, wanted to make number theory an objective and general science, that is a science whose subject matter is an unchanging world of unchanging objects quite separate from the physical collections which ordinarily it is our business to number.[4] Thus

---

[4] Though, according to Annas ([1976], p. 28), Aristotle clearly distanced himself from this 'separate entity' or 'Platonic' view of number. For some (crude) remarks on the relation between Cantor's and Aristotle's positions, see §3.4.

One of the great Greek contributions to the very concept of mathematics was the conscious recognition and emphasis of the fact that mathematical entities, numbers, and geometrical figures are abstractions, ideas entertained by the mind and sharply distinguished from physical objects or pictures. (Kline [1972], p. 29)

It seems that to achieve this with the theory of number the Greeks took the primitive tally mark system and gave it an idealist (or Platonist) twist. In any collection to be counted each of the elements is replaced by the so-called pure *unit* or *monad* and the collection or heap of repeated monads is then itself the number sought after. Clearly this is just an abstract reconstruction of the tally mark system; elements are now counted by reference to an *ideal* tally mark, the pure monad or pure unit, instead of a physical tally or a written stroke '|'. And now numbers are collections or assemblages of abstract units instead of assemblages of symbols.

Cantor's theory of 'ones' follows directly in this line. His use of something like the Greek ideal units avoids any purely formal or material element. And although the origin of the theory is the notion and process of *counting*, that is, the successive enumeration or assignment of marks to the collection to be numbered, it seems (as Cantor after 1883 wanted) that the much more Platonic idea of instantaneous replacement avoids dependence on what Cantor calls 'the subjective [or one can also say successive] numbering process'. Thus, it seems that part of his desire to avoid subjective elements has been achieved. Cantor's position can be summed up in some remarks he made against the views of number put forward by Helmholtz and Kronecker which (if Cantor's characterization is fair) goes *back* towards the formal tally mark system. He says:

. . . both scholars take numbers in the first place as *signs*, but not signs for concepts which relate to sets, but *signs for the individual things counted in the subjective numbering process*. (Cantor [1887–8], p. 382)

But Cantor says clearly that this 'stands opposed to my position': in this respect he aligns himself indeed with Euclid.

Actually Cantor does not successfully avoid the 'subjective number process' or dependence on literal *successive* counting. For we must not forget that his theory, unlike the Greek, is intended to apply to infinite number as well, and in this respect it still conceals well-ordering assumptions which Cantor never succeeded in divorcing from the 'subjective' or 'successive' elements (though his successors did). But this is jumping ahead a little. To bring us towards this we must first consider the question: are the *units* or monads different? Or is there only one unit?

3.3. The theory of 'ones' sensibly construed

It seems clear that in Plato's conception there is a *unique* 'one'. (See, for example, Szabó [1978], p. 259, where he quotes in particular from *The Republic*.) Thus we can also assume that in Euclid's conception, where the numbers are multitudes of units, the units are taken to be the same unit repeated. Cantor's own position is not clear. When dealing with cardinal numbers Cantor always speaks of its 'ones' [*Einsen*] in the plural, thus suggesting a plurality of different 'ones'. But this should not be taken too seriously, since Euclid also speaks of units in the plural while meaning 'sameness repeated'. And Cantor allies himself sufficiently closely to Euclid to suggest that he also means 'sameness repeated'. In the case of the ordinal numbers Cantor states clearly that the 'ones' *are* all the same:

To the individual elements E, E', E'', ... *of* the set M there correspond in its ordinal type $\overline{M}$ only *ones* $e = 1$, $e' = 1$, $e'' = 1$. . . . , which as such are all identical but are differentiated from one another . . . through their position inside the order-type $\overline{M}$; . . . (Cantor [1887–8], p. 423; the passage stems from 1884)

This seems ungainsayable. And it is also quite clear from Cantor's accounts that a cardinal number will arise from an ordinal number (or type) by 'abstracting' *once* from the subsisting order. This would leave unstructured, repeated, *identical* ones. Thus the corresponding cardinal would consist of unstructured, and therefore undifferentiated, identical ones—and arguably (in view of the 1883 ordinal theory of cardinality) all cardinals can be explained in this way. It should also be noted that in [1883*b*] in reference to the natural numbers Cantor definitely says that the 'unities' [*Einheiten*] of the numbers are to be 'looked upon as equal' (see §3.2, p. 131). But Cantor is not quite textually pure. For against this, in one place he describes the 'ones' of an order-type as 'conceptually differentiated' ([1887–8], p. 380). And he does once ([1887–8], p. 418, n.1) speak of the 'different ones' in a number (at the particular place cited he means finite cardinal number). Perhaps the *best* we can say is that Cantor is never clear on this point. Nevertheless, it seems from the historical and some of the textual evidence that we have at least to face up to the problem raised earlier of sets, extensionality, and coalescence.

Can we assume, perhaps, that Cantor would attempt to solve the problem of the coalescence of the single repeated 'one' by giving up the axiom of extensionality, that is, by giving up the insistence that numbers be sets in the normal sense? In this connection it is interesting to note that Weierstrass regarded *real* numbers as certain collections of rational numbers in which finitely many repetitions are allowed. (One of these collections defines a real number if the sum of any finite number of its elements is less than some fixed rational bound: see van Dantscher [1908], or Jourdain [1909].) It is not surprising therefore that some of Weierstrass's followers (probably relaying Weierstrass's views) were happy to regard natural numbers as collections of 'ones' repeated (see Frege [1903], pp. 149–53). But there are two reasons why Cantor could not have followed Weierstrass in this respect. First, it seems impossible to give up extensionality for collections without relying on a prior notion of cardinal number. For one would have to provide a criterion of identity for non-extensional collections, and surely non-extensional collections will be different just in case they do not contain the same elements *with the same cardinal number of re-petitions*. Secondly, Cantor *insisted* that extensionality holds for sets. For example, in [1887–8], p. 387, he states that a set 'consists of clearly differ-entiated, conceptually separated [*wohlunterschiedenen, begrifflich getrennten*] elements'. And later he writes in a similar vein that a set 'is a bringing together into a whole of definite, clearly differentiated [*bestimmten, wohlunter-schiedenen*] objects' (Cantor [1895], p. 282; see also Dedekind [1888], p. 1). And, as we have seen, Cantor insisted that numbers are sets. So this does suggest that extensionality must hold for the numbers also.

There are a few short passages where Cantor seems to suggest that numbers may not be normal sets, that they are, so to speak, 'more than sets'. For ex-ample:

The cardinal numbers as well as the order-types are *simple* conceptual forma-tions: each of them is a *true unity* (*monas*), because in it a multiplicity and manifold of *ones* is uniformly joined.
The elements of the set M we are given are to be imagined as separated. In its intellectual image $\bar{M}$, which I call its order-type, the ones [*Einsen*] are on the contrary *united to one organism* [my italics]. In a certain sense, each order type can be looked on as a composition of *matter* and *form*. The conceptually differentiated ones it contains furnish the *matter*, while the order subsisting among them is the corresponding *form*. (Cantor [1887–8], p. 380)

Similarly Cantor says:

Every number according to its nature is a *simple* concept, in which a manifold of ones is put together organically-uniformly in a *special* way. (Cantor [1887–8], p. 418, n.1)

The terms used in respect to the 'ones', 'uniformly joined' and 'united to one organism', seem strange in connection with the statement that in M the elements 'are to be imagined as separated'. It *perhaps* suggests that $\bar{M}$ is a different *kind* of set from the 'normal' set M. If this is so then it seems by interpreting Cantor as taking numbers to be sets in the normal sense we are *mis*interpreting him. Cantor gives really no further indication of what these special *organic* sets are. In one other place where he mentions the peculiarity of the numbers, he seems to point to their being *intellectually conceived* as their crucial feature:

Are not *a set* and *the cardinal number belonging to it* quite different things? Does not the *first* stand to us as an object, whereas the second is an abstract image of this object *in our* intellect? (Cantor [1887–8], p. 416)

This does not take us very far. For one thing, Cantor sometimes says that sets are also objects or entities 'arrived at' in the mind or the intellect. Thus, recall, in [1883*b*], p. 204, n.1, he says: 'By a "manifold" or "set" I understand in general any many which can be thought of as a one.' (See also the discussion in §1.3.)

But to come back to the notion of 'organic unity', this does not seem to provide a sufficient separation from sets either. For Cantor *always* takes a set as a collection which is 'united to a whole' ([1895], p. 282), or 'one thing'

([1883*b*], p. 204) or 'a thing itself' ([1887-8], p. 411). Indeed it was one of the most important features of Cantor's conception that sets be taken as single objects. And it is hard to see what difference there could be between objectness, thingness or wholeness and 'organic unity'. Thus it seems that if we are to take the above remarks as indicating a serious difference between numbers and ordinary sets we are, in effect, back with the 'primitive entity thesis'. This seems unacceptable. It may be held that to take numbers as primitive entities is quite natural for someone, like Cantor, for whom numbers were of such crucial structural importance. But this overlooks how important the justification of the existence of (transfinite) numbers was for Cantor *and* the argument he gives for this justification. As explained above, it is based on the assumption that mathematics must accept the existence of infinite sets. Numbers are then in some way or other tied to these. But the argument becomes much more compelling if numbers are also construed as sets, that is *things of the same kind*. This would then mean that accepting numbers does not commit one to anything qualitatively different from the entities *already* accepted. (This argument, if pushed hard enough, would eventually work in favour of the number-class interpretation of cardinal number. For example, if mathematics must accept the existence of the set of natural numbers *and* this set is taken *as its own cardinal number* then mathematics has to accept at least one infinite cardinal.)

We are thus in the following position. If one insists on the non-differentiability of 'ones' in the Cantor cardinal numbers then we cannot make sense of the theory without going outside set theory. Perhaps in the end we shall be forced to say that Cantor's theory cannot be made sense of. But first let us examine the thesis that in a cardinal number the 'ones' *are* differentiated. After all, there is no doubt that Cantor was aware of the difficulties that arise if they are not. For one thing he had read Frege. But for another, in his characterization of order-types he says that the *equal* 'ones' are differentiated by their ordering, thus implicitly recognizing the danger of collapse if they are not differentiated. It is inconceivable that he did not recognize the same danger with cardinal numbers also. And after all there are one or two passages, as we saw, where he says that 'ones' are 'conceptually differentiated' (by attachment to their base elements in *M* perhaps?) or 'different'.

We have seen clearly that the Euclidean theory, so similar to Cantor's, was in effect and in origin a *representational* theory (i.e. subject to the demand that $card(x) \sim x$). With hindsight we know that the only way to obtain a representational theory (i.e. satisfying demands (*i*)-(*iv*), p. 88) is to appeal to some strong *ordinal* notions. In its origin, the theory presented by Euclid was a counting theory, that is to say an ordinal theory, for the ability to count a collection presupposes at least that the collection be well-orderable. It seems that the Greek idealists, and later Cantor, wanted to expel the counting notion from the theory. For the Greeks this was probably because of emphasis on

Being rather than change (see Szabó [1978]); for Cantor it was clearly because it smacks too much of human capacities and is too closely related to the actual assignment of signs or tally marks. But while insistence on counting itself is dispensed with, this does not mean that all ordinal notions have been eradicated. Indeed closer analysis of Cantor's operation with, and arguments concerning, his own 'ones' theory show that indeed strong ordinal notions are still secreted in it.

If the 'ones' or 'units' are differentiated then either there is only one single 'one' which is used with a system of indexing, *or*, what amounts to the same thing, there is a multiplicity of ones. This seems quite reasonable even in the original stroke mark system. After all, in that system although we might feel that the *same* mark is being used over again, that is not strictly the case. Either we are using successively different fingers or chalk marks or matchsticks. *Or* we are using one mark E (or marks which at a glance are indistinguishable) with some spatial or temporal indexing denoted by the complex (E, i), and these *complex* marks are different. Indeed, in the sophisticated formal approach (represented, say, in the Babyloanian symbolism) successive elements are marked by successive and *different* symbols (I), (II), (III), ... Intuitively what is important is not that a new mark should be the *same* as the old, but that it should be, while different, as little different as possible from the old. (It should be noted that the use of the successive ordinal numbers captures this beautifully with the principle that a new counting mark is the immediate successor of the one used immediately before.) The assertion that the 'ones' are different from one another, therefore, fits perfectly well even with the original primitive tally mark system. The interesting historical question here is why in giving this theory an abstract or idealist twist the Greeks should have adopted the view that there is only *one* ideal unit or 'one' and not a genuine differentiated plurality. One plausible answer is furnished by Szabó's hypothesis that the Euclidean/Platonic definition of number came from the Pythagoreans via the Parmenidean (Eleatic) doctrine of the 'One'. The central thesis of the Eleatics was that only the 'One' existed, that there is no plurality and thus no divisibility. The Pythagoreans, according to Szabó's interpretation, wished to maintain plurality (which is the essence of arithmetic) without wholly giving up the doctrine of the 'One'. (As Frege remarked: 'If abstraction caused all differences to disappear, it would do away with the possibility of counting.' [1906], p. 125.) This was done by considering number as something *abstract* and non-sensual (like the 'One' or 'Being'), and by construing numbers (i.e. from 2 onwards—presumably 1 is no number because it is 'the One' itself) as multiplicities or repetitions of the 'One'. Fractions could then be represented as ratios without it being necessary to assert that the 'One' *actually becomes divided* (see Szabó [1978], pp. 257-65, especially pp. 263-4).

To come back to the thesis that Cantor's 'ones' *are* differentiated, the interesting question is: what properties must the collection of 'ones' have? The first to notice is that it cannot be the case that each object in the universe gives rise to a 'one' or 'monad' special to itself. For suppose this were the case, that is to say if $a \neq b$ then a and b give rise to different 'ones' 1_a and 1_b. Then $card(\{a\}) = \{1_a\} \neq \{1_b\} = card(\{b\})$, contradicting the condition that $x \sim y$ iff $card(x) = card(y)$. Indeed, there must be a unique 'one' or 'monad' 1^* chosen from all the 'ones' to form the cardinal number $\{1^*\}$ of $\{a\}$ *whatever* object a is. Now apply similar considerations to arbitrary pairs. Again there must be two special 'ones', $1'$ and $1''$ which are used to form the cardinal number of $\{a, b\}$ for *any* a and b. Moreover, Cantor insists ([1887–8], p. 418) that each number should contain its predecessors, or at least the material out of which they are formed, as 'virtual components' ('component' being one of Cantor's terms for 'subset'):

Each number [*Zahl*] according to its nature is a *simple concept* in which a manifold of ones [*Einsen*] is put together organically-uniformly in a *special way*, so that it contains as virtual components both the different ones [*die verschiedenen Einsen*] and the numbers which arise from the partial collection of these ones. (Cantor [1887–8], p. 418, n.1)

Cantor stresses that, for example, the concept 5 does not contain the concept 3 'as a real part'. But it is clear from the above that the 'ones' of 5 must contain among themselves the 'ones' which go to make up 3. This is enough for us to conclude that one of the 'ones' used for forming the cardinality of pairs must surely be 1^*. Hence, the other 'one' must be a unique 'one' distinguished from among all 'ones' other than 1^*. Call this 1^{**}. By similar reasoning one can argue that there must be an ω-sequence of ideal 'ones' $1^*, 1^{**}, 1^{***}, \ldots$ etc. from which all the finite cardinal numbers are composed. Thus it would seem that in some sense Cantor's approach already presupposes the existence of the collection of natural numbers.

The argument here strongly resembles Cantor's own presentation of the natural numbers in his [1895]:

To a single thing e_0, if we subsume it under the concept of a set $E_0 = (e_0)$, there corresponds as cardinal number that which we call 'one' [*Eins*] and denote by 1. We have

$$1 = \overline{\overline{E}}_0. \tag{1}$$

If one joins another thing e_1 to E_0, and calls the union set E_1 then

$$E_1 = (E_0, e_1) = (e_0, e_1). \tag{2}$$

The cardinal number of E_1 is called 'two' and is denoted 2:

$$2 = \overline{\overline{E}}_1. \tag{3}$$

Through the addition of new elements we obtain the series of sets

$$E_2 = (E_1, e_2), \ E_3 = (E_2, e_3), \ldots$$

which provide us in unlimited succession with the other so-called *finite cardinal numbers*, denoted 3, 4, 5 . . . The auxiliary application of the same numbers here as indices is justified because a number is only used in this sense after it is defined as a cardinal number. (Cantor [1895], pp. 289–90)

And, of course, according to Cantor's remarks earlier in the same paper, $1 = \bar{\bar{E}}_0$, $2 = \bar{\bar{E}}_1$, etc. should themselves be sets of 'ones'. With regard to this whole conception of natural number and particularly that numbers virtually *contain* their predecessors, it is worth noting what Cantor wrote in a letter to Peano, 14 September 1895: '. . . our definition of the finite cardinal numbers is such that each finite cardinal number v is *connected chainlike to all earlier numbers*.' (Cantor *Nachlass VII*, p. 180.)

I suggest, however, that not only does Cantor's approach in some sense presuppose the natural numbers, but also much more. It is difficult to see *exactly* how far the above line of argument will take us. For instance, it is not immediately clear that the *whole* collection of 'ones' or 'monads' must be well-ordered. Nevertheless, there is some evidence that this must be at the heart of Cantor's conception. For example, consider Cantor's proof that if $M \sim N$ then $\bar{\bar{M}} = \bar{\bar{N}}$, given in [1887-8], p. 413, and [1895], pp. 283-4. The proof rests on the following assertion. If an element m of M is replaced by its counterpart $f(m)$ in N (f being a map which establishes the equivalence of M and N) then the cardinality of the resulting set M' ($= (M - \{m\}) \cup \{f(m)\}$) is the same as that of M, i.e. $\bar{\bar{M}}' = \bar{\bar{M}}$. Given Cantor's claim that for any A, $A \sim \bar{A}$, it is trivial to conclude that $\bar{\bar{M}} \sim \bar{\bar{M}}'$. But to conclude that $\bar{\bar{M}} = \bar{\bar{M}}'$ one surely has to argue that the elements of M give rise to exactly the same 'ones' when they appear in M' as they did when they appeared in M, *and* that $f(m)$ gives rise to the same 'one' as m did. But I can see no way of arguing this except by using a *place* or *ordering* argument, i.e. an argument which says that each element of M has a fixed place in M which is not disturbed in the shift from M to M', and that the 'ones' are assigned not so much to elements but to places, a particular place always being canonically assigned the same 'one'. Then one could argue, of course, that $f(m)$ takes the *place* of m and hence takes the 'one' previously assigned to m; all the other elements of M keep their places and hence their 'ones'. But this argument is obviously founded not only on the well-orderability of arbitrary sets, but also on the assumed arrangement of the whole stock of 'ones' in well-ordered form *guaranteeing* that assignments of 'ones' to elements is always canonical. Without this there seems no way of establishing the fundamental condition $x \sim y \rightarrow [card(x) = card(y)]$. (A similar kind of 'place argument' must also lie behind Cantor's assertion ([1887-8], pp. 415-16) that every infinite set M possesses a proper subset M_1 such that $\bar{\bar{M}} = \bar{\bar{M}}_1$.)

The implicit well-ordering assumption is perhaps even clearer in Cantor's [1887-8] presentation of the argument:

The cardinal number $\bar{\bar{M}}$ of a set M remains unchanged according [to the abstraction idea], if other things are substituted in place of the elements m, m', m'', . . . of M. If now $M \sim M_1$, then there is a correspondence by which the elements m, m', m'' . . . of M correspond to the elements m_1, m_1', m_1'', . . . of M_1. One can think of the elements m_1, m_1', m_1'' . . . of M_1 as substituted for the elements

$m, m', m'', \ldots$ all at once. Thereby the set M is transformed into the set M_1, and since with this transformation nothing in the cardinal number is altered then $\bar{\bar{M}} = \bar{\bar{M}}_1$. (Cantor [1887–8], p. 413, n. 1)

Here, no doubt unwittingly, Cantor has even begun to write out the sets M and M_1 in well-ordered form.

Thus, it seems, it is exactly here in these arguments that Cantor pays the price for a representational theory which we now know *must* be paid. If, therefore, we take the Cantor 'ones' to be differentiated (and to make clear sense of the theory we *must*) then this 'abstraction' theory of number is nothing other than a disguised *ordinal* theory.

What is striking about this analysis is the similarity between this reconstruction and the ordinal number-class theory of cardinality which Cantor introduced in his [1883b] and which he took as basic from that time on. (For recall that in virtually all the texts where Cantor appeals to the abstraction theory of number one can also find a passage where he explicitly says that the number-classes represent all infinite powers.) The central difference is that there the ordinal assumptions are quite open and clear, whereas in the later accounts they are wrapped in the mysterious cloak of abstraction and the theory of 'ones'.

In his notes on Cantor's [1895] explanation of abstraction Zermelo wrote:

The attempt seemingly to elucidate the abstraction process leading to 'cardinal number' by conceiving the cardinal number as 'a set put together our of nothing but ones' was certainly not a happy one. Since the 'ones', if they are all different, as indeed they must be, are now nothing more than the elements of a newly introduced set equivalent to the first, we have not come one step further forward in the required abstraction. (Zermelo [1932], p. 351)

But while one can sympathise with Zermelo's frustration, his conclusion does not seem to me to be right. As we have seen simply by looking at Cantor's arguments, and *not* by dragging in sophisticated modern set theory, the replacement of the given set is not replacement by just *any* set. It must be *canonical* replacement. And the demands of such canonical replacement do certainly lead us forward (or in Cantor's case we might say back!) conceptually or heuristically. For it becomes clear what kind of assumptions are required if such replacement is always to be possible.

If one accepts the parallelism between the number-class theory and Cantorian abstraction then one is brought back to the thesis that in his mathematically explicit ordinal theory of cardinality the number-classes themselves are the natural candidates for the cardinal numbers. In the abstraction theory the cardinals are taken to be actually sets of 'ones'. Thus, the natural reading of the parallel ordinal theory is that the cardinals are canonical sets of ordinals, in other words (if they are to be the same for all sets of the same power, independent of the particularity of an ordering) the *number-classes*. For example, according to the [1895] theory the number $\aleph_1$ is a set of pure 'ones' which is cardinally equivalent to the number-class (II). But if $\aleph_1$ must be a well-ordered set of ones starting from 1* and equivalent to (II) then it may just as well *be* (II) (or (I) $\cup$ (II)).

The parallelism would also conversely provide an explanation of Cantor's insistence that a cardinal number is achieved by 'double abstraction'. The first level of 'abstraction' is in fact (see §3.2) replacement of the elements of the set by 'pure ones' or 'ordinal numbers'. (This fits very much with the Greek theory of 'ones' as interpreted by Aristotle, for he saw the replacement of the objects by 'ones' as prior to the actual counting process. See Klein's discussion of Aristotle on number in his [1968].) The second level of abstraction would now be the 'collapse' of the set of ordinals or 'ones' to the appropriate number-class (or cumulative class). This second level abstraction is quite consistent with Cantor's claim that in forming a cardinal number one abstracts 'from the order in which they [i.e. the elements] are given' ([1887-8], p. 379, also [1895], p. 282) or from 'the order which may rule among them' ([1887-8], p. 387). For while in this process some ordering is imposed on the given set, in general this ordering will be quite different from any *given* ordering. Moreover, once the 'collapse' to the number-class is made (and the result of this collapse is the same for *all* equivalent sets) any imposed ordering on the original set can be forgotten. Here it is important to recall that the theory of 'ones' was itself part of Cantor's attempt to explain the notion of 'abstraction'.

By this analysis and construal of the theory of 'ones' I do not mean to claim that it was Cantor's intention to present a disguised ordinal theory of cardinality. Indeed, as I said in §3.1, it seems plausible that Cantor was attempting to *avoid* reliance on the ordinal explanation of cardinality. And there is certainly one point where my analysis does not fit with Cantor's actual statements. For in one (though only one) of his accounts of the 'ones' theory of cardinality, Cantor says:

... the elements [of the cardinal number] as so-called ones [*Einsen*] have in a certain way grown together organically into one another to a uniform whole in such a way that none of them has a privileged relation of rank to the others. (Cantor [1887-8], p. 379)

And this conflicts with the conclusion that the 'ones' if different must constitute a natural well-ordered totality. Rather, all I have claimed is that to make sense of Cantor's abstractionism as a *set* theory of number, one is forced to bring in strong ordinal assumptions. Thus, if one imagines oneself in discussion with Cantor, one would say: 'You have two choices. *Either* give up the abstractionist account of number altogether, and go back to the number-class explanation of cardinality which at least has the merit of relative clarity. *Or* regard the abstractionist account as an intuitive informal explanation of cardinal number, but stress that mathematically speaking it leads in exactly the same direction as the number-class account.' Whatever Cantor may have *intended*, the abstractionist account does not loosen the ties between the general theory of cardinality and well-ordering. Indeed, if Cantor was actually trying to shift away from dependence on sweeping ordinal assumptions, it is all the more remarkable that his attempt, after a rather complicated detour, leads back to the starting point.[5]

3.4. Order-types

As I have mentioned already, Cantor relied on abstraction in a very similar way in his later explanation of order-types in general and, therefore, ordinals in particular. How does my analysis of the theory of 'ones' as essentially an *ordinal* theory affect this? One question certainly arises: if the 'ones' are properly construed as 'ordinals in disguise' is not Cantor's explanation of ordinals as sets of 'ones' circular or, at least, somewhat hollow? Before coming to this, though, let me briefly review what Cantor says about order-types.

In most of the places where Cantor gives an abstract 'definition' of cardinal number he also gives an account of order-types. The existence of an order-type for a simply ordered set is again taken to be demonstrated by *abstraction*, but this time *only* from the nature of the elements and *not* from their order. Cantor variously describes the result again as a 'general concept', a 'universal', an 'ideal number' or the 'ideal paradigm' of the given ordered set. (For these accounts see [1887-8], p. 379, pp. 422-3 and [1895], p. 297.) Again the abstraction is explained in terms of the mysterious 'ones', that is to say each element is replaced by or 'becomes' a 'one', and the 'ones' are then 'organically' united togther to a 'form or structure [*Bildung*]', an 'intellectual image [*intellektuelles Bild*]' of the given set (see [1887-8], pp. 379, 423). The order-type of M is denoted $\bar{M}$ to signify that only one level of abstraction has been effected, as opposed to the two levels in the case of a cardinal. In analogy with the cardinals, Cantor also claims that $\bar{M} = \bar{N}$ if and only if M and N are isomorphic as ordered sets (which he denotes by $M \simeq N$). This is asserted in [1887-8], pp. 379, 424, and again in [1895], p. 295. He gives no proof, but he asserts:

Both parts of this double proposition follow easily from the concepts of order-type and of similarity of ordered sets in a way analogous to that by which we proved that two sets have the same cardinal number when and only when they are equivalent. (Cantor [1887-8], p. 424)

It seems implied in all this, particularly in this last assertion, that the order-types $\bar{M}$ are themselves taken to be ordered sets of 'ones'. And this is indeed explicitly asserted in [1895]:

---

[5] I add the following suggestion, footnotes being comfortable places to stretch points. If (unlikely) Cantor knew that his 'ones' theory must be based on ordinal assumptions, his commitment to the similar number-class theory *may* explain why he ignored Frege's criticism of 'ones'.

Accordingly the order-type $\bar{M}$ *itself* is *an ordered set* whose elements are nothing but ones [*lauter Einsen*], and which have the same rank ordering among themselves as the corresponding elements of M, out of which they arise by abstraction. (Cantor [1895], p. 297)

To summarize this, Cantor lays down four conditions on order-types which are analogous to the four conditions on cardinals. These are:

(*i*)′ every simply ordered set $(x, \leqslant)$ has a definite order-type $ordtyp(x, \leqslant)$;

(*ii*)′ for every $(x, \leqslant)$, $(y, \leqslant^*)$, $ordtyp(x, \leqslant) = ordtyp(y, \leqslant^*)$ if and only if $(x, \leqslant) \simeq (y, \leqslant^*)$;

(*iii*)′ for every $(x, \leqslant)$, $ordtyp(x, \leqslant)$ is a set;

(*iv*)′ for every $(x, \leqslant)$, $ordtyp(x, \leqslant) \simeq (x, \leqslant)$.

(So (*iv*)′ implies that $ordtyp(x, \leqslant)$ is a simply ordered set.) Thus, what Cantor is proposing here is an interior representation theory of order-type very similar to the interior theory for cardinals. Let me first add some remarks about these conditions.

We have seen that in the cardinal case, fulfilment of the four conditions requires some strong ordinal assumptions, namely appeal to the well-ordering theorem and to the existence of a sufficiently rich class of ordinal numbers. The well-ordering theorem tells us that within any given cardinal equivalence class C there will be sets of ordinals; in particular C will generally contain many sets x *determined* by an ordinal, that is to say of the form $\{\alpha : \alpha < \beta\}$ for some ordinal β. (In the modern conception, β therefore *itself* belongs to C.) Thus we can canonically choose representatives for the classes C by choosing for each that set x in C determined by the smallest ordinal. (For infinite cardinalities these will be just the cumulative number classes, or, in the von Neumann conception, the initial ordinals.) In other words, focusing on sets of ordinals is already enough. In the case of order-types in general this is *not* enough; one needs something much stronger than the well-ordering theorem. In fact, it seems that one needs to assume a well-ordering of the whole universe of sets (or entities), or a selection term defined over proper classes of sets.

Why we need this is not difficult to see. Cantor assumes that an order-type is actually itself an ordered set of 'ones'. Now the 'ones' should be very much like (if not actually be) ordinals. This construal is not forced on us just by reference back to the analysis of cardinal numbers. For, as mentioned, Cantor appeals to the same kind of argument for the proposition '$\bar{M} = \bar{N}$ if and only if $M \simeq N$', as he does for '$\bar{\bar{M}} = \bar{\bar{N}}$ if and only if $M \sim N$', and exact analysis of this would show that *again* appeal must be made to 'place' assumptions and the existence of a canonical well-ordered stock of 'ones'. Thus an order-type must be tantamount to a simply ordered set of ordinals, and the assumptions behind the theory so construed are *at least* as strong as in the cardinal case. As in the cardinal case, these assumptions [*mutatis mutandis*] allows us to conclude that any isomorphism class of simply ordered sets will contain an ordered set whose field is a set of ordinals. For example, if $(x, \leqslant)$ is an ordered set in an isomorphism class C and ω_α (to use the modern conception for a moment) is $card(x)$, then there must be members of C with field ω_α, say $(\omega_\alpha, \leqslant^*)$. But fixing on these 'canonical' ordinals ω_α is of no great help. For, while we have narrowed down the whole equivalence class C to a set (the set of all ordered sets in C with field ω_α), there will be (infinitely) many ordered sets in this subset of C. Thus, to choose canonically representatives for all isomorphism classes one is faced with the problem (in modern terminology) of choosing simultaneously from a *proper-class* of non-empty sets. The only known way of doing this is to assume as given a well-ordering of the universe (or, equivalently, the so-called global axiom of choice). And this, of course, is a much stronger assumption than merely that of the well-orderability of all sets (see §4.1). We shall see (§4.1) that Cantor made implicit assumptions of similar strength in his later theory of absolute and non-absolute collections.

Thus, Cantor's notion of order-type can, like that of cardinal, be made sense of, though only at the expense of rather strong ordering assumptions. (John Bell has pointed out to me that the problem of defining order-types which accord with Cantor's conditions is an instance of the problem of finding *skeletal* categories. A *skeleton* for a category C is a (full) sub-category D such that each object of C is isomorphic to a unique object of D. Thus, to take two examples, (*a*) the category of ordinals is skeletal for the category of well-ordered sets and (*b*) the category of cardinals (i.e. natural numbers and initial ordinals) is skeletal for the category of sets. In example (*b*) construction of the skeleton requires the local axiom of choice and transfinite recursion (therefore replacement), while (*a*) only requires recursion. However, it seems that proving the existence of skeletal categories for other natural large categories, like the category of all groups, or—to take the Cantor case—the category of all simply ordered sets, requires the assumption that the universe of sets (or the initial category) is well-ordered. For it seems that the only way of uniformly selecting representatives for groups, say, is to well-order the class of all groups and for each group G choose the least element in this well-ordering isomorphic to G. This would be the natural method in the Cantor case also if one is not constrained by the condition that the field of the underlying representative set must be a set of ordinals.)

Let us now briefly turn to the specific question of the structure of the ordinal numbers. Since Cantor says very little about the 'ones', we have as little idea in the case of the ordinals as in the case of the cardinals as to what the 'ones' are like. But if, following the above analyses of cardinals and types, the 'ones' effectively must be ordinals, then an ordinal number on Cantor's construal must be a *set* of ordinal numbers. Is there not a circularity here? Or, again, if the 'ones' are not actually the ordinals, they form a collection which has *exactly* the properties that the collection of ordinals should have. Thus the further introduction of ordinals appears redundant.

Certainly, if one attempted to give a general definition of an ordinal in terms of ordinals, this would be circular. But as I have pointed out, it is perhaps much more helpful to view Cantor's abstractionist statements about number as explanations rather than definitions, a step on the way to understanding what properties ordinals as sets should have. And if one says that ordinals must be sets of ordinals there *need* not be either redundancy or circularity. For one can here simply point to the von Neumann conception of ordinals where ordinals are just that, sets of ordinals (though not *defined* in this way). Now this only makes a point about consistency or coherence, and is not *at all* meant to impute to Cantor ideas which came only later. But it is perhaps worthwhile making one further observation here.

According to von Neumann an ordinal is *exactly* the set of its predecessors. Thus, a 'new' ordinal is recursively determined by the 'old' ordinals. (This, of course, does not amount to a definition of ordinal. But recognition that ordinals do or should have this form no doubt led to the isolation of properties which do serve to define them.) Now there is one place where Cantor deals explicitly with the characterization of specific numbers according to his theory of abstraction. This is where he deals with the finite cardinal numbers ([1895], pp. 289–90; see pp. 138-9). And here the numbers *are* explained recursively. This is no great surprise. But if in addition one takes seriously the explanation of the 'ones' as very like ordinals (and the point about the canonical assignment of these 'ones'), then these finite cardinal numbers can only be something very much like the von Neumann finite ordinals. Identify the finite cardinals with the finite ordinals (this is no great step), and one is well on the way to the modern conception. Note here also Cantor's characterization of the finite ordinals in [1883*b*], p. 195, as the (successive) uniting together of 'successive laying down of unities', which is quite close to 'taking the set of predecessors' (see pp. 52-4). Note also that in this passage ω is explained as the 'expression' of the fact that the totality (set) of finite ordinals is given 'in natural succession'. This quasi-recursive approach is then extended to $\omega + 1$ etc. To this I would only add: look at how *simply* Mirimanoff arrived at the von Neumann notion of ordinal by reflecting on the demands of Cantor's idea that an ordinal must be a genuine set representative of the set it numbers (see §8.1).

The point here can be simply summarized. If one concludes that the 'ones' can only be sensibly interpreted as 'hidden' ordinals, there is still the possibility of avoiding circularity and redundancy. (Note here that Cantor himself, in the passage from [1895] cited above, is sensitive to the point about circularity, and the recursive procedure he adopts certainly avoids it.) And, as we have seen, there is some evidence that the ordinal interpretation of 'ones' does capture a good deal of what lies behind Cantor's conception of numbers.

3.5. Cantor and well-ordering

I pointed out in §3.1 that the motivation behind Cantor's abstractionist account was probably twofold: (a) to *divorce* (cardinal) number from well-ordering; (b) to construe number as far as possible as something objective. The analysis of abstraction shows that neither (a) nor (b) was satisfactorily achieved: to understand the theory of 'ones' it seems that all numbers must ultimately depend on well-ordering, and appeal to 'acts of abstraction' does not obviously avoid all *subjective* elements. Yet one might argue that the failure of (a) through the intrusion of well-ordering at least now allows (b) to be achieved. If 'abstraction' really means 'replacement by "ones" (ordinals)' then surely the assertion that the various numbers exist can simply be construed as the claim that certain well-orderings and well-ordered sets exist. Cantor's use of psychological terminology could then be taken (as already suggested) only as part of an attempt to explain how we grasp or conceive what the numbers are like. But is this really so? Certainly with hindsight one could conceive the role of well-ordering here in this way. But this raises the question of what Cantor's own conception of well-ordering was and whether *that* is divorced from all subjective elements. This question is already in effect raised by the number-class theory of cardinal, so my examination will not be especially dependent on the analysis of the abstractionist account.

Let us go back to Cantor's treatment of well-ordering in his [1883b]. There the statement that a set is well-orderable is phrased as: 'it is possible to *bring* it into the form of a well-ordered set' ([1883b], p. 169). Moreover, the characterization of a well-ordering given there refers to the definite 'succession' of the elements ([1883b], p. 168). The dependence on succession appears too in the conception of the ordinal numbers, which are understood to be the representatives or reflections of well-orderings. The finite ordinals are achieved by 'successive laying down of unities', with the implication that this is extendable to the transfinite numbers. Ordinal numbers are taken to enumerate or count sets, and there is a strong suggestion that Cantor takes 'counting' here in a more or less literal sense. For the key verb is *abzählen*, which literally means to count off, as in *abzählen an die Fingern*—to count off on one's fingers. Moreover, the idea of ordinals as counting numbers was actually *crucial* in the original discovery of the notion of well-ordering (see §2.1). All these elements are brought to the fore in the following passage from [1883b]. Referring to Aristotle's conclusion that 'countings [*Zählungen*]' are only possible on finite sets, Cantor goes on:

However, I believe I have proved above, and it will be shown even more clearly in the rest of this work, that definite countings can be effected [*vorgenommen*] both on finite and on infinite sets, assuming that one gives a definite law according to which they become *well-ordered* sets. That without such a lawlike succession [*gesetzmässige Sukzession*] of the elements of a set no counting with it can be effected lies in the nature of the concept *counting*. (Cantor [1883b], p. 174)

The terminology in these references suggests (though perhaps not in a very definite way) *either* appeal to our actual mental capacity or ability (say, to arrange a set in well-ordered form, or to count through a set), *or* to some kind of abstract extension of our actual ability to perform a construction (to count, or arrange, or specify a law), what one might call loosely 'constructive capacity'. The intrusion of these 'constructive elements' introduces considerable tension in Cantor's system, certainly at this [1883*b*] stage. Now in some of his later papers Cantor attempted (as we have seen) to divorce himself from some of the subjective/constructive aspects of the [1883*b*] theory. For example, he rejects 'the subjective numbering process' and the 'successive laying down of unities'. And even in [1883*b*] (see §2.1, and the passage just quoted) there is a clear attempt to reduce the notion of counting (ordinal) number to that of a well-ordered set. But it remained the case that in all important instances where he attempts or requires to establish that a set can be well-ordered, the 'subjective' or 'constructive' element creeps back in. One example is his characterization of finite set given in his [1887-8]:

By a *finite* set we understand a set M which arises out of an original element through the successive addition of elements in such a way that also the original element can be achieved *backwards* out of M through *successive removal* of elements in reverse order. (Cantor [1887–8], p. 415)

(This definition was heavily criticized by Frege ([1892], pp. 164–5) precisely because it contains a subjective element.) Reliance on *succession* is later clearly extended to the treatment of infinities. For example, Cantor's proof ([1895], p. 293) of the claim that every infinite set has a subset of power $\aleph_0$ is proved by a *successive* selection argument. And apparently he believed that one could establish the well-ordering theorem in general in just this way: take any set, select an arbitrary element as first, select as second an arbitrary element from the remainder, and so on, until the set is exhausted. (See Zermelo [1908*a*], p. 193, and also §4.1 below.) Zermelo remarks that Cantor '. . . apparently had reservations about accepting [this method] as a *proof*.' But the point is that Cantor apparently had available no *other* approach to proving the existence of a well-ordering. Thus, despite Cantor's intentions, if number is based on well-orderability, then the 'psychological' or 'constructive' element is certainly by no means yet avoided.

Russell later remarked that the psychological (perhaps one could also say 'constructive') aspects of counting *can* be dispensed with if we concentrate only on well-ordering:

What is meant by counting? To this question we usually get only some irrelevant psychological answer . . . [But] counting has, in fact, a good meaning which is not psychological. But this meaning is only applicable to classes which can be well-ordered. (Russell [1903], p. 144)

But in Cantor's case we see this does not get us very far. Via the idea of immediate replacement by 'ones' some of the problems associated with actual counting of sets *are* dispensed with. But in so far as this explanation itself must rely on well-ordering, the 'psychological' or constructive elements still remain, though now shifted onto the notion of well-orderability.

It is perhaps helpful to try to pin down more exactly where the tension in Cantor's system lies by tracing the heuristic development in the idea of a counting theory of cardinal size. We can distinguish (somewhat crudely) three stages.

(*i*)   The existence of a definite size for a collection is directly connected to our actual ability to count the collection.

(*ii*)   Acceptance that there are arbitrarily large counting (size) numbers generated from 1 by the operation 'addition of 1'; thus *actual* ability to count has been transcended.

(*iii*)   The existence of a counting number (and therefore size) is tied *only* to the *existence* of a well-ordering.

(*i*) is somewhat vague, but it is based on the idea that existence of a number is inextricably bound up with capacity to *determine* or *name* number. This belongs to the earliest conception of number, and the connections between existence and determination are nicely illustrated in Archimedes's *Sand-Reckoner*. The problem Archimedes addresses is this. Imagine the universe (assumed closed and finite) filled with grains of sand; does it make sense to say that this collection of grains has a number? Archimedes (assuming Aristarchus's estimate of the diameter of the universe) shows that it does by devising a nomenclature for numbers which enables him to name or exhibit numbers sufficiently large to put a bound on the collection of grains. (See Heath [1921], pp. 19–20, 327–30.) Here existence of number is directly linked with capacity of the human agent to determine number, the implication being that only if there are enough nameable counting numbers available can we say that the collection has a size. Stage (*ii*) is in effect the concept of natural number as we have it today, where what is important is the existence of a function which 'generates' an ω-sequence. It is possible here to interpret 'generate' in a quasi-literal sense, and thus connect (*ii*) to (*i*) by saying that ability to determine in actual practice is replaced by ability to determine only in principle. (One might say too that this step is really already potentially present in Archimedes. For it is clear that the collection given in the problem is too large to be *actually* counted. Yet Archimedes attempts to show that there are enough numbers available for the count to be made in *principle*. In (*ii*), the successor function makes available infinitely many numbers.) (*ii*) is thus somewhat ambiguous philosophically: it can be interpreted as based just on an existence claim, or as a theory about constructive possibilities. (*iii*) is really the modern set-theoretic position due to Zermelo and von Neumann. As Russell suggests, *one* feature of counting has been singled out—its connection with well-ordering—and this is treated existentially, i.e. the existence of a counting number for a set depends only on the *existence of* a well-ordering on it. Here the role of actual step-by-step counting has been taken over completely by the abstract, non-constructive notion of definition by transfinite recursion. Thus (*iii*) generalizes (*ii*) in so far as (*ii*) is interpreted non-constructively.

Cantor sits somewhere between stages (*ii*) and (*iii*). His position with respect to (*ii*) is not altogether clear. Sometimes he is clear that it is only the independent existence of the natural numbers as a whole that is important. But sometimes, as in his definition of finite sets quoted above, Cantor seems to take the view that it is succession, construction, and determination that matters. He himself refers to such a dichotomy in a letter to Veronese, 7 September 1890:

With the numbers [*Zahlen*], one must distinguish between them as they are *in and for themselves, and in and for the Absolute intelligence*, and how these same numbers appear in our restricted, discursive comprehension and are differently defined by us for systematic or pedagogical purposes.
For systematic purposes, I let the numbers arise successively (the greater out of the immediately preceeding), just as you do. However, when it is a question of saying what the numbers are in and for themselves, it must be recognized, that each number is a simple concept and a unity, just as much a unity as the one itself. Taken absolutely, the smaller numbers are only virtually contained in the larger. Taken absolutely, they are all independent of one another, all equally good and all equally necessary metaphysically. (Cantor *Nachlass VII*, p. 3)

And again Cantor wrote to Hermite, 30 November 1895:

You say beautifully in your letter of 27th Nov.: 'The (natural) numbers seem to me to constitute a mode of realities which exist outside us, and this with the same character of absolute necessity as the realities of that nature, knowledge of which is given to us by our senses, etc.'
Permit me to remark, however, that the reality and absolute uniformity of the whole numbers seems to be *much stronger* than that of the world of senses. That this is so has a single and quite simple ground, namely the whole numbers both separately and in their actual infinite totality exist in that highest kind of reality as eternal ideas in the Divine intellect. I expressed a similar idea to yours in 1869 in my *Habilitationsschrift* . . . Of the three theses which I publicly defended then, the third read as follows: 'numeros integros *simili modo atque corpora coelestia totum quoddam* legibus et relationibus compositum efficere'. [Because there are laws and relations among them the natural numbers, *like the heavenly bodies, form a certain whole.*] (Cantor *Nachlass VIII*, pp. 48–9; partially reproduced in Meschkowski [1967], pp. 262–3)

These statements appear to make Cantor's position with respect to natural numbers clear: the 'successive' (or constructive) presentation of numbers is only for didactic purposes. But *in fact* it is not so clear. For, given his decision to base numbers on sets, to demonstrate the existence of finite numbers he has to use the notion of a finite set, and his characterization of finite set is, as we have seen, definitely tied to some notion of constructive capacity. But Cantor also takes the crucial step away from (*ii*) of applying the notion of a counting number to *infinite* collections—the essential step behind the number-class

theory of power and thus the most important heuristic step in the creation
of the modern set-theoretic theory of number. (It is already present in the
indexing of derived sets *beyond* the finite.) Cantor says, as with the finite
numbers, that the 'new' transfinite counting numbers exist independently of
any determination, indeed any set which is well-ordered 'gives rise to' or possesses such a number. Thus, we could interpret any reference he makes to
'counting' or literal enumeration again as only a didactic device (as it was
originally a *heuristic* device). (See the letter to Harnack, 3 November 1886,
Cantor *Nachlass VI*, p. 86, quoted in §1.2, p. 28.) Here we *seem* to obtain
the full transfer to position (*iii*), with the singling out of well-ordering as the
key feature of counting. But again, to argue for the existence of transfinite
ordinals (and such an argument was very important for Cantor), Cantor has
to appeal to the existence of well-ordered sets. The obvious candidate for
one such existence proof is the set of natural numbers. But since the natural
numbers are infected with the 'psychologism of well-ordering' so is the set
of them. Indeed, the infection spreads to all the other usual mathematical
domains which are defined in terms of these. But in any case, to base the notion
of cardinal finally on ordinal he needs the well-ordering theorem. And here
again his proof idea was bound up with 'successive enumeration' or 'constructive capacity', or determination through a given law. Thus despite intentions
Cantor moves only half the way towards position (*iii*), for the difficulties associated with counting are in effect all stored in the notion of 'arrangement in a
well-ordering'. Thus, unlike with the later (*iii*), there is no clear divorce from
constructive determination, and, also unlike (*iii*), Cantor does not make it
clear exactly how the decision to extend the notion of counting to infinities
fits coherently with the intuitive notion of counting which lies behind the
decision.

Note that it is not *necessarily* the association of 'constructive determination'
and 'counting or numbering' infinities which creates tension. There are constructive theories which do extend to infinite sets. But these theories must
not only be very careful with regard to the notion of construction, but also
rather restrictive with regard to the notion of *set* which it is applied to. But
Cantor, as we have seen, wants the notion of set to be as broad as possible,
and he seems to want to apply counting in some sense to all of these, for example to the classically conceived continuum, the set of all real functions,
and so on. In his [1879], he applied the term 'countable into the infinite'
[*ins Unendliche abzählbar*] to point-sets

... which have the same power as the natural number series: 1, 2, 3,... v, ...
and thus can be imagined in the form of a simply infinite series [sequence],...
(Cantor [1879, p. 142)

The term 'countable into the infinite' here obviously implies that one can
count, but that the count never properly ends. But in [1883*b*] he applies
'countable' to *all* and arbitrary infinite sets. And *this* is where the tension

really lies. Cantor remarks ([1883*b*], p. 168) that the *only* essential difference between finite and infinite sets is that the latter can be enumerated (counted) in various ways while the former can be enumerated in just *one* way. Surely if one takes counting (or arrangement) seriously one would expect the essential difference to be stated: (most) infinite sets cannot be counted *at all*. The heuristic slogan behind Cantor's theory might well be formulated as follows: the finite and the infinite must be treated in the same way; the finite can (in principle) always be counted, the infinite not; but let us *ignore* this unpleasant feature of infinity, and proceed to develop a general theory of counting numbers for infinite collections and a notion of cardinality based on this.

Comparison with Frege on some of these points is instructive. He too believed that a theory of number should apply equally well to the finite *and* the infinite. Indeed, he praises Cantor for attempting to extend number theory to the infinite ([1884], p. 97). But Frege's own approach was quite different. In effect, *he* says: the finite and infinite numbers should be treated in the same way, but we should therefore ignore the fact that (some) finite sets can be actually counted and develop a theory of cardinality which has nothing to do with determination of number or counting at all.

Frege's position is that the extension of a concept must have a definite size or cardinal number regardless of whether we are in a position to determine the number or not, thus in particular regardless of whether it is infinite or not. Concepts like 'x is a sun-like star' or 'x is an atom of hydrogen' have associated numbers even though there is no hope of determining them, and even though we have no certain idea of whether there are finitely or infinitely many stars or hydrogen atoms. The result Frege comes up with is a theory which breaks right out of the sequence of counting theories (*i*)–(*iii*) sketched above. Of course, the resulting theory of finite cardinal number can be interpreted as an ordinal theory at stage (*ii*), since the cardinals have all the correct formal properties. But the point is that, unlike Cantor, Frege does not attempt to build a theory which in spirit generalizes (*i*). Determination of size through counting may have been the origin of the notion of number. But Frege takes this as irrelevant. (He remarks in the introduction to his [1884] (p. vi): 'Never again let us take a description of the origin of an idea for a definition, . . .') What this means is that ability to count and therefore ability actually to arrange in an order must be irrelevant for the existence of a cardinal number. This position Frege shares with the modern ordinal theory (*iii*). But interestingly since Frege is not attempting to generalize (*i*), he goes further and claims that order has nothing at all to do with cardinal number. This can be seen first in his discussion of the 'monad' or 'ones' theory of number in his *Grundlagen*.

Frege considers the idea of using some ordering to distinguish elements in a collection which is to be numbered *prior* to the construction of the set of monads or 'ones'. He considers first the idea of using spatial or temporal ordering (either or both of which are involved in the primitive theory of counting), and here he is severely critical. Because we want to number collections which are not naturally given in space or time, he concludes that this will involve unnecessary imposition of structure on the given collections. And indeed he says natural orderings in space and time which some collections have are quite irrelevant for their number; they are merely accidental arrangements as far as number is concerned:

All these are relationships which have absolutely nothing to do with number as such. Pervading them all is an admixture of some special element, which number in its general form leaves far behind. (Frege [1884], p. 53)

Frege then goes on to dismiss *any* attempt to use serial order:

Another way out is to invoke instead of spatial or temporal order a more generalized concept of series, but this too fails of its object; for their positions in the series cannot be the basis on which we distinguish the objects, since they must already have been distinguished somehow or other, for us to have been able to arrange them in a series. Any such arrangement always presupposes relations between the objects, whether spatial or temporal or logical relations, or relations of pitch or what not, which serve to lead us on from one object to the next and which are necessarily bound up with distinguishing between them. (Frege [1884], p. 54)

Of course, here Frege is specially attacking attempts to build a theory of number based on a theory of monads, and 'abstraction' to monads. But he would doubtless apply the same point against any attempt at a characterization of number based on an order on the underlying collections. For here I take it that Frege is saying, correctly, that in general collections will not be given in ordered form, since concepts and not their extensions are primary and most concepts do not naturally impose an order on the objects which fall under them. *And* he is saying, also correctly, that we cannot always (or even very often) hope literally to *impose* or arrange order on extensions which are not given to us in some serial form. Thus, Frege as a realist is distancing himself from the involvement of subjective capacities; he is thus careful always to observe distinctions where Cantor is not, although the latter seems *sometimes* to be aware of them. Frege certainly does *not* confuse 'subjective capacity to arrange in an order' with 'existence of an order'. As he notes, for example, in his criticism of Cantor's definition of finite set:

What the author wishes to say with the word 'successive', I have defined precisely in my *Begriffschrift* and again in my *Grundlagen* without mixing it up with time. And there the connection with complete induction is recognizable. (Frege [1892], p. 165)

And indeed, we find the following very clear statement in the *Grundlagen*. Given a relation ϕ, Frege defines 'y follows in the ϕ-series after x':

First, since the relation ϕ has been left indefinite, the series is not necessarily to be conceived in the form of a spatial and temporal arrangement, although these cases are not excluded.

Next, there may be those who will prefer some other definition as being more natural, as for example the following: if starting from x we transfer our attention continually from one object to another to which it stands in the relation ϕ, and if by this procedure we can finally reach y, then we say that y follows in the ϕ-series after x.

Now this describes a way of discovering that y follows, it does not define what is meant by y's following. Whether, as our attention shifts, we reach y may depend on all sorts of subjective contributory factors, for example, on the amount of time at our disposal, or on the extent of our familiarity with the things concerned. Whether y follows in the ϕ-series after x has in general absolutely nothing to do with our attention and the circumstances in which we transfer it; on the contrary, it is a question of fact, just as much as it is a fact that a green leaf reflects light rays of certain wave lengths whether or not these fall into my eye and give rise to a sensation, . . .

My definition lifts the matter onto a new plane; it is no longer a question of what is subjectively possible but of what is objectively so. (Frege [1884], pp. 92-3)

What Frege shows is that it is quite possible to view the existence of an order relation (and a well-ordering at that) in a completely objective way. Thus, one might well ask: since he shows order *can* be objectively treated, why does he rule out a theory of cardinal number *based* on orderings? If 'existence of ordering' is confused with 'ability to arrange' then it would be quite natural for a realist like Frege to avoid orderings altogether: numbers exist regardless of our ability to manipulate collections in a certain way. But we have just seen that Frege does not conflate 'existence' with 'subjective capacity to arrange'.

Russell later was much clearer and more explicit on these matters than Frege. Russell asserts, adopting the existential position (both Frege and Russell adopt this position with regard to number, for the numbers are demonstrated to exist by appeal to a pure existence principle), that if a set exists, then so do all the orderings on that set:

The correct analysis of ordinals has been prevented hitherto by the prevailing prejudice against relations. People speak of a series as consisting of certain terms *taken* in a certain order, and in this idea there is commonly a psychological element. All sets of terms have, apart from psychological considerations, all orders of which they are capable; that is, there are serial relations, whose fields are a given set of terms, which arrange those terms in any possible order. In some cases, one or more serial relations are specially prominent, either on account of their simplicity, or of their importance. Thus the order of magnitude among numbers, or of before and after among instants, seems emphatically the *natural* order, and any other seems to be artifically introduced by our arbitrary choice. But this is a sheer error. Omnipotence itself cannot give terms an order which they do not possess already: all that is psychological is the *consideration* of such and such an order. Thus, when it is said that we can arrange a set of terms in any order we please, what is really meant is, that we can consider any of the serial relations whose field is the given set, and that these serial relations will give between them any combination of before and after that are compatible with transitiveness and connection. From this it results that an order is not, properly speaking, a property of a given set of terms, but of a serial relation whose field is the given set. (Russell [1903], p. 242)

When the divorce from our capacity to order is put this sharply, the possibility clearly arises that in the existent collection of orderings on a given set, there may *not* be one of the appropriate kind. This was certainly Russell's view with respect to the well-orderings: it is logically possible that such orderings may not exist for all sets. Perhaps if Frege had devoted any attention to Cantor's ordinal theory of cardinality he would have taken a similar view, or at least would have said that the existence of a well-ordering has to be proved.

It should be pointed out that the mistake of conflating 'arranging a set in an order' or of 'defining an order' and 'proving the *existence* of an order' was not special to Cantor. Indeed, in respect of well-orderings it was not really sorted out until the debate that followed Zermelo's proof of the well-ordering theorem. The conflation is well illustrated in Hilbert's comments on the continuum problem in his famous 'Mathematical Problems' lecture. Hilbert first connects the continuum hypothesis with the well-ordering question, and rightly points to a proof of well-orderability of the continuum as a prerequisite or a 'key' in a proof of the continuum hypothesis. But the possible well-orderability of the continuum is formulated as follows:

The question now arises whether the totality of all [real] numbers may not be arranged in another manner so that every partial assemblage may have a first element, i.e., whether the continuum cannot be considered as a well-ordered assemblage—a question which Cantor thinks must be answered in the affirmative. It appears to me most desirable to obtain a direct proof of this remarkable statement of Cantor's, perhaps by actually giving an arrangement of numbers such that in every partial system a first number can be pointed out. (Hilbert [1900a], pp. 8–9)

This, of course, strongly recalls Cantor's 'bringing into the form of a well-ordered set' or 'giving a definite law' (p. 146).

However, while Cantor's treatment of well-ordering was logically and philosophically sloppy, at least in comparison with Russell's and Frege's, the discussion cannot be confined to points of philosophical rigour. Cantor's principle of treating the infinite as little differently as possible from the finite may have been applied 'in the wrong direction' in that it attributes to arbitrary infinite sets properties which strictly speaking only finite sets possess. But the heuristic benefits of this move were enormous. In the first place, the (philosophically 'wrong-headed') application of the intuitive notion of counting to infinite sets created an extremely important set-theoretic theory of number. Secondly, it shifted the difficulties with counting onto the notion of well-ordering, and thus focused attention on the well-ordering problem. Zermelo's solution of this problem introduced the axiom of choice specifically to avoid successive selections (see §4.1). In the present context this had two crucially important consequences. It separated finally the question of the existence of well-orderings (or choice functions), and in effect also the existence of *sets*, from our ability to construct or exhibit them. Thus, by concentrating on existence it relieves the tension in Cantor's system along realist lines. The Fregean realist and anti-psychologistic objections are fully met. In addition, investigation of the consequences of the axiom of choice eventually showed *contra* Frege not only that ordinal notions can be coherently involved in a theory of cardinality, but that they are *indispensable* for a sensible theory of infinite cardinal number. Cantor's theory of cardinal number now turns out to be the simplest, because it makes the best use of necessary well-ordering assumptions (see §2.3(*b*)). The reconstructed (von Neumann) theory, while an ordinal theory of cardinal, completely agrees with Frege that *determination* of number is quite separate from *existence* of number. Though here one must repeat: concentrating on the method of determining number (counting) led to the isolation of the key features of cardinal number in the Cantor–von Neumann theory. During the initial growth of a mathematical theory philosophical naivety is perhaps no bad thing.

I come back finally to Cantor's attitude to the *status* of the well-ordering theorem itself. In [1883*b*] he wrote:

The concept of *well-ordered set* is fundamental for the whole theory of manifolds. It is a basic law of thought [*Denkgesetz*], rich in consequences and particularly remarkable for its general validity, that it is always possible to bring any *well-defined* set into the form of a *well-ordered* set. I will return to this law in a later memoir. (Cantor [1883*b*], p. 169)

Here it is not altogether clear what Cantor means by a set being 'well-defined', and it is not at all clear what he means by 'law of thought'. And unfortunately he did *not* come back to it in a later paper. Indeed, as I mentioned in §2.2 it is possible, even likely, that later Cantor had some doubts about the validity of the well-ordering hypothesis, at least during 1884 when he thought he had proved that the continuum is equivalent to *no* number-class. However, Cantor's belief in the number-class or aleph theorem in its full generality reasserted itself, certainly by 1884–5, indicating that his belief in the correctness of the well-ordering theorem was also restored. For instance, Cantor clearly affirms the number-class theory of powers in a section of his [1887–8] which stems from February 1884 (pp. 387–8; the passage is quoted on p. 65). And in his further work on point-sets in his [1885*b*], p. 271, Cantor freely assumes that every power is a number-class power. Indeed, the conviction of the correctness of the number-class/aleph theorem can only come through or together with a conviction that the well-ordering theorem is correct. Whether or not he continued to hold that it is a 'law of thought' is simply not known. Certainly he seems never to have taken the view that the well-ordering hypothesis is sufficiently indisputable that it can be openly adopted as an unproven assumption.

But let us focus a little more closely on the term 'law of thought'. In view of later developments, the claim that the well-ordering proposition is a 'law of thought' *could* be construed as meaning that it is a *logical* principle. It is doubtful that Cantor ever took it as such, for otherwise why not openly present it so and make as much explicit use of it as possible? Moreover, in analysing Cantor's work this is not a very fruitful path to pursue, for he wrote practically nothing about logic, and, as is made clear in Chapter 1, he saw himself as dealing not with logical possibility but with *reality*. (However, the logic question becomes a little more interesting again with the discovery of the axiom of choice. This is discussed further below.) There are, though, two other aspects of the 'law of thought' claim which deserve mention. First, 'law of thought' could be taken as meaning objectively that it is 'a proposition basic to the kind of entities considered', in this case sets. Secondly, that the well-ordering proposition is a 'law of thought' could be taken more literally, that is as an assertion about our actual mental conception of sets. Indeed the foregoing discussion shows that Cantor's approach to the well-ordering proposition involves just such a subjective element, ability to arrange in an order. (However, it should be clear from Cantor's treatment of existence and existence arguments discussed in Chapter 1 that the subjective and objective elements are not totally separate. Thus it is no great surprise that they are mixed here.) But let us consider the objective element first.

Certainly Cantor's *philosophical* views lead to the position that the possibility of well-ordering is basic to sets. In one indirect sense it is basic because the well-ordering assumption turns out to be essential to many of the propositions about sets and cardinality which Cantor instinctively regarded as correct. Most important here was his fundamental conviction from the beginning that all sets are comparable, though one might also mention the proposition that the union of denumerably many denumerable sets is denumerable and the proposition that the natural numbers represent the smallest infinite power, i.e. that every infinite set has a subset equivalent to the natural numbers. All of these go back to Cantor's earliest dealings with powers. But more importantly and more directly, the well-ordering proposition was fundamental to Cantor's approach to cardinal number and cardinality. Here above all is the crucial assumption that countability is the route to cardinality even for infinite sets. And the well-ordering theorem, as first construed by Cantor, says just that countability is a property of all sets.

It is important to be clear that it is Cantor's philosophical doctrines which support this approach, for it is based on the view that transfinite sets are like finite sets in their fundamental properties. It is not wholly clear which properties must be 'fundamental', but Cantor's metaphysical and theological views strongly suggest that one thing finite and infinite sets must share is enumerability, countability. We saw in Chapter 1 that Cantor held that all sets are 'made finite to God', to use Augustine's phrase. Thus, since countability (well-orderability) is essential to *our* finite, it is also essential to *His* finite. This would be enough to convince

Cantor of the correctness of the well-ordering hypothesis, and that it is part of the *nature* of sets, finite *and* transfinite, that they can be counted and well ordered. And since this means that 'countings' for transfinite sets will exist 'as ideas in the mind of God', Cantor would allow mathematics to assume that such 'countings' actually exist, that statements about 'counting' can be taken as statements about objective reality, and not about *our* subjective capacity. The appeal to God thus relieves a little of the tension created by applying counting to infinite sets.

But, as I pointed out, the subjective element itself was also very important in Cantor's treatment. For his operation both with sets and with proofs indicates strongly that he regarded it as always possible for *us* to conceive of a set as arranged in a well-ordering. This is explicitly shown in his use of successive selection arguments for infinite sets. But it is also revealed to some extent in the way he refers to sets extensionally, for Cantor often uses $m, m', m'', \ldots$ as a notation for the elements in an *arbitrary* set M. This occurs notably in his operation with the 'ones' theory of number (see [1887-8], pp. 387, 413, 420), and thus reveals an interesting connection with the Aristotelian forerunner of this theory. In these extensional contexts Cantor usually speaks of the elements of a set as being 'conceptually separated' (see §§3.2-3.3, pp. 134-5). Now in an extensional set, the elements are certainly 'separate' in the sense of being 'distinct', 'different'. But more than this, when one looks away from the *unity* of a set, one is left with a mere collection of individuals. Such individualization was important to Cantor in numbering contexts. For example, he says that in comparison with $\bar{M}$ (the order-type), 'The elements of the set M we are given are to be imagined as separated: . . .' (Cantor [1887-8], p. 380). In this respect, Cantor appears to be following Aristotle, for in the latter's account of abstraction to the collection of 'ones', part of the process is first seeing the objects to be counted as individualized. (I follow here Klein's account; see Klein [1968], Chapter 8, especially pp. 102-4.) Cantor's approach seems to be similar. Of course, Cantor is dealing with sets, where the objects are somehow 'welded together' into a unity. But when one is making the conceptual shift to the number one must, at least temporarily, 'discretize' the objects as things *apart* from the set of them. Now this 'picturing' or 'conceiving' the elements as discrete is not the same as saying that they can be conceived as well-ordered, but it goes a certain way towards it. All the more so if one imagines the discretization as something spatial or linear, or if one bases the picture on the finite counting process, where for example the 'ones' theory has its origin. Here the elements are presumably imagined or pictured as discretely separated like apples spread out on a table, and although no well- or linear-ordering is at *first* apparent, we are at liberty to begin choosing from among them, and so begin to *impose* an ordering. It may be that this conceptual picture is what lies behind Cantor's successive selection proofs. If this chain of ideas is a reasonable reconstruction of Cantor's thinking, it shows again how

important his principle of finitism is. (Emmy Noether apparently told the story
that while Dedekind imagined a set as being like a bag full of objects, Cantor
imagined a set as like 'an abyss [*Abgrund*]' (see Becker [1964], p. 316). Here
it looks as if I am attributing to Cantor the 'bag' thesis rather than the 'abyss'
thesis (whatever that might be). This need not be so, however. In sketching this
picture I stress that the elements are conceived as being *apart* from the set that
unifies them. A collection of elements *apart* from their sethood may well be
bag-like.)

The connection between the Cantorian ('ones') theory of number and the
Aristotelian theory is a complicated matter. Cantor once ([1883*b*], p. 204)
calls his notion of set Platonic, and since numbers for him are sets, this means
at least that Cantorian numbers should be separate independent objects. (Indeed,
he often calls numbers 'universals'.) But Cantor also says that in respect of
number he represents 'Aristotelian realism' ([1887-8], p. 382). But Aristotle
severely attacked Plato's theory of number, particularly his claim that numbers
have a pure independent existence apart from the collections which are to be
numbered (see Annas [1976], and especially Klein [1968], Chapter 8). Certainly in this respect Cantor was much more Platonic than Aristotelian. But
Cantor's approach to number *is* Aristotelian in so far as his attempt to explain
or establish the existence of number is based on abstraction from the underlying
sets. For Aristotle seems to have held that numbers arise by being 'lifted off'
from the collections numbered. Abstracting from the elements counted produces
heaps of monads which are now taken as the numbers (Klein [1968], p. 107).
This conceptual 'separation' of numbers from what is numbered is for Aristotle
the basis of arithmetical thought and study, though strictly speaking he allows
no separate realm of number objects. The 'separateness' of numbers is rather
just a useful posit, which for example, enables us to conclude in one sweep
that three apples and two apples give five apples, three sheep and two sheep
gives five sheep, etc., by concluding that three plus two is five, even though
numbers do not really exist as separate objects apart from the apples, the
sheep, or whatever else can be counted. Cantor's argument that given a set M
(finite or infinite) there must be a cardinal number belonging to M is somewhat
reminiscent of this Aristotelian account of how an arithmetical science is possible. In both accounts numbers are 'tied' to collections and collections are the
necessary foundation for a theory of number. Indeed, Cantor often speaks
in a rather Aristotelian way of numbers 'arising' by abstraction, suggesting a
kind of later 'separation', and thus that the numbers are not among the pre-
existent objects. But this suggestion is misleading. For Cantor, abstraction is
actually either an existence argument for numbers or an explanation of their
structure; either way, for Cantor numbers do have independent existence.

The solution of the well-ordering problem by Zermelo led away from the
question of how it is *we* actually conceive sets and actually perform operations
on them to the axiom of choice—a pure existence principle. Now one must

not make the mistake of assuming that choice was anticipated by Cantor, or
that the well-ordering theorem as formulated by Zermelo has the same meaning
as Cantor's well-ordering hypothesis. The discussion on pp. 146–54 is enough
warning against that. Nevertheless we have no reason to suppose that Cantor
did not accept Zermelo's solution. I do not want here to enter on a long historical discussion of the arguments surrounding the axiom of choice. But it is
at least worth paying attention to Zermelo's own defence of the axiom, since
this connects at some key points with the analysis of the Cantorian position
just given.

Zermelo's central statement justifying the axiom points to its 'self-evidence'
and its 'necessity for science', which means in this context its repeated appearance in reasoning with sets. In an attack on Peano's dismissal of the axiom, he
raises the question of how one decides what are fundamental principles in
mathematics:

Evidently by analyzing the modes of inference that in the course of history have
come to be recognized as valid and by pointing out that the principles are
intuitively evident and necessary for science—considerations that can all be
urged equally well in favour of the disputed principle. That this axiom although
it was never formulated in textbook style, has frequently been used, and successfully at that, in the most diverse fields of mathematics, especially in set theory,
by Dedekind, Cantor, F. Bernstein, Schoenflies, J. König, and others is an
indisputable fact, which is only corroborated by the opposition that, at one
time or another, some logical purists directed against it. Such an extensive use
of a principle can be explained only by its *self-evidence*, which of course must
not be confused with its provability. No matter if this self-evidence is to a
certain degree subjective—it is surely a necessary source of mathematical
principles, even if it is not a tool of mathematical proofs. . . . (Zermelo [1908a],
p. 187)

There are certainly some confusions in Zermelo's statement, not least a
certain distortion of history. For one thing, the axiom of choice was *not* used
by the predecessors Zermelo mentions. What were *implicitly* used were various
assumptions (or 'modes of inference' as he calls them) which arguably, when
analyzed and realistically construed, turn out to be equivalent to the axiom of
choice. The first to use the axiom was Zermelo himself. Moreover, how can an
unformulated, unknown principle be used because of its self-evidence? In fact, it
does not seem at all helpful to speak of 'self-evidence' here. The point Zermelo
is making could be put more forcibly by repeating that the well-ordering assumptions he refers to and the successive selection arguments on which they were
usually based are intrinsic to the operation with sets, and that the axiom of
choice is a plausible reconstruction of these. By *itself* the axiom cannot be
evident; but on reflection it may well be *intrinsic* to the surrounding subject
matter.

This is any case, I suggest, is what Zermelo means by his argument that the
axiom is 'necessary for science'. To support this he points to various theorems

whose proof requires something like the axiom of choice and which were previously accepted as correct either with proof or without. Here above all he points to various propositions connected with cardinality, among them the denumerable sum theorem, the equivalence of finiteness and Dedekind finiteness, the non-emptiness of products in general, and the proposition that 'the sums of equivalent sets are again equivalent, provided all terms are mutually disjoint' (Zermelo [1903a], p. 188). (Presumably here Zermelo means the theorem, mentioned in §2.3(b), that cardinal summing is independent of the representative sets chosen.) Again it does not seem of much help to appeal to the 'obviousness' of *these* propositions as a justification for the axiom of choice. For if up to a certain time one accepts proposition A as obvious and *then* one discovers that A is equivalent to (or depends on) B, one can now *either* accept B too as correct, *or* if one decides that B is not 'obvious' one can say: 'We were simply mistaken: we must now accept that A is not obvious. There is much more to this issue than we previously assumed.' Thus appeal to 'obviousness' or 'intuition' can cut both ways,[6] and will in the end reduce to a fight between conflicting intuitions. (Indeed the axiom of choice turned out to have consequences which arguably are 'obviously wrong' or 'intuitively incorrect'.) Zermelo does not make the mistake of appealing to obviousness here. Rather he says we are dealing with '. . . a number of elementary and fundamental theorems and problems that, in my opinion, could not be dealt with at all without the principle of choice' ([1908a], p. 188). Indeed, the core of Zermelo's argument seems to be that the propositions he cites are basic to the notion of set or cardinality. Thus the sum theorem mentioned above is '. . . a theorem on which the entire calculus of cardinalities rests' (Zermelo [1908a], p. 188). And, as Zermelo says, the theories of denumerable sets and of the ordinals of the second number-class (both of which by this time had already been widely and successfully applied in analysis) 'rest' on the denumerable sum theorem. But if these are basic propositions, then so too is the axiom of choice a basic proposition. For in concluding his argument Zermelo writes: 'Cantor's theory of cardinalities, therefore, certainly requires our postulate, and so does Dedekind's theory of finite sets, . . .' (Zermelo [1908a], p. 189). What this argument says, then, is that if the theory of sets or cardinality makes no sense without these propositions, then it makes no sense without the axiom of choice.

To a certain extent this position fits with the Cantorian position on well-ordering outlined above. But again it should be emphasized that if one accepts as basic full cardinal comparability or that a cardinal sum is independent of

---

[6] Thus, to take one example, while Borel had implicitly used well-ordering assumptions *before* Zermelo's proof (for instance in his first proof of the Heine–Borel theorem; see Hallett [1979a], p. 21), after seeing the proof and therefore what well-ordering assumptions rest on (are equivalent to) he rejected both the axiom and the previous well-ordering assumptions.

the representatives, this can only be because one is extending to the transfinite case propositions which are basic in the finite case.

The 'similarity of finite and infinite' as a justification of the axiom of choice can be framed in a rather clear and strong way. Zermelo, again in reply to Peano, notes that the axiom of choice for finite sets (of non-empty sets) follows from one of Peano's *logical* axioms; thus *his* (Zermelo's) axiom is an extension of Peano's axiom to the infinite case. So, in the first place, the axiom of choice is clearly related to a logical principle. Indeed, in his original article on the well-ordering theorem, Zermelo calls the axiom of choice itself a 'logical principle' ([1904], p. 141). But if the logical principle, and thus the existence of choice sets for finite sets, is accepted then 'finitism' should allow *also* the existence of choice sets for infinite sets (of non-empty sets). Choice is nothing more than the generalization of a logical principle under the Cantorian principle of finitism and thus arguably a Cantorian axiom.

The kernel of this argument, which Zermelo indirectly hints at in his remarks, is also found later in Hilbert's [1923]. As Hilbert says, most of the objections raised against Zermelo's proofs of the well-ordering theorem were directed against the principle of choice, and suggest

. . . that the permissibility of the choice principle is doubtful, while the other methods of argument . . . are not singled out for objection in the same way. This view I consider quite wrong. Rather, logical analysis, as it is carried out in my proof theory, shows that the essential idea [*Gedanke*] lying behind the choice principle is a general logical principle, which is already necessary and indispensable for the most elementary level of mathematical argument. (Hilbert [1923], p. 179)

Later in the paper, Hilbert gives an argument for the choice principle which I shall come back to later. Of course the kind of argument for choice being adopted here partly depends on the acceptance of certain means of 'collecting together' the choices made. (In the case of finite sets this is done just by finite repetition of the application of existential specification and forming singleton sets, followed by the union of the singletons.) This has an interesting connection with another axiom which might with justice be called Cantorian, the axiom of replacement. (It is Cantorian at least in the sense that it is a necessary addition to the Zermelo axioms if one wants to reconstruct axiomatically the Cantorian theory of ordinals. This is discussed further in Chapter 8.) Fraenkel believed at one time that the addition of the axiom of replacement allows one to *prove* the axiom of choice. The reasoning was as follows: for any given set of non-empty sets, selections from all the elements exist without using choice, but we cannot assume that these selections are gathered into a set; replacement allows us to assume this. Thus:

The axiom of choice . . . could be dispensed with on the basis of the new [replacement] axiom. . . . The possibility of the choice acts themselves is a consequence of the other axioms, whereas it is not a consequence of them that the totality of selected objects forms a set. (Fraenkel [1921], p. 98)

And:

> With the aid of the replacement axiom introduced in I, the existence of a set S_1 [a selection set for $T = \{M, N, \ldots\}$], and thereby the axiom of choice, is apparently *provable*. For according to that axiom [i.e. replacement], there must exist a set S_1 which arises out of T in such a way that each of the (non-empty) sets $M, N, \ldots$ is replaced by one of its elements. (Fraenkel [1922b], p.233)

Fraenkel does not mention the principle of existential specification, but it is clear that this lies behind his argument. For in attempting to dispel 'misunderstanding' over the choice axiom, he writes:

> One can express the axiom of choice as follows: it is always possible to select out of each of the elements $M, N, R, \ldots$ of T a single element $m, n, r, \ldots$ and unite all these elements to a set S_1.
> This last proposition has often been interpreted as saying that the kernel of the axiom lies in the claim that it is possible to "choose a distinguished element" from each of the sets $M, N, \ldots$ or in the claim that it is possible to make a "simultaneous choice" from them all. Both views could be interpreted as incorrect. For each single set M, or otherwise stated, for the case where T possesses only a single element M, the choice required by the axiom is *provable*. For if a is some element of the set M (assumed to be non-empty) then the set $\{a\}$ exists according to the Axiom II [Zermelo's axiom of elementary sets], and it has the required property. Therefore the simultaneous choice for all sets of T is naturally possible, since the choice like every mathematical operation is to be looked upon as something timeless. (Fraenkel, [1922b], p. 232)

Fraenkel goes on to say that, by dint of this argument the axiom of choice does not really concern 'choices' (these are already made), but the *existence* of the *set* of the choices. And this Zermelo also assumed: 'For, once we accept the above mentioned principle of Peano, which permits a choice from a single set, there is no longer any limit to its repeated application.' ([1908a], p. 193.)

Fraenkel, of course, is wrong about the deducibility of choice from replacement, since choice is *independent* of the Zermelo-Fraenkel axioms. But the mistake is an interesting one, for two reasons. First, it seems quite clear from the above passages that Fraenkel regards the extension of the idea of single selection to arbitrary *transfinite* selections as perfectly legitimate, indeed as basic to mathematics and set theory. This conviction is again nothing other than the conviction of the legitimacy of extending to the infinite what holds in the finite. Secondly, he focuses on replacement as the axiom required to extend the means of collecting the arbitrary selections. Although Fraenkel did not favour adopting replacement as a new axiom (at least until a long time after it was in fact adopted) its importance is *precisely* that it acts as a transfinite collecting principle.

The reason that choice for infinite sets does not follow simply from existential specification via replacement is that an infinite number of applications of specification would be required and this is not allowed by the logical axioms.

Specifically, if a is a given set, then the replacement schema allows the existence of the set $\{y : \exists x \in a[\phi(x, y)]\}$ where $\phi(x, y)$ is any relation which acts *functionally* on a (i.e. $\phi(x, y) \wedge \phi(x, y') \rightarrow y = y'$). If we try to form a condition ϕ using only the existential quantifier to say that y is some element of x, say $\phi(x, y) \equiv \exists u \in x[y = u]$, then clearly ϕ is not functional. To make it functional we would first have to apply existential specification (i.e. $\exists u[u \in x] \rightarrow u_x \in x$) for *each* member x of a, and then form $\phi(x, y) \equiv y = u_x$. Since this would (if possible) *fix* elements of each x, $\phi(x, y)$ *would* be functional. But we cannot do this in the normal predicate calculus, because if a is infinite it would require infinitely many applications of existential specification. However, to take this story a little further, Hilbert introduced a device, the so-called ε-operator, which, so to speak, set-theoretizes the quantifiers and allows the kind of selection required here. In set theory, formulated in the predicate calculus together with Hilbert's symbol (and the axiom schema governing this term), the ε-operator acts as a *selector*; that is, for every set x, $\varepsilon(x)$ is also a set, and when x is non-empty $\varepsilon(x) \in x$. If one now stipulates that the axiom schema of replacement covers also formulas which include ε-terms, then the axiom of choice is provable. The reason is simply that if a is a given set (of non-empty disjoint sets), then the formula we attempted above can be written $\phi(x, y) \equiv [x \in a \wedge y = \varepsilon(x)]$, which is a perfectly good functional relation on a and correlates to each x in a the member of x chosen by the ε term. Thus, by replacement (extended to cover formulas containing ε) $\{y : \exists x \in a[\phi(x, y)]\}$ is a perfectly good set, and crucially a *choice set* for a. (All this is pointed out in Leisenring [1969], pp. 106–7.) (This result is certainly compatible with the famous Second ε-Theorem of Hilbert and Bernays on the eliminability of the ε-symbol. This latter says that if σ is deducible from τ in the ε-calculus, and σ and τ *contain no occurrences* of ε, then σ is already deducible from τ in the ordinary predicate calculus. Thus, if we base *ZF* on the ε-calculus *without* allowing ε-terms in the replacement schema, then the axiom of choice is not deducible.)

Hilbert's procedure in [1923] goes half way along this line of argument. A predecessor of the ε-symbol, a τ-symbol, is introduced, and Hilbert shows how this τ-term can be used to correlate to every set of subsets of real numbers a selected element, in effect producing the formula $\phi(x, y)$ above. He then *assumes* that the selections will themselves form a set. This, says Hilbert, proves the Zermelo axiom for sets of sets of reals. The key point about the 'collecting together' of the selections is brushed over. (See Hilbert [1923], pp. 190–1.)

Hilbert does not actually call the axiom of choice a *logical* axiom, but as we saw he does say that it is *based* on a logical principle, implying that it is the generalization to the transfinite of this principle (Hilbert [1923], p. 183). Indeed, he calls the basic axiom governing the τ-term the *transfinite axiom*. By this he means that it is actually a substitute for the quantifiers, which in a sense allow infinity into the logic. But the τ-term is used in his justification of the choice axiom, and he does say that: 'The transfinite axiom is the fundamental source for all transfinite concepts, principles and axioms.' It is interesting to note here Russell's view ([1911], p. 173) that although he regards the axiom of choice as probably *false*, he believes it has a *logical character*, i.e. the whole question of its correctness or incorrectness is a logical matter.

4

THE ORIGIN OF THE LIMITATION OF SIZE IDEA

4.1. The Absolute and limitation of size

In §1.4 I made the point that Cantor's principle (c) with its type distinction between the Absolute and the transfinite gave his system considerable *prior* immunity against the shock of the set-theoretic contradictions. It is time now to expand this contention and to introduce the theme that forms the core of Part 2, the influence of Cantor's doctrine of the Absolute on modern axiomatic set theory. In Chapters 2 and 3 we saw how deeply embedded ordinality and well-ordering assumptions were in Cantor's thinking about sets. The crucial importance of these was re-emphasized in his 1899 attempt finally to prove the aleph theorem while steering clear of the obvious difficulties of Burali-Forti style contradictions. This reworking of the theory was in effect a transformation of Cantor's earlier theory of the Absolute into a theory of 'absolute collections', a transformation which gave birth to a notion of 'limitation of size'.

The crucial principle of the ordinal theory of cardinality is that all sets (or forms) in the extended realm of the finite and transfinite are capable of being ordinally numbered or 'counted'. From this it seems to follow that the universe of forms—the Absolute universe of forms—is reflected in the extended sequence of ordinal numbers (see §1.4). As Cantor himself states: 'The absolutely infinite sequence of numbers therefore seems to me in a certain sense a proper symbol of the Absolute.' ([1883b], p. 205, n. 2.) (The whole remarkable passage from which this comes is quoted on p. 42.) If we now apply principle (c), it follows that neither the aleph sequence nor the ordinal sequence can be assigned an ordinal number. For both sequences 'stand for' the Absolute and since the Absolute is 'beyond mathematical determination' it cannot be numbered. Cantor himself stated quite explicitly in 1885 that 'the Absolute [which he calls here 'the Actual Infinite in God'] falls wholly outside number-theory'. (Cantor, letter to Carbonelle, 28 November 1885, *Nachlass VI,* p. 34; see also my analysis of Cantor's criticism of Frege [1884] in §3.1.) This would mean that for Cantor, Burali-Forti type paradoxes are quite out of the question. For in these cases, contradiction is derived only after one has agreed to assign an ordinal to the whole sequence of ordinals or alephs.

This is already witness of *some* immunity (as early as 1883) to the difficulties later presented (from 1895) by the paradoxes (see §2.2, p. 74). So stated, of course, it is certainly not an explicit theory indicating how to avoid the paradoxes altogether. It says only that *the Absolute* cannot be numbered, cannot be taken as an element, etc. But it gives no general characterization or criterion of 'Absoluteness', nor does it explicitly tie mathematical determination of the Absolute to contradictoriness. Nevertheless, when Cantor later dealt expressly with the Burali-Forti contradiction, he attempted to make more precise the rather vague theory of the Absolute, and with it to give a more precise version of his theory of sets. What one sees here is the beginnings of a genuine type distinction between totalities, a reworking of his old type distinction between the finite and transfinite on the one hand and the Absolute on the other. 'Absoluteness' now appears as something which can do genuine mathematical work. (Though it is important to realize that, whatever the mathematical transformation that Cantor worked on his theory of the Absolute, he certainly seems in this later period to have held to the 'theological' or 'mystical' aspects of Absoluteness that were present earlier. The best example of this is his letter to Pater Ignatius Jeiler, 13 October 1895 (Cantor *Nachlass VII*, pp. 194-7, published in Meschkowski [1967], pp. 257-9). See also the letter to Pater Thomas Esser, 1 and 15 February 1896 (Cantor *Nachlass VIII*, pp. 133-7, published in Meschkowski [1965], pp. 510-13).)

The 1899 theory was (briefly) expounded in a series of letters to Dedekind. Cantor's first important step was to shift from the notion of 'the Absolute' to a new notion of 'absolutely infinite collections':

If we start from the notion of a definite multiplicity (a system, a totality) of things, it became clear to me that we must necessarily distinguish between two kinds of multiplicity (by this I always mean *definite* multiplicities).

For on the one hand a multiplicity can be such that the assumption that *all* of its elements 'are together' leads to a contradiction, so that it is impossible to conceive of the multiplicity as a unity, as 'one finished thing'. Such multiplicities I call *absolutely infinite* or *inconsistent multiplicities*.

As one easily sees, the 'totality of everything thinkable', for example, is such a multiplicity; later still other examples will present themselves.

When on the other hand the totality of elements of a multiplicity can be thought without contradiction as 'being together', so that their collection into '*one* thing' is possible I call it a *consistent multiplicity* or a *set*.[1]

Two things stand out here. One is the connection now drawn between *absolute* and *inconsistent*. The second is the clear reference in the formulation of the notion of absoluteness to one of the main uses of principle (*b*), i.e. the treatment of totalities as single objects. In my remarks on Cantor's criticism of Frege's [1884] I suggested that what lies behind the criticism is the doctrine that there are collections which cannot be treated as single objects (§3.1). This doctrine is now expressly admitted, bringing thus a restriction on the application of principle (*b*), or, in effect, a distinction of type among collections.

---

[1] Letter to Dedekind, dated 28 July 1899, in Cantor [1932], p. 443 (my translation). The published translation is in van Heijenoort [1967], p. 114. According to Grattan-Guinness [1974], this 'letter' is really an amalgam of several letters put together without any accompanying explanation by Zermelo in editing Cantor [1932].

Cantor's intention, in the spirit of the old theory of the Absolute and now following also the Burali-Forti argument, is that this distinction carries with it a clear mathematical differentiation. Above all, absolute collections do not carry numbers.

This position does not come immediately, however. For it is *not* clear from the above statement exactly where the separation in type comes. To begin with, there is a problem again with 'collecting' or 'conceiving'. But even leaving this aside, there seems to be a problem with the connection between absoluteness and inconsistency. No inconsistency seems to arise just from the assumption that the whole universe or the whole ordinal sequence form single wholes (though this is not the case with the Russell set, discovered later). A direct connection between the old theory of the Absolute and absolute collections might imply that it is wrong in principle to treat these collections as single objects. But to derive actual contradictions it is necessary to perform some mathematical operations. It is not clear in the passage quoted above whether Cantor was relying on such a connection to the original notion of Absoluteness or on the actual derivation of contradiction. For instance, the question is left quite open when he says 'one can easily see' that the 'totality of everything thinkable' is an inconsistent multiplicity. But later in the same letter (Cantor [1932], p. 445; van Heijenoort [1967], p. 115) the inconsistency of the ordinal sequence is proved by a Burali-Forti argument relying on mathematical operations. And in a later letter to Dedekind (31 August 1899, Cantor [1932], p. 448), Cantor proves the inconsistency of the 'system S of all thinkable classes' by an argument based on his theorem of [1891] that for any set there is a cardinally larger set. Thus, it would seem that whatever the underlying principle, in practice (as it would have to be) 'inability to conceive of a totality without contradiction' actually means 'a contradiction can be subsequently *derived* if the totality is treated as a single object'.

But in any case 'derivation of contradiction' is of no great theoretical help in building a two-sorted theory of collections. In general there is no effective way of telling in advance whether or not a contradiction will result if the extension of a given property is taken as a set; the best that can be done is to wait and see whether a contradiction appears. Thus we cannot do directly what ideally we would like to do, namely to rule out from sethood all and only the contradictory collections. This situation was well summed up a little later by Hessenberg:

We come now to a serious defect affecting our results in their complete generality. It consists in the lack of clarity of the set concept itself. We know infinite sets to which we can apply our results without any contradiction arising. But hitherto we have no knowledge of those conditions which must be imposed on the definition of set so as to guarantee freedom from contradiction. (Hessenberg [1906], p. 627)

We noted in §1.3 how unclear Cantor's conception of set is, and how unclear it appeared to set-theorists like Dedekind and Frege. The contradictions make the philosophical need for clarity a little more immediate. As Hessenberg also notes:

As yet there has been no profound analysis of the concept of set . . . Hence, when we have established that it is not permissible to put all the ordinals together as a set, we are in no position to *prove* this from the concept of 'set' and 'ordinal number'; we must be content solely with the fact of the contradiction. (Hessenberg [1906], p. 634; see also pp. 633–6)

Cantor's work of 1899, unlike his earlier work, seems to tackle just this problem. Although there is little chance of characterizing non-contradictory collections *directly*, there is still the possibility of isolating the strange (absolute or inconsistent) collections indirectly by singling out a general family of collections which is known (or better, *assumed*) to contain all the absolute collections. It was this that Cantor attempted, and it is just here that the old notion of Absoluteness begins to play a role. For Cantor establishes a connection between the Absolute and absoluteness by using the 'appropriate symbol' for the Absolute, namely the ordinal number sequence.

What Cantor in effect establishes in his 1899 letter is that

(*) X is not a set (is an absolute collection) if and only if it has a sub-collection equivalent to the ordinal number sequence.

This now allows the ordinal sequence, so to speak, to measure absoluteness. Let us now attempt to follow the rather complicated path to (*).

Using a Burali-Forti argument, Cantor takes the ordinal sequence itself as absolute or inconsistent. (Recall that this sequence could also be classified as Absolute.) Thus:

(1)   The ordinal numbers do not form a set, but an absolute collection.

Cantor also assumes a form of replacement 'axiom', namely:

(2)   Two equivalent [i.e. cardinally equivalent] multiplicities either are both 'sets' or are both inconsistent. (Cantor [1932], p. 444, or van Heijenoort [1966], p. 114)

And there is also a clear subset principle:

(3)   Every submultiplicity of a set is a set. (Cantor [1932], p. 444, or van Heijenoort [1967], p. 114)

From (3) and (1) it follows immediately that if X contains the whole ordinal sequence then X is itself absolute; thus from (1), (2), and (3) it follows

(4)   If X contains a part equivalent to the ordinal sequence then X is absolute (not a set).

This is half of (*). Cantor now puts (4) to work, together with other interesting assumptions, to prove:

(5) Every cardinal number is an aleph.

Recall that such a proof was one of the crucial things missing from his 1883 and later accounts of cardinality. Let us, then, follow its assumptions through rather carefully. I begin by quoting Cantor's proof in full (Cantor [1932], p. 447). Here Ω stands for the sequence of ordinal numbers.

Proof. If we take a definite multiplicity V and assume that *no aleph* corresponds to it as its cardinal number we conclude that V must be *inconsistent*.

For we readily see that, on the assumption made, the whole system Ω is projectible into the multiplicity V, that is, there must be a submultiplicity V' of V that is equivalent to the system Ω.

V' is *inconsistent* because Ω is, and the same must therefore be asserted of V.

Accordingly, every transfinite *consistent multiplicity*, that is every transfinite set, must have a *definite aleph* as its cardinal number. Hence:

C. The system $\daleth$ of all alephs is nothing but the system of all transfinite cardinal numbers.

All sets, and in particular all "*continua*" are therefore "*countable*" in *an extended sense.*

(The passage is quoted from the translation in van Heijenoort [1967], pp. 116–7, except that I put 'countable' in place of 'denumerable' in the last sentence. Cantor later outlined the same proof in a letter to Jourdain, 4 November 1903 (Grattan-Guinness [1971], p. 116), and, according to this letter, also in a letter to Hilbert in 1896.) Note the assumption here that only sets have cardinal numbers, for it is only this which allows the step from 'every set has a *definite aleph* as its power' to the conclusion that *all* cardinals are alephs. This, of course, is perfectly natural if one takes the origin of the notion of 'absolute collection' as being Cantor's earlier notion of the Absolute. For the assumption is then simply guaranteed by principle (*c*). This is certainly the most natural interpretation here. The second interesting assumption is that if V has no aleph as cardinal number, then 'the whole system Ω is projectible into V', for this is what permits (via (4)) the conclusion that V is absolute and thus has no cardinal at all. The basis of the 'projection assumption' was described by Zermelo in a footnote to his 1932 edition of Cantor's works:

Cantor apparently thinks that successive and arbitrary elements of V are assigned to the numbers of Ω in such a way that every element of V is used only *once*. *Either* this procedure would of necessity come to an end once all elements of V had been exhausted, and then V would be mapped onto a *segment* of the number sequence and its cardinality would be an aleph contrary to assumption, or V would remain inexhaustible, and hence contain a constituent part that is equivalent to all of Ω and therefore inconsistent. (Zermelo [1932, p. 451, n. 1)

This dependence on the legitimacy of 'successive and arbitrary' choices is in effect an extension of Cantor's proof of his theorem A ([1895], p. 293) that every infinite set has a subset of power $\aleph_0$. Of this proof Zermelo noted:

The 'proof' of theorem A, which is purely intuitive and logically unsatisfactory, recalls the well-known primitive attempt to arrive at a *well-ordering* of a given set through successive extractions of arbitrary elements. (Zermelo [1932], p. 352, n.6)

And Zermelo's criticism of the 1899 proof is focused exactly on these successive choices:

It is precisely at this point that the weakness of the proof sketched here lies. That the whole number series Ω must be 'projectible' into every multiplicity V that has no aleph as its cardinal number has *not* been proved. (Zermelo [1932], p. 451, n.1)

He then remarks that 'the intuition of time is applied here to a process that goes beyond all intuition, and a fictitious entity is posited of which it is assumed that it could make *successive* arbitrary choices.' (Zermelo [1932], p. 451, n.1.) Zermelo presumably means that transfinite time is required for the 'fictitious being' to carry through the selection process. Now although Cantor makes no mention of any 'fictitious' being, his willingness to accept the reasoning here at least as an heuristic argument might well stem from the theological element in his ontology: 'God can do it, so assume it done!' (see §3.5). And Zermelo may be indirectly referring to this in his criticism. In any case, Zermelo regards the intrusion of time or subjective ability as quite unacceptable. The only way to avoid this, as Zermelo states, is to assume some mechanism of *simultaneous*, not successive, choice. It was precisely to make this shift, Zermelo tell us, that he proposed his famous axiom of choice (*AC*). In his 1883 theory, Cantor had founded cardinality on the well-ordering postulate. Now he uses a choice assumption. Thus the 1899 argument might be seen as a step away from dependence on well-ordering towards the discovery of *AC*. However, one must be a little careful here, since the assumption of successive choices is very close to the assumption of well-orderability (see §3.5). In all this, too, one must remember that Cantor never published his 1899 proof, which may indicate some dissatisfaction with the proof method. Interestingly, in one of his late letters to Dedekind, after Dedekind had sent him a proof of the Schröder–Bernstein theorem, Cantor asks Dedekind if he could not also provide a proof of *full* comparability. Cantor notes that he lacks a proof with *simple* methods and 'I can only prove it *indirectly*.' (Cantor to Dedekind, 30 August 1899, Cantor [1932], p. 45.)

However, some principle of choice is involved in Cantor's proof. So far in our proof analysis, it seems that a choice assumption together with 'only sets have cardinals' and (4) (which is inspired directly by the Burali-Forti argument) when united yield the desired enumeration theorem (5). So far then, it looks as if we have reconstructed something very similar to the modern proof of (5), a proof due essentially to von Neumann. (See von Neumann [1929]. This was really von Neumann's *second* proof. His first proof proves much more. See Chapter 8.) What this modern proof does is to take any set *a* and map the ordinals 'step by step' onto distinct members of *a*; it is then shown that one *must* reach an ordinal where *all* of *a* is exhausted. This last step is established by appeal to an application of replacement reminiscent of Cantor's (2), and a prior proof that the ordinals cannot be a set (i.e. a Burali-Forti argument). (See, for example, Levy [1979], pp. 160-1, or Drake [1974].) One immediately clear difference between the Cantor argument and the modern reconstruction should be pointed out. In the von Neumann proof the 'selections' from *a* are arranged against the ordinals by a process of defining a function by transfinite recursion. Although the definition of the particular function involved here is simple, the legitimation of such definitions in general is not. Indeed, it was one of von Neumann's great achievements in set theory to recognize the importance of legitimizing such procedures and to provide the axiomatic means to do so. Cantor, of course, has no such explicit procedure. It could perhaps be said that when one uses *successive*, not simultaneous, choices the arrangement of the selections goes hand in glove with the selections themselves. But in any case, there is in this respect a considerable gap between the Cantor proof and its axiomatic successor. Nevertheless there are clear senses in which von Neumann's proof can be regarded as a reconstruction of Cantor's, whereas Zermelo's proof of the theorem 'every set can be well-ordered', which shuns the ordinals (and which therefore depends on *AC* but *not* on an axiom of replacement) cannot. This matter is taken up again in Chapters 7 and 8.

But comparison with von Neumann's *second* proof does not quite capture the power of Cantor's method or assumptions, for it ignores the important 'projection postulate' which Cantor uses. We shall see that Cantor's assumptions amount to much more than a simple choice assumption. Indeed they come closer to the axioms behind von Neumann's *first* proof of the enumeration theorem, which comes as a corollary to the theorem that the universe of sets is cardinally equivalent to the class of all ordinals.

Cantor starts from the premise

(6)   *X* is a set iff *X* has a cardinal number

and via his proof transforms it (for infinite *X*) into

(7)   *X* is a set iff *X* has an aleph as cardinal.

Contraposing (7), by making use of the fact that there are two categories of collections (sets and absolute collections) we obtain

(8)   *X* is absolute iff *X* has no aleph as cardinal number.

Now Cantor's projection assumption states that if *X* has no aleph as cardinal it has a part equivalent to the whole ordinal sequence. Since the ordinal sequence has no aleph, then certainly if *X* contains a part equivalent to the ordinals it too can have no aleph. Thus we obtain our (*) of p. 168, a criterion of absoluteness:

(*) X is absolute iff X has a part equivalent to the ordinal number sequence.

It may seem that (*) does not really go much beyond the enumeration theorem (7), or if so, it only does so *qualitatively* by mentioning absolute collections instead of sets. But this is not so.

To argue this, we have to resort again to axiomatic reconstruction of the argument. Since Cantor uses arbitrary multiplicities, the nearest modern axiomatic system would be a set theory with classes, say the system *VNB* after von Neumann and Bernays. (I follow the notation of Fraenkel, Bar-Hillel, and Levy [1973], pp. 119 ff., although it is rather inappropriate. The system *VNB* is quite different from von Neumann's original system both in strength and spirit, and the designation *VNB* ignores Gödel's contribution. However, it is quite consistent with what seems to be a tradition of misascription in the naming of systems of set theory! See Chapter 8.) Assume that *VNB* includes a class comprehension axiom, the so-called axiom of predicative comprehension (the extension of any property not containing class quantifiers is a class), an extensionality axiom for classes (assuming, *contra* Bernays, that sets are also classes), and the usual *ZF* axioms, including foundation and with replacement given now in a single sentence. Assume that *AC* is omitted. The analogue of what Cantor calls an 'absolute or inconsistent collection' is now a proper class. And the analogue of (*) is

(9) X is a proper class (i.e. not a set) iff X has a part equivalent to *Ord* (the proper class of all ordinals).

We assumed that Cantor has some principle of choice available in proving (7). Following Zermelo, let us assume that this is a principle of simultaneous choice, which for the sake of argument we take to be *AC*. Now the result is surprising, for it is known that (9) and *AC* together imply that the universe can be well-ordered. Indeed it is known that if we put

(10)   The universe can be well ordered

then

(11) $VNB \vdash (10) \leftrightarrow [(9) \wedge AC]$.

(This equivalence holds only in the presence of the axiom of foundation.) Thus, we can say (providing we accept the information provided by this 'modern' reconstruction) that Cantor's instincts lead to a very strong result indeed. And one can show that not only is the combination of (9) and *AC* much stronger than (7) but that both parts are necessary here.

(10) itself is equivalent in *VNB* to the so-called 'global' axiom of choice *GC*, i.e.

(12) $VNB \vdash GC \leftrightarrow (10)$

where GC asserts the existence of a function F (a class) such that $\forall x\,[x \neq \emptyset \to F(x) \in x]$. (For (12) see Levy [1979], p. 176, and the accompanying discussion.) But it is known that

(13) $VNB \nvdash AC \to GC$

a result due to Easton. (See Fraenkel, Bar-Hillel, and Levy [1973], p. 134, or for a proof see Felgner [1976].) Yet

(14) $VNB \vdash AC \leftrightarrow (7)$.

Thus, Cantor's assumptions are much stronger than are needed to prove the (7) he was aiming at (or depended on) in the 1883 version of his theory. Moreover (13) shows that AC and replacement are not enough to yield the principle (9) which incorporates Cantor's 'projection postulate'. For if so we would have $VNB \vdash AC \to (9)$ and thus from (11) and (12) $VNB \vdash AC \to GC$ contradicting (13). In addition, (9) is fully independent of AC, since

(15) $VNB \nvdash (9) \to AC$.

Thus since Cantor wanted at least to achieve (7), the 'projection postulate' while it adds considerable power to the choice axiom, is not enough alone; it has to be combined with some choice assumption. (I am grateful to John Bell and Ulrich Felgner for pointing out to me and clarifying the various relations between AC and GC and (9) with respect to VNB, in particular the non-derivability of AC from (9), and equivalence (11) and the essential use of the axiom of foundation here.)

I do not mean to suggest with this discussion that the way to understand Cantor's argument is by a modern axiomatic reconstruction. Indeed not. Nor do I mean to suggest that Cantor knew precisely or at all that his argument could yield something so strong as the well-orderability of the universe. But the reconstruction does provide further evidence, not just of the power and sweep of Cantor's assumptions, but that Cantor's thinking about sets is deeply, perhaps *inevitably*, connected with well-ordering. Involved in this, too, is the structural role given to the ordinal numbers. This was crucial, as we saw, in the Cantor theory of cardinality, particularly since Cantor's attempt at an abstractionist and set-reductionist explanation of cardinality seems to lead back to the ordinals. But in the 1899 argument the ordinals are given a further structural role—in effect to act in totality to gauge which collections form sets. In this sense, they form a key scale of 'size' within the universe of objects. The VNB reconstruction shows that (local) well-ordering together with this structural assumption lead directly to the well-orderability of the universe. But while Cantor almost certainly did not know this, it would not have been such a surprise to him. To begin with, as I pointed out earlier, the 1899 argument with its successive choice assumption and the projection postulate is directly inspired by Cantor's theory of the Absolute, and not least by the notion that the ordinal number sequence *reflects* the Absolute. But since the Absolute was itself partly an expression for the totality of created abstract objects (see §1.4), it may well be (by a kind of 'transitivity of reflection' argument) that Cantor intuitively viewed the whole universe of mathematical objects as well ordered.

More important for the subsequent development of set theory is that Cantor's 1899 theory gives a mathematical transformation (or a translation into mathematical language) of the doctrine of the Absolute. And this clearly presages a theory of 'limitation of size'. For (*) now apparently provides a working criterion of type differentiation between absolute collections and sets specifically on grounds of size. A collection is *absolute* (not a set) just in case it is as large as or bigger than the sequence of ordinal numbers. The original doctrine of the Absolute and Absoluteness strongly suggested that Absoluteness is connected with size. For while all transfinites according to Cantor can and should be numbered (numbered size being increasable), the Absolute, which Cantor sometimes called an 'Absolute maximum', is *beyond* numbering (unincreasable). This might easily be construed as saying that the Absolute is *too large* to be numbered. In any case, this is exactly what the 1899 theory says about the absolute collections. For the term 'equivalent' in Cantor's proof and in (*) means 'equivalent in size' in exactly Cantor's original sense.

At any rate, Cantor's 1899 argument marks the birth of the limitation of size idea, an idea which, in various forms, was to be of continuing importance. However, three particular difficulties with Cantor's suggestions should be pointed out here, for they form standing points of discussion in the rest of this book. The first problem concerns principle (c) of Chapter 1. As has just been made clear, this principle is heavily involved in Cantor's 1899 argument, and indeed is directly and mathematically productive. But is the end result in some sense incompatible with principle (c)? It is clear in the resulting theory that while the absolute collections are not treated as single objects they do play a direct mathematical role in the proof of (7). This might be taken as counter to the earlier theological doctrine of the Absolute. Later set theorists did not at all preserve the connection between 'absoluteness' and *the* Absolute or God, and this might have made direct operation with absolute collections and a two-sorted class theory rather easier. But even so, there was a good deal of confusion about the precise status of absolute collections. Zermelo, for instance, took a strong line against them, while von Neumann developed a theory in which principle (c) is again *prima facie* challenged, since he treats absolute collections as mathematical objects of a certain kind which can even have a cardinality. One of the difficulties von Neumann faced was that of how to treat *properties* and *relations*. Von Neumann considered that to get the full strength of Cantorian theory one sometimes needs to use relations and functions which cannot be non-absolute Cantorian sets. But even before the recognition of this difficulty, there was the question of whether one *should*, and if so *how* one should, treat absolute collections as genuine objects. The difficulties here are shown up by the manifold confusions in Jourdain's treatment of the subject which I come to in §4.2. Actually, I shall suggest in §8.3 that von Neumann's theory does not essentially violate Cantor's principle (c). When a heuristic principle evolves, its consequences may change or become more clearly delineated. This is what happened in this case. Nevertheless, the von Neumann theory preserves the spirit of principle (c).

The second difficulty concerns Cantor's claim, following his proof, that 'all continua are countable in an extended sense'. But this would follow only if one could show that they (particularly the linear continuum) are *sets*. But Cantor does not argue for this. No doubt he would have regarded it as obvious by the extendability and cardinality arguments mentioned in §1.3. But we shall see that the incorporation of the continuum (more generaly, the power-set operation) into limitation of size frameworks was one of the trickiest points of this approach to set theory. In particular, it is extremely difficult to make the notion of 'extendability' *a priori* clear. This difficulty over the continuum indicates a more general problem, our third here. Cantor's (*) yields a criterion of sethood. And one might then want to take it not as a derivative fact about the Cantorian universe of sets but rather as a starting point for deciding what is *in* the universe. But (*) by itself does not yield much information about what sets there are. Indeed it indicates only what are sets *relative* to the ordinal number sequence. But how many ordinals are there? This was a difficulty Cantor himself implicitly recognized. He put it in the following form:

One ought to raise the question: how do I know that the well-ordered multiplicities or sequences which are assigned the cardinal numbers

$$\aleph_0, \aleph_1, \ldots, \aleph_{\omega_0}, \ldots, \aleph_{\omega_1}, \ldots$$

are actually 'sets' in the sense of the word which I have explained, that is to say, 'consistent multiplicities'? Is it not conceivable that these multiplicities are already 'inconsistent' but that the contradiction which results from the assumption that 'all their elements can be taken together' has not yet been noticed? (Letter to Dedekind, dated 28 August 1899; Cantor [1932], p. 447)

Cantor admits that in the end the question can only be resolved by postulation. The difficulty is to state as precisely as possible what postulates one is relying on. This Cantor does not do. Instead, in the same letter he contents himself with the following declaration of faith:

. . . the fact of the 'consistency' of finite multiplicities is a simple, undemonstrable truth, it is '*the axiom* of arithmetic' (in the ancient sense of the word). And likewise the 'consistency' of multiplicities to which I assign the alephs as cardinal numbers is 'the axiom of the extended, of the transfinite arithmetic'. (Letter to Dedekind, dated 28 August 1899; Cantor [1932], p. 447)

There are certain passages in Cantor's 1899 letters which seem to suggest an axiomatic approach to the question of set existence, particularly the statements

(2) and (3) quoted on p. 168. But one should not be misled by these. At best one can say that Cantor's 1899 theory *foreshadows* an axiomatic approach in only *one* clear sense (though see also §1.3). The theory is an attempt, through the construction of a proof, to isolate principles which will lead to the desired enumeration theorem, while at the same time avoiding the Burali-Forti contradiction. Later it was careful proof analysis and careful proof building directed to almost the same end (though for the well-ordering theorem, not the enumeration theorem) which gave Zermelo his axiom system for set theory. And similar considerations directed von Neumann, though now with Cantor's enumeration theorem back in prime position.

This part of the axiomatization story is pursued again in the later chapters of Part 2. For the time being I want to follow the limitation of size idea more closely. The idea was first expounded in print by Jourdain and Russell in a form quite close to Cantor's own suggestions. Their attempts begin to show the difficulties of proceeding with such a theory in a non-axiomatic way. In particular, the acute criticisms of Russell point perceptively to the reason why modern set theory, while strongly influenced by 'limitation of size', is *not,* despite many claims to the contrary, based *solely* on limitation of size principles.

4.2. Jourdain's limitation of size theory

Cantor drew a connection between absoluteness and inconsistency, and then connected absoluteness with size. If one then also definitely connects the 'contradictory' collections revealed by the paradoxes with the 'inconsistent' collections, as Cantor certainly did, then one has apparently a basis for building a set theory which avoids these paradoxes. For suppose one conjectures now a direct connection between size and contradictoriness by a *limitation of size hypothesis*

LSH: All contradictory collections are too big (in some sense of 'bigness' to be specified), 'contradictory' collection here meaning any collection such that a contradiction can be derived from assuming it to be a set.

Then *LSH* suggests that any system in which all sets are small (not too big) will be free at least from the traditional source of contradiction. As I say, the proposal is a spiritual descendant of Cantor's way of thinking represented in his 1899 correspondence. But in published form *LSH* and its use as a starting point for building a contradiction-free set theory stems from Russell and Jourdain.

Jourdain's aims were effectively the same as Cantor's: the desire to prove that every cardinal number is an aleph and simultaneously to avoid the Burali-Forti contradiction. His procedure presented in his [1904*a*] and [1905] was also remarkably similar to Cantor's. (See also Grattan-Guinness [1977], which is a useful source of information about Jourdain's work, his life, and his relations with Russell.) Jourdain starts from Hardy's theorem (Hardy [1904]) that every cardinal number is either an aleph or greater than all alephs. The keynote of the argument, which Jourdain does not challenge, is again successive selection. Indeed Hardy links his own proof directly to Cantor's proof of [1895] that every infinite set has a subset of power $\aleph_0$:

These considerations must, I imagine, have been familiar to Cantor. But he confines himself to showing that there is no cardinal between α_0 and α_1 [Hardy's notation for $\aleph_0$ and $\aleph_1$], i.e. $> \alpha_0$ and $< \alpha_1$, and never explicitly rejects the possibility contemplated by Mr. Russell. (Hardy [1904], p. 88)

The 'possibility raised by Russell' is that of strong incomparability of cardinals. Russell stated:

We do not know that of any two different cardinal numbers one must be the greater, and it may be that 2^{α_0} is neither greater nor less than α_1 and α_2 and their successors, which may be called well-ordered cardinals because they apply to well-ordered classes. (Russell [1903], p. 323)

Hardy does not address himself to the well-ordering problem, but clearly saw his result as refuting Russell's 'possibility'. Indeed, as part of this Hardy 'constructs' a subset of real numbers of power $\aleph_1$, thus showing $2^{\aleph_0} \geqslant \aleph_1$. About this result, he added, and one might say with considerable foresight, that it

. . . may be of some interest as throwing some light (though of course a very partial one) on one of the most fundamentally important and apparently hopeless questions in the whole range of pure mathematics. (Hardy [1904], p. 87)

What follows from Hardy's 'successive selection' is that a collection is either equivalent to some initial segment of the ordinal numbers, *or* has a part equivalent to the whole sequence. Jourdain now reckons with the Burali-Forti contradiction in much the same way as Cantor did, namely by denying that the sequence of ordinals is like other ordered sets, and in particular denying that it has an ordinal type, or indeed a cardinal number. (If the sequence has a cardinality then, since it is well ordered, it must be an aleph, say $\aleph_\beta$. Thus the ordinals would be isomorphic to all ordinals less than ω_β, so ω_β would be a type for the ordinals, contrary to assumption.) This, of course, is not enough, for if the collection of ordinals has no cardinality or type, neither can any aggregate containing them or a part equivalent to them. Jourdain thus decided to rule out *all* such aggregates from possessing types or cardinal numbers. He did this by distinguishing between 'consistent' and 'inconsistent' aggregates, these being the aggregates which can have neither cardinality nor type. The 'inconsistent' aggregates are described, very much as Cantor did, as those 'which it is impossible to think of as a whole without contradiction'. But Jourdain adds, again like Cantor's (*) above:

. . . for formal purposes, I use the following definition: An 'inconsistent' aggregate is an aggregate such that there is a part of it which is equivalent to W [the ordinal number sequence]. (Jourdain [1904a], p. 67)

This principle, like Cantor's a 'limitation of size' principle, now guarantees that there can be no cardinal number greater than all alephs, and thus, in conjunction with Hardy's theorem, that all cardinals are alephs. It is quite clear that, despite the remarkable similarity to Cantor's procedure, Jourdain discovered his proof independently. He wrote to Cantor about it on 29 October 1903, and received a reply (4 November 1903) in which Cantor pointed out the similarity. Cantor encouraged Jourdain to publish. (See Grattan-Guinness [1971], p. 116.)

This limitation of size solution appears at first sight to be quite natural and straightforward. But the difficulties are in fact considerable. The most obvious concerns the characterization of inconsistency itself. Jourdain wanted 'M is consistent' to appear as an hypothesis in the definitions of 'the cardinal number of M' and 'the ordinal type of M'. However, as long as W, the collection of all ordinal numbers, appears in the characterization of inconsistency/consistency, he regarded this as suspect. For the collection of (ordinal) numbers is then assumed to exist before the notions of number have been defined. Thus, as Jourdain explained to Russell (letter of 24 May 1904): 'But, of course, since this criterion of inconsistency should be prior to all talk of ordinals, we must define an aggregate similar to W in another way.' (See Grattan-Guinness [1977], p. 34, and also Jourdain [1906], pp. 56-7.) This he had done (in his [1904a]) by replacing W in the characterization of inconsistency by a well-ordered aggregate which he termed 'the well-ordered aggregate of which every well-ordered aggregate is a segment' (Jourdain [1906], p. 67). But this attempt to get around the problem is hopelessly confused. Russell, in a letter of 28 April 1905, noted:

You ought to speak of a series such that etc. One has 'the series of ordinals in order of magnitude'; but your new series is not unique: e.g. it still fulfills your definition if any two of its elements are interchanged . . . You do not show the existence theorem in the present case, but by using the for a your conceal the need of it. (Grattan–Guinness [1977], p. 45)

Russell puts his finger on a crucial point here, for what is to act as the crucial measure of size or inconsistency? I shall return to this very important point below.

However, the *central* difficulty was that Jourdain overlooked the effect that his limitation of size theory has on Russell's comprehension principle (every propositional function determines a class), the key definitional tool of the set theory of *The Principles of Mathematics*. Jourdain wished, for obvious reasons (convenience, simplicity), to preserve this principle. Indeed, he seems to have regarded his theory as a simple modification of the set theory of *The Principles*.

But he seems not to have asked himself seriously whether the preservation of comprehension is actually feasible. The problem arises when one asks: what is the status of the 'inconsistent' aggregates; are they or are they not legitimate mathematical objects? The impression given by Jourdain's [1904a] is that they are *not,* for an inconsistent aggregate is one 'which it is impossible to think of as a whole without contradiction' (p. 67). In this case comprehension would *obviously* fail. Yet this impression is hastily corrected in [1905], where Jourdain says that 'absolutely infinite' would have been a better term than 'inconsistent': '. . . because an "inconsistent" aggregate is not itself contradictory (it exists in the mathematical sense of the word), but a cardinal number or type of it does not exist.' (Jourdain [1905], p. 54.) It seems from this that the inconsistent classes are to be retained as objects, though objects of a different sort from ordinary sets, since they are not subject to all the set operations. (In particular, they will not be legitimate arguments for the terms *card*(x) and *ord*(x).) Jourdain clearly thought that this would enable him to retain Russell's comprehension principle. For in a letter to Russell of 17 March 1904 he noted: 'The concept of Cls [class] I take from you ($x \ni \phi(x)$) [i.e. the collection $\{x : \phi(x)\}$].' (Grattan-Guinness [1977], p. 28.) He made clear also that for consistent classes he followed Russell's definition of number, and repeated: 'I think that W is not contradictory, but the class of classes similar to W (Nc W) [i.e. the Russell cardinal of W] is.' (Grattan-Guinness [1977], p. 28.)

The advantage of adopting this position is obvious, for it would, if coherent, enable Russell's class theory of [1903] to be taken over virtually unhindered, except in denying numbers to the large classes. Moreover, since W and its like are legitimate objects, they can be put to solid mathematical work as they are in the Cantor–Jourdain proof of the aleph theorem.

But unfortunately the comprehension principle is just *not* defensible in this theory, as Russell was quick to notice. In Jourdain's system we can legitimately apply comprehension to obtain $Ord = \{x : Ord\ (x)\}$ which is thus a perfectly good class. But then, as Russell pointed out, we can apply comprehension again to obtain $\{v : v \text{ sim } Ord\}$. ('$v$ sim u' is Russell's notation, in his correspondence with Jourdain, for 'v is equinumerous with u'.) Hence the Russell cardinal number of the ordinals *does* exist. And he noted in a letter of 12 April 1904:

. . . it seems to me plain that, whatever sort of class u may be, $v \ni (v \text{ sim } u)$ [i.e. $\{v : v \text{ sim } u\}$] is a good class; thus by refusing to call this a number, we get out of no difficulties. (Grattan-Guinness [1977], p. 29)

Russell saw this as a direct attack on the comprehension principle. We certainly know that if *Ord* and *card*(*Ord*) both exist we have a contradiction. He argues that if *Ord* exists, so does *card*(*Ord*) by a simple application of comprehension. Thus *neither* can exist, so comprehension *must* fail. Dropping the Russell definition of number would not help greatly, for one would still have objects available which could act as the numbers of inconsistent aggregates. And, quite possibly, the cardinal and ordinal contradictions would still be reproducible, though in a slightly different form. As Russell says, Jourdain's theory seems 'merely a proposal not to talk about the awkward cases' (letter of 12 April 1904; Grattan-Guinness [1977], p. 29).

That it is the principle of comprehension which is at fault here, and *not* the Russell definition of number, was hammered home by Russell in his [1906a]. The core of Russell's analysis of the paradoxes there is his exhibition of a 'generalized contradiction' which shows that the various contradictions all have a common form. Russell's generalization is rather simple. Assume that ϕ is a predicate (or propositional function), and f a function of sets; then if ϕ and f satisfy the following condition:

(*) $\forall u \ [\forall x(x \in u \to \phi(x)) \to (f(u) \notin u \land \phi(f(u)))]$

then $w = \{x : \phi(x)\}$ is contradictory. This covers both Russell's and Burali-Forti's contradictions. To obtain Russell's contradiction put $\phi(x) \equiv x \notin x$ and $f(x) = x$; to obtain Burali-Forti's contradiction put $\phi(x) = Ord(x)$ and define $f(x)$ to be the least ordinal number bigger than all ordinals in x. (Cf. Russell [1906a], pp. 141-3; also [1906b], p. 199.) By using an early form of definition by transfinite induction Russell then showed that given a ϕ and f satisfying (*) it is possible to generate a series of sets with property ϕ similar to the ordinal number sequence. In effect he constructed a correlation F between the ordinals and some ϕ sets by the following procedure

$$F(\alpha) = f(\{F(\beta) : \beta < \alpha\})$$

where $F(0)$ is some arbitrarily chosen ϕ set (see [1906a], p. 143). So if $F(0) = a$, then

$$F(1) = f(\{a\}), F(2) = f(\{a, f(a)\})$$

and so on. In the particular case corresponding to Russell's paradox, f is the identity map. Hence here

$$F(0) = a, F(1) = \{a\}, F(2) = \{a, \{a\}\}, \text{ etc.}$$

This clearly foreshadows the construction of the von Neumann ordinals: just put $F(0) = \emptyset$ instead of just any non-self-membered set. Thus Russell, with Mirimanoff and Zermelo, was another who had potentially solved the problem of generating set objects to match the ordinals before von Neumann. (Though see §8.1 and Hallett [1981], p. 398.)

In one sense, this result was a reason for optimism, for it shows that the Russell set (i.e. $w = \{x : x \notin x\}$) is also too large, or inconsistent, on Jourdain's criterion. This strongly suggests that Jourdain's theory had isolated an important *general* feature of the 'queer' sets. And this generality certainly impressed Russell (see below, pp. 182-3). But what it *also* shows, and this is important here, is that Jourdain's theory cannot consistently maintain the comprehension axiom. For, when the predicate ϕ and the set operation f satisfy the conditions specified, to avoid contradictions one can either deny that $w = \{x : \phi(x)\}$ exists, or deny that $f(w)$ exists (Russell [1906a], p. 153). In the Burali-Forti case Jourdain appears to have the freedom to accept the existence of $w = \{x : Ord(x)\}$ and to deny the existence of $f(w) = ord(w)$. Yet in the case of Russell's contradiction *there is no such choice*: comprehension must be abandoned since $f(w) = w$. This was a strong argument against Jourdain's position on comprehension, for it is arrived at precisely by extending Jourdain's own procedure, namely treating the Burali-Forti contradiction as the basic contradiction (it is this which gives rise to the Jourdain theory) and generalizing it. The denial of comprehension then follows from solving the Russell contradiction *by exactly the same method* as Jourdain solves the Burali-Forti contradiction.

It would seem that Jourdain did not immediately take the point (see Jourdain to Russell, 28 December 1905, Grattan-Guinness [1977], p. 66). Indeed when Russell in his [1906a] took the Jourdain theory to deny uniformly the existence of the ws (because of the above analysis) Jourdain complained that Russell had misrepresented him (see letters of 28 December 1905 and 21 March 1906, Grattan-Guinness [1977], pp. 66, 84). He still accepted that the inconsistent collections are indeed classes. However, by the time he wrote his [1907] he had apparently accepted that comprehension must fail. For in that paper (submitted on 31 March 1906) Jourdain states:

In particular, Russell and I were led independently to the observation that certain propositional functions do not define classes: he, from working on his (cardinal) contradiction; I from working on the ordinal contradiction of Burali-Forti. (Jourdain [1907], p. 282)

(The passage is from a note added on 30 April 1906 (see also pp. 268-9).) Despite what Jourdain says here, it is fairly clear that the point was Russell's; Jourdain's working with the Burali-Forti contradiction alone did not lead him to the rejection of comprehension. Jourdain did claim (in his [1905], i.e. before Russell's [1906] appeared) that his theory *does* avoid Russell's contradiction. However, as the remark in the previous quote indicates he was clearly confusing this contradiction with the version of the greatest cardinal contradiction derived from applying Cantor's theorem to the universal class. Possibly he misunderstood Russell's explanation of how he (Russell) was led to his own contradiction by just such an application (see Russell [1903], p. 101). Anyway Jourdain's analysis of the contradiction clearly did not lead him to reject comprehension. He could 'solve' the greatest cardinal contradiction by denying the universal class a number. In any case, Russell's instinct here was much more finely tuned than Jourdain's. For example, in a letter of 28 April 1905 Russell noted:

I think both you and other mathematicians do not realise how any limitation such as you propose involves a modification of the fundamentals of logic, since these, as commonly accepted, exclude any such limitation. When the right limitation has been found, it is likely to exclude more than might be wished. (Grattan-Guinness [1977], p. 46)

And, in the same letter:

I have at present no view on Burali-Forti's contradiction; but I am sure the solution lies in the foundations of logic, and not in any late technical mathematics.

Jourdain, though, remained deeply confused about the status of the 'inconsistent classes', sometimes holding that they are genuine classes, sometimes not. Strangely, the confusion is at its height in his [1907] *despite* the apparently unequivocal statement recently quoted. For example, he holds that the Russell numbers (p. 283) and the Russell 'set' (pp. 268-9) are non-entities, while the collection of ordinals is a genuine class (p. 282). Yet all these collections are 'inconsistent' or 'too big' according to his theory. This is quite *ad hoc*, and it makes nonsense of distinguishing between 'consistent' and 'inconsistent' on grounds of size. Jourdain did later, apparently, seek ways to purge his theory of this *adhoc*ness, specifically by looking for new definitions of number. For an account of this, I refer any remaining readers to Hallett [1981], section 4.

4.3. Modifying comprehension by limitation of size

From the beginning, from his discovery of his contradiction, Russell, unlike Jourdain, was convinced that the comprehension principle was at fault. In his famous letter to Frege of 1902, in which he explained the contradiction he had just discovered, he remarked:

From this [i.e. the contradiction] I conclude that under certain circumstances a definable collection does not form a totality. (Russell [1902], p. 125)

And more explicitly in *The Principles of Mathematics*:

We took it as axiomatic that the class as one is to be found wherever there is a class as many; but this axiom need not be universally admitted, and appears to have been the source of the contradiction. By denying it, therefore, the whole difficulty will be overcome. (Russell [1903], p. 104; cf. also pp. 102-3)

And in his [1906a] he states clearly:

What is demonstrated by the contradictions we have considered is broadly this: 'A propositional function of one variable does not always determine a class.' (Russell [1906a], pp. 144-5)

But having identified the comprehension principle as the source of the contradictions the 'whole difficulty' is only shifted. The problem is now that of identifying the 'certain circumstances' under which a 'definable collection does not form a totality'. This problem was much more difficult, as Russell recognized:

. . . the question arises: which propositional functions define classes which are single terms as well as many, and which do not? And with this question our real difficulties begin. (Russell [1903], p. 103)

And for Russell, of course, the difficulty over comprehension was very serious since the system of set theory he uses in *The Principles* was based on it.

It is here that Jourdain's work, the notion of *bigness* and the observation that all the known contradictory collections are too big, becomes interesting. In a sense the idea was already in Russell's *Principles*, for there he remarked that the difficulty with the contradictions '. . . does not concern the infinite as such but only certain very large infinite classes.' (Russell [1903], p. 362.) But in any case Jourdain's and his own clarification of size and of the size of the contradictory collections apparently made possible an attempt to modify comprehension along limitation of size lines. And Russell sketched a proposal for such an attempt in his [1906*a*].

The generalized Russell contradiction we encountered in §4.2 shows that the known contradictions share the 'largeness' property of the ordinals. Russell took this as showing that excessive size (as characterized by Jourdain) might well be a key to the whole mystery. He noted:

This generalization is important because it covers all the contradictions that hitherto emerged in this subject. (Russell [1906*a*], p. 142)

And then

It is probable in view of the above general form for all known contradictions, that if ϕ is any demonstrably non-predicative property [i.e. if a contradiction results from taking its extension as a set] we can actually construct a series, ordinally similar to the series of all ordinals, entirely composed of terms having the property. (Russell [1906*a*], p. 144)

This conjecture is just the Limitation of Size Hypothesis (*LSH*) noted at the beginning of §4.2, now using Jourdain's definition of excessive size. If this *LSH* is correct, it is natural to try to rehabilitate the comprehension principle by building in a size condition:

We still have the distinction of predicative and non-predicative functions [i.e. propositional functions which respectively do and do not define sets]; but the test of predicativeness is no longer simplicity of form [as in the 'zig-zag theory': [1906*a*], pp. 145–41] but is a certain limitation of size. (Russell [1906*a*], p. 152)

Modified comprehension would then read: given any predicate (propositional function) ϕ, the collection $\{x : \phi(x)\}$ is a set if and only if it is not too big, 'bigness' being understood in Jourdain's sense. All the contradictions are met in a unified way, and (for Russell at least) the contradictory collections are all eschewed as 'non-classes' or 'non-entities': '. . . any supposed class which reaches or surpasses this limit [of size] is an improper class, i.e. is a non-entity.' (Russell [1906*a*], p. 152.)

The proposal, however, is unworkable for reasons strongly reminiscent of the difficulties raised with Cantor's 'limitation of size' theory. First, if the collection of all ordinals is not a good class, then it cannot be used straightforwardly either in the characterization of inconsistency or in Jourdain's proof of the aleph theorem. Moreover, as with Cantor, there is no guarantee that the continuum is a consistent aggregate and therefore an entity. This is a serious threat for a supporter of classical Cantorian mathematics like Jourdain, particularly when one considers that for Cantor the central purpose of the aleph theorem was to pave the way for a proof of the continuum hypothesis. However, the central difficulty with the proposal concerns the notion of number itself, as Russell again was well aware.

The characterization of inconsistency depends on the availability of the ordinals. Here Jourdain, as we have seen, initially accepted Russell's definitions of number. But these definitions are based on the availability of the full comprehension principle, and indeed are *not* available if the size-modified principle is adopted. This is because, as Russell pointed out to Jourdain, all his numbers, except 0, are equinumerous with the entire universe (see Jourdain's letter to Russell of 28 December 1905, and Russell's reply of 1 January 1906; Grattan-Guinness [1977], pp. 66-7). Thus, they are 'inconsistent' and so non-classes, or non-entities. This does not, of course, entirely rule out the 'size'-modified comprehension principle; one can try to find a new definition of the numbers, not suffering from this difficulty. But this move already reveals the relative weakness of the modified comprehension principle. If numbers are not taken as sets, then we need extra principles governing the introduction and existence of numbers. And if we demand that numbers be sets (Jourdain and Russell were both reductionists, i.e. against taking numbers as primitives), then the search for an adequate new theory of number is already a search for a partial replacement for comprehension since it will involve set existence assumptions guaranteeing enough cardinals (sets) for ordinary mathematical work (including, perhaps, a proof of the aleph theorem). All this must be prior to, or at least alongside and independent of, any use of the modified comprehension principle. To put it another way, if numbers are sets, modified comprehension will characterize the legitimacy (or sethood) of the extension of a propositional function or predicate according to a measurement against a collection of *sets*. Thus there is a prior need to have available a collection of sets (and principles guaranteeing their existence) to do the measuring. Size-modified comprehension thus cannot be autonomous. It is worth recalling (see p. 178) that Jourdain had already suggested replacing the ordinals by some other well-ordered sequence of sets in the characterization of 'inconsistency'. But again the question arises: what sets, and what principles justify their existence? (Note again Russell's remark quoted on p. 178. Note also that Jourdain did attempt to find set definitions of ordinal and cardinal number alternative to Russell's; see Hallett [1981], §4.)

Russell, I think , was aware of this central problem with modified comprehension, if only implicitly:

This theory has, at first sight, a great plausibility and simplicity, and I am not prepared to deny that it is the true solution. But the plausibility and simplicity tend rather to disappear on examination . . . A great difficulty of this theory is that it does not tell us how far up the series of ordinals it is legitimate to go . . . We need further axioms before we can tell where the series begins to be illegitimate. (Russell [1906*a*], pp. 152–3)

Russell does not go into the difficulties of defining number. But he does say that if α is a legitimate ordinal then the class of smaller ordinals must be a legitimate one (i.e. a set); thus the required 'further axioms' governing the legitimacy of ordinals are already, in effect, axioms governing the existence of sets.

This failure of the size-modified comprehension proposal is what really makes Jourdain's theory interesting. For the difficulties it faces encapsulate the reasons why set theory cannot be formulated *wholly* according to principles of limitation of size. Since nobody has been able to specify any means of recognizing whether a collection is intrinsically small (safe) or not, the only alternative is to specify an external measure of size via a measure collection as Jourdain did. Thus the existence of some collection of objects (sets) must be guaranteed independently of considerations of size. This does not preclude the use of limitation of size principles, but it means that they cannot be the sole arbiters of set existence. This brings us neatly and precisely to Part 2 which is concerned with the attempts of Zermelo and von Neumann, following a quite different course from Jourdain and Russell, to axiomatize set theory. The 'limitation of size argument' claims in the spirit of *LSH* that the resulting axiom systems work (in the sense that they do not succumb to the traditional paradoxes) because the axioms they use (which are, in effect, instances of comprehension) *successfully limit size*. There are clear and important 'size-limiting' axioms in these systems, and I shall argue later that the limitation of size idea was of key importance in the formation, even the formulation, of these axioms. But they are crucially still *relative* size principles, like Jourdain–Russell comprehension. They depend for their effect and their power on the existence of other sets whose existence is guaranteed by principles not obviously connected with 'restricted size'. Indeed, it turns out that many of these sets are intuitively and intrinsically *big*.

4.4. Mirimanoff

Before looking at the limitation of size argument and the Zermelo and von Neumann systems, I want to jump ahead of Zermelo briefly to the work of Mirimanoff ([1917*a*], [1917*b*] and [1921]). (There is a short description of Mirimanoff's work in Wang [1974], pp. 215–8.) Although he had certainly read Zermelo's work, Mirimanoff cannot be seen as a continuation of Zermelo. For one thing he is not an axiomatizer in Zermelo's sense, nor is he strictly speaking a reductionist. Rather he appears more as a successor of Cantor-Jourdain, attempting to characterize sets directly in terms of size. Like them, Mirimanoff concentrates on the Burali-Forti paradox, and like Russell's analysis before, Mirimanoff shows how, in terms of size, the Burali-Forti paradox is basic and that if we solve this the other paradoxes will be solved too. But Mirimanoff's analysis is both clearer and more profound than Russell–Jourdain. It is clearer because he is careful to set out the various postulates about sets which are needed for his argument to go through. For while his basic conception is that of 'limitation of size', it is a *relative* limitation of size theory which must rely on explicit auxiliary postulates. It is also more profound at least in the sense that it provides a much deeper analysis than Russell did of the structure of sets which can mimic the ordinal numbers. (While in many ways Mirimanoff's ideas are close to those of Cantor in the 1899 letters and to Russell [1906a], there is no indication in his papers that he knew of any of this work.)

Mirimanoff's key idea is to give a set-theoretic form to the Burali-Forti paradox, which of course originally involved ordinals taken as primitives. Russell's paradox is then tied to this set-theoretic Burali-Forti paradox. Mirimanoff begins by showing that Russell's paradox is actually a special case of more general paradoxes. For example, he introduces a concept of isomorphism by describing it as follows ([1917a], pp. 40-2): a set x is isomorphic to y if there is a function f mapping x one-to-one onto y and

(a) $\forall z \in x$, z is an urelement iff $f(z)$ is
(b) $\forall z \in x$, if z is a set, then z and $f(z)$ are isomorphic.

Now let R' be the set of sets which are isomorphic to none of their elements. Then Mirimanoff points out that $R' \subseteq R$ (the Russell set of all sets not containing themselves as elements) and that R' has the same inductive property as R, namely if $\forall x \in y[x \in R'] \rightarrow y \in R'$. Mirimanoff now introduces a new concept, that of 'ordinary set', which corresponds to the later notion of well-founded set. He calls a set x ordinary if there are only finite descending membership chains stemming from x, i.e. if . . . $x_3 \in x_2 \in x_1 \in x$ must always end in an urelement (or 'indecomposable element') after finitely many steps. (If one assumes the axiom of choice ordinary sets coincide with well-founded sets where a well-founded set is defined as a set x such that $\exists y \in x[y \cap x = \emptyset]$ (see Fraenkel, Bar-Hillel and Levy [1973]). This notion of well-foundedness was introduced by von Neumann [1928a], p. 498. In his axiom of foundation in [1930] Zermelo mentions both finite descent and well-foundedness, claiming that they are equivalent.) Call the set of all ordinary sets V, then V also has the inductive property, and $V \subseteq R'$. Because V, R' and R all have the inductive property, the assumption of their existence gives rise to a contradiction: for example, since no member of R' is isomorphic to any of its elements, R' must have this property. Thus $R' \in R'$ which shows that R' is isomorphic to one of its elements; thus contradiction. This shows, says Mirimanoff, that none of V,

R', R can exist ([1971a] pp. 40, 42, 43). If one now assumes that 'the existence of a set entails that of each of its subsets' (Mirimanoff [1917a], p. 44), then it is enough to show that V does not exist in order to conclude the non-existence of R' and R.

But Mirimanoff takes the chain of subsets of R further. If x is an ordinary set, then all membership chains starting from x end in urelements. Mirimanoff calls these *nodes* of x. Let V'_x be the set of all ordinary sets whose nodes belong to a given set x of nodes. V'_x like V generates a contradiction; in particular any V'_e which stems from a *single* node e generates a contradiction. But what Mirimanoff shows crucially is that any well-ordered set can be represented by a set well-ordered under inclusion (or membership) having only a single fixed node e. (These representatives are in effect the von Neumann ordinals. Mirimanoff's analysis of these, and the explanation of how he arrives at them, is discussed in §8.1.) These sets, which Mirimanoff calls S sets, form a totality $W' \subseteq V'_e$ for the fixed e. (This e corresponds to the 'empty segment' of a well-ordered set. Thus, it is a privileged urelement analogous to the empty set: see §8.1, p. 273.) Mirimanoff does not argue directly that W' does not exist, as he could easily have done by showing that if it did it would itself be an S set and thus one of its own members. (Mirimanoff gave a general characterization of S sets as transitive, well-founded (ordinary) sets connected by $\in$—in other words, the standard modern definition of set-ordinals (see §8.1, p. 274). It is easy to show that, if it is admitted as a set, W' satisfies these conditions.) Rather he argues that W' does not exist by connecting it with the Burali-Forti antinomy. If we assume that there are ordinal numbers α with no corresponding S set α_S then, argues Mirimanoff, W' certainly cannot exist. Let π be the least ordinal such that π_S does not exist; since W' would satisfy the conditions for being such a π_S, W' cannot exist either ([1917a], p. 47). Mirimanoff actually rejects this assumption, that is he assumes that to every α there is an α_S, and proceeds to show that even so W' cannot exist. First W' is actually cardinally equivalent to W, the set of all ordinal numbers. Applying the standard Burali-Forti argument, the set W of all ordinals does not exist. The key is now to assume also that no existent set contains a subset equivalent to W. This means that the Burali-Forti paradox is now sufficient to show that W', V'_e (or any of the V'_x), V, R' and R cannot exist. In other words, given this cardinality or equivalence connection between W and W', the Burali-Forti paradox is revealed as the basic paradox: solve it and the others are solved.

What is now required is to develop a theory in which W does not exist, and according to which none of the other paradoxical sets can appear either. Mirimanoff states the central problem as follows ([1917a], p. 38): 'What are the necessary and sufficient conditions for a set of individuals to exist?' His answer to this problem is motivated by clear limitation of size considerations:

The study of the various antinomies we have encountered up till now has shown the following: in each of our examples it is possible to form more and more comprehensive sets, but the set of *all* the individuals does not exist; whatever the set envisaged (provided it exists), new individuals appear and a more comprehensive set necessarily appears also; one is thus confronted with an indefinite extension which carries neither stopping point nor bound. In what follows I will try to make more precise this rather vague notion of bound and absence of bound. (Mirimanoff [1917*a*], p. 48)

(Note the similarity between this passage and that from Russell [1906*a*], p. 152: see p. 223.) In a sense, Mirimanoff now works backwards. W is outlawed, and then various postulates are introduced which guarantee that none of the more comprehensive paradoxical collections can appear either. The result, the safety of sets in terms of size (at least for ordinary sets, for Mirimanoff restricts himself to these), is summarized in a theorem for ordinary sets:

(**)   . . . it is necessary and sufficient [for a collection of sets to be a set] that the ranks of these sets have a Cantorian bound. (Mirimanoff [1917*a*], p. 51)

That is to say (ranks being ordinal numbers) that these ranks have an upper bound. The notion of rank Mirimanoff uses here is exactly the modern notion:

Definition (*r*)—the rank of an ordinary set is the least ordinal number greater than the ranks of its elements. The rank of a node [urelement] is zero. (Mirimanoff [1917*a*], p. 51)

And he then shows that every ordinary set has a rank. The proof for the boundedness criterion is now carried through using various postulates which Mirimanoff states explicitly. The first two postulates we have met already:

1.·   Any sub-collection of a set is a set.
2.   No set has a subset cardinally equivalent to W.

(These are called Properties 1 and 2, [1917*a*], pp. 44–5.) But Mirimanoff requires more than just these. First, as I said, he assumes that he is working with only ordinary sets and that the urelements of nodes themselves form a set. Thus, we have the two working hypotheses:

3.   The postulate of ordinariness (in effect the foundation 'axiom').
4.   The urelements form a set.

There is also a postulate of 'distinction':

5.   The elements of every set are distinct.

(Properties 3–5 are found on p. 48 of [1917*a*].) In addition there are three set postulates (Mirimanoff's Postulqtes 1–3, [1917*a*], p. 49):

6.   If a set exists, so does the set of all its subsets.
7.   If a set exists, so does its union set.
8.   If a collection is a set so is every collection (cardinally) equivalent to it.

(Given 1 and 8, 2 can be simplified; see p. 192.) And one must add here also the assumption mentioned above:

9. To every ordinal α there is a corresponding S set α_S (i.e. an S set whose order type is α).

(In his [1917b], p. 210, Mirimanoff explicitly adds an extensionality postulate.)

I shall come back to the limitation of size aspect of these postulates later. For the moment, let us trace the proof of (**). Mirimanoff first shows that (**) holds for S sets. For if E is a collection of S sets whose ranks have an upper bound π, then all the elements of E belong to π_S (using 9), and so E is a set by 1. Suppose that the ranks of a collection E of S sets have no upper bound. Mirimanoff now argues as follows. Suppose π_S is an element of E. Let $A(\pi) = \{\alpha_S : \alpha_S \in E$ and $\mathrm{rank}(\alpha_S) < \pi\}$ and let $B(\pi)$ be the collection of all S sets β_S with rank larger than any in $A(\pi)$ and not greater than π. The $B(\pi)$ thus serve to fill in any gaps in the unbounded sequence of elements of E. $B(\pi)$ is certainly always a legitimate set, and indeed is a subset of $(\pi + 1)_S$. If E is a set so is the collection of all $B(\pi)$ for $\pi_S \in E$ by the replacement assumption 8. By 7 the union of these $B(\pi)$ is also a set, but this must be just W' which is forbidden by 2. Hence E is no set. The converse, that if the ranks of E have a bound, E is a set is immediate. The next step is to extend (**) to all ordinary collections. Let E be a collection of ordinary sets whose ranks have no upper bound. Then. the function $f(\alpha) = \alpha_S$ on the ranks yields a collection of S sets those ranks have no bound; this collection is no set and therefore E is no set either. Now let E be a collection of ordinary sets whose ranks do have an upper bound π. Let O_α be the set of all sets of rank α, so $A = \{O_\alpha : \alpha < \pi\}$ is clearly a set by 8, and so is the union of A by 7. But every element of E is in this union and so E is a set by 1. Thus, the criterion is established (Mirimanoff [1917a], p. 52).

Mirimanoff proves explicitly that the O_α must be sets. Each element of an O_π is a set of sets of smaller rank and thus a subset of the union of O_α for $\alpha < \pi$. Thus O_π is a sub-collection of the power set of this union, and therefore must be a set (using 8, 7, 6, and 1 respectively). So if O_α are sets for $\alpha < \pi$ then so is O_π. But since O_0 is just the collection of urelements, thus by 4 a set, *all* O_α must be sets (Mirimanoff [1917a], pp. 51-2. Thus, note that the power-set principle is explicitly used in showing how we climb up through the hierarchy.

What strikes one above all about Mirimanoff's whole conception is its modernity and its relation to the now standard idea of the von Neumann cumulative hierarchy R_α of well-founded sets. Mirimanoff does not show that his ordinary sets form such a hierarchy, but clearly each O_α (the set of sets of rank α and presumably of no lower rank) is just $R_{\alpha+1} - R_\alpha$. Whether his ideas had a direct influence on later developments is not clear. In his [1925], p. 404, n. 11, von Neumann mentions Mirimanoff's [1917a] in connection with the idea that there *can* be sets with descending membership chains. However, in his [1929], where he first introduces the notion of rank and the cumulative hierarchy (for the purpose of a relative consistency proof), he does *not* mention Mirimanoff. (Von Neumann proves that if his system without his 'limitation of size' axiom IV2 (see §8.3) but with a global axiom of choice and a replacement axiom is consistent so is the original system with IV2. To do this he shows the consistency of the weakened system with the axiom of foundation, and uses the cumulative hierarchy to show that any proper class is equivalent to the ordinals. See von Neumann [1929], pp. 503–5, or Bernays [1948].)

But in any case one should be wary of seeing too much modernity in Mirimanoff. Let us look first at his foundation assumption 3. What is clear is that Mirimanoff only uses this as a working hypothesis, that is as a means of producing a *partial* solution to the problem of the antinomies. He does *not* adopt it as an axiom. (It is suggested by von Neumann in his [1925], p. 412, but appears first as an explicit axiom only in Zermelo [1930], and then in Gödel [1940] and Bernays [1941].) Mirimanoff does in [1921] (pp. 30–1) make some remarks to the effect that the notion of rank of a set and the idea of sets being ranked is implicitly used by Cantor, Dedekind, and Hessenberg, which suggests the view that he is only extending and making explicit a principle natural to the Cantorian notion of set. But he gives no serious analysis of the Cantorian notion which would support this. And in any case the attitude in [1917*b*] is rather *against* 3, that is, it suggests that the ordinary sets are only *some* of the sets. And on pp. 212–13 he gives as an example of a non-well-founded set the cover of a book which has a picture of two children admiring what seems to be the same book, which has a similar picture and so on to infinity. One might question whether what Mirimanoff gives here is a set. But the example clearly indicates that non-well-foundedness is perfectly conceivable. Gödel, too, clearly suggests that Mirimanoff rejects the foundation principle as an axiom, for in [1944] (p. 140, n. 30) he says that Mirimanoff is working towards a theory of sets in which Russell's vicious circle principle (in the form 'no class can contain members *involving* this totality') does not apply. (Recall that Gödel distinguishes three versions of Russell's principle according to whether one focuses on Russell's terms 'definable only in terms of', 'involving' or 'presupposing' (see Gödel [1944], p. 133).) Gödel's remark deserves some discussion. It is, of course not at all clear what Russell meant by 'involving'. But certainly there seems to be a connection between this form of the vicious circle principle and well-foundedness. Certainly one can say with some plausibility that if $a \in a$, or $a \in b \in a$, then a clearly has members which *involve* a. The same could be said for any finite circle $a \in a_n \in \ldots \in a_1 \in a$ provided that we accept the transitivity of 'involvement'. Thus if such 'circular' sets exist, then the vicious circle principle mentioned by Gödel is violated. But infinite descending chains need not necessarily involve the initial set again at all. This is the case with Mirimanoff's book analogy. The book does *not* reappear, for each 'reappearance' is only the appearance of a non-identical isomorphic copy of the book. And here there seems at first sight no violation of the vicious circle principle. Nevertheless, let us examine Gödel's comment a little more closely. He says:

I even think there exist interpretations of the term "class" (namely as a certain kind of structures [*sic*]), where it [the vicious circle principle] does not apply in the second ['involvement form'] either. (Gödel [1944], p. 140)

And the footnote reference to Mirimanoff is added here. The mention of 'structure' is the interesting thing. For 'isomorphism structures' are just what Mirimanoff has been discussing before the remarks on those particular non-well-founded sets which contain isomorphic copies of themselves, isomorphism being used in the sense introduced in [1917*a*] (see p. 186 above). Mirimanoff writes:

What is common to isomorphic sets is their structure or their composition. If one abstracts from the particular properties which distinguish a set from its isomorphs, and if one retains only the details of its structure, one arrives at a new concept which one could call the structure type of this set, a concept closely related to Cantor's notion of power and ordinal number. In fact, it coincides with the notion of power or cardinal number when one abstracts from the structure of the elements of the set, which amounts to regarding these elements as urelements [indecomposable elements]. On the other hand, the order-types of well-ordered sets, Cantor's ordinal numbers, are derived directly from the structure types of certain special sets which I have called S sets. (Mirimanoff [1917*b*], p. 212)

Suppose now we assume that Mirimanoff intends these 'structure types' to be *internal* representatives of the isomorphism equivalence classes. (This is a natural assumption. For one thing this is exactly how he develops the S sets two pages later; see §8.1 where these are more fully discussed. For another, in [1917*a*], p. 43, Mirimanoff describes the process of representing isomorphic sets in exactly this way.) And suppose that J is a set which is isomorphic to one of its members J', and thus one of the special kinds of sets with infinite descending chains which Mirimanoff singles out. J corresponds to Mirimanoff's book, which as he says has as elements two children e_1, e_2, the background f, and J', the isomorphic image of J; likewise J' contains e'_1, e'_2, f', J'', etc. Thus $\ldots \in J'' \in J' \in J$ and all members in this sequence are isomorphic. If now T_J is the isomorphism type of J, then T_J must contain an element isomorphic to J', which must contain an element isomorphic to J'', and so on. But J, J', J'', $\ldots$ are all *mutually* isomorphic, so we must have $\ldots T_J \in T_J \in T_J$. Thus we arrive at a self-membered set, and thus a set which clearly has members 'involving' themselves. Therefore, if we admit the existence of Mirimanoff's isomorphism structures, then the existence of non-well-founded sets certainly *does* imply, as Gödel suggests, the violation of the vicious circle principle even in the 'second form'.

Mirimanoff does not treat the other postulates as genuine axioms either.

Rather, 'Although these postulates have been frequently used in the study of problems in the theory of sets, they are far from being evident, . . .' (Mirimanoff [1917a], p. 52). He discusses them further in [1921], but the discussion is unclear and certainly inconclusive. This tentative approach need not mean much; it may by itself just indicate a degree of caution. But unlike Zermelo and von Neumann, Mirimanoff does *not* treat the postulate system as a basis from which to begin the development of a pure set theory. Of course, he uses his postulates to show that the troublesome sets W, W', V_x', V, R', and R are excluded. And one might say, since his system has much in common with Zermelo's, that he could rely on Zermelo's work. But Mirimanoff allows ordinals as primitives, and is apparently not interested in showing that or how they can be treated purely set-theoretically, *even though* his results show quite clearly that the S sets can do all the work of the Cantorian ordinals. Indeed, Mirimanoff could not only simplify this theory, but also his solution to the paradoxes, by dispensing with ordinal primitives in favour of S sets. But this he manifestly does not do. But, more important, the postulate system is built up around and used solely for the solution to the paradoxes for ordinary sets and, as we have seen Mirimanoff does not accept the crucial postulate of ordinariness (postulate 3). It is not at all clear how much of this 'postulate solution' Mirimanoff would carry over to the non-ordinary sets.

Nevertheless, Mirimanoff's partial solution *is* a theory of limitation of size, and does warrant comparison with Cantor's theory of 1899 discussed in §4.1. In what sense is Mirimanoff's theory a limitation of size theory? The theory excludes W (the ordinals) and (via postulate 1) any more comprehensive class. Thus, in one sense, the class of ordinals acts as a kind of upper limit of comprehension. In this sense, 1 is clearly a size principle (see Chapter 7). But 2 and 8 strengthen this into a theory of limitation of *cardinal* size, by bringing in the notion of cardinal equivalence. And now the similarities with Cantor are quite evident. As Mirimanoff says ([1917a], p. 45), by virtue of 1, 2 can be replaced by:

2'.   No set is cardinally equivalent to the set W.

And indeed, since he adds 8, 2' can be further weakened to simply:

2". W is not a set.

So far this follows Cantor's arrangement since Cantor also has 2", 1 and 8 (his (1), (3), and (2) respectively in §4.1). Moreover, both Cantor and Mirimanoff proceed to establish criteria of sethood in terms of ordinals, Cantor directly and Mirimanoff indirectly. But now the difference of emphasis begins to emerge. In the Cantorian theory the alephs (or the ordinals) can be used to express the legitimacy of sets in terms of size because they are intended to capture everything (in the infinite) which is numerable or small, everything which falls under the 'Absolute maximum' of the Absolute. Although the theorem that every infinite cardinal is an aleph is not at all obvious, nor easy to prove, the result for Cantor is, so to speak, tautologous. Mirimanoff's result, on the other hand, gives no direct expression of legitimacy in terms of Cantorian size. Rather he shows that sets are arranged in levels according to rank, and legitimacy is expressed as a matter of being somewhere in the hierarchy. There is no indication that the individual sets within a given level are small or have a size which is numerically expressible at all. There is no concern in particular for the *breadth* of the ranks. Mirimanoff's criterion is a limitation of size one in the sense that to be a set, a collection can only gather elements from within a proper segment of the universe, that is within a limited part of the universe. Thus despite the presence of *cardinality* principles (that is 2 and 8), Mirimanoff's theory really expresses a limited comprehension view of sets—sets can be formed only through limited selection in the universe. And from Mirimanoff's postulates it can be established that these 'limited parts' of the universe are always sets. So far described the theory says nothing about how many levels, ranks (or ordinals) there are, that is how long the universe really is. Nor is there any indication about how broad or how big the levels are. Mirimanoff does use his power-set principle in describing why each level is actually a set. Thus, we know that potentially the levels could be very broad, at least cardinally bigger than any of the levels coming before. And by using power-set again, Mirimanoff *could* have shown that the universe is long too. For repeated use of power-set would enable the generation of (cardinally) longer and longer S sets. And since these correspond to ordinals, there must be correspondingly many ordinals too. But, as I have said, there is no concern for a scale of size with Mirimanoff, for the size of sets within levels.

Cantor's theory, as we have seen, fails to be explicit enough about the size of all sets or the length of the ordinal sequence, partly because of the intractability of the continuum problem and partly because, unlike Mirimanoff, Cantor lacks clear set existence principles. But there is a sense in which specifying cardinal size is the main goal of the theory, which is why Cantor and Jourdain focused on the aleph and well-ordering theorems as well as the Burali-Forti paradox. Indeed, their solution of the paradox was simply one component of the attempt to achieve the aleph theorem. Thus, one consequence of Mirimanoff's different focus with regard to limitation of size is the absence of any well-ordering or choice argument and, this is the reason why he cannot sharpen his 2 into the (*) of §4.1, namely:

A collection is not a set if and only if it has a subset equivalent to W.

For the converse of 2 cannot be established with just Mirimanoff's postulates. Suppose a collection X is not a set. Then the ranks of X have no upper bound by Mirimanoff's criterion (**). To prove the converse of 2 we must set up a map which takes a subset of X onto the whole ordinal sequence W. This we cannot do. To make the problem simpler, assume that $\forall \alpha \exists x \in X[\text{rank}(x) = \alpha]$. If we could choose an x with $\text{rank}(x) = \alpha$ for each α then the theorem would be established. But as long as infinitely many of the subsets Y_α of X, where $Y_\alpha = \{x \in X : \text{rank}(x) = \alpha\}$, have more than one element we need to use a well-ordering or choice assumption to make the selection. (In fact it is clear that we would need a version of the global axiom of choice.)

In modern set theory (with both the axioms of foundation and choice) both Cantor's aleph theorem and Mirimanoff's rank theorem are established. Thus, with respect to the universe as a whole there is both a Cantorian bound on the cardinal size of a set, and a bound on the 'spread' of its elements, that is, on how broadly they can be chosen from within the whole hierarchy. Such a dual characterization, or limitation of sets stems first explicitly from Zermelo [1930]. Cantor's solution in terms of the aleph theorem has the advantage that it does not depend on the postulate of ordinariness (foundation) which Mirimanoff could not accept. Nevertheless, use of this, as Mirimanoff first showed, makes the universe very tidy, indeed gives the ordinals a further role in the theory of limitation of size. But Mirimanoff is not in a direct line from Cantor to Zermelo [1930]. Rather he stands off to one side with a different view of limitation of size. Perhaps most important is that Mirimanoff's theory represented a substantial gain in clarity over the Cantorian theory of Jourdain and Russell in that it is quite clear in specifying the auxiliary postulates required for his limitation of size conception to be carried through. The dependence of limitation of size on auxiliary principles is the subject of Chapter 5.

PART 2

THE LIMITATION OF SIZE ARGUMENT
AND AXIOMATIC SET THEORY

INTRODUCTION TO PART 2

In part 1, I devoted considerable attention to the philosophical and metaphysical doctrines associated with the Cantorian conception and development of set theory. These doctrines were intended to justify, explain, and delimit the use of actual infinities in mathematics, acting both as a defence of actual infinity and a source of theoretical ideas about them. Cantor's own discursive treatment of infinity was intimately connected with his theological views, though I maintained that the core can be separated out and summarized in the three principles (*a*)-(*c*) (p. 7). The importance of these ideas in shaping the theory was profound, particularly in respect of the tendency towards set-theoretic reductionism, the structuring of the theory of number, and the attempt to circumscribe the domain of set-theoretic mathematics. In tracing the development of set theory, it is thus pertinent to ask how far Cantor's successors continued to rely on these same or similar ideas. Certainly the theological element in Cantor's thought has subsequently played no role. But the core principles (*a*)-(*c*) were and are of continuing importance (see, for example, §2.3(*c*)). Above all it is interesting to see how far the main axiomatizations of set theory, at first sight so different from Cantor's approach, used or absorbed these principles.

In this respect, principle (*a*) presents no great problem. As was explained in §1.2, (*a*) ('potential infinities are based on actual infinities') was a crucial principle in the 'arithmetization' of analysis well before Cantor began his detailed development of a theory of the transfinite. In his hands the successful reduction of potential to actual infinity in analysis was used both as an argument for the mathematical treatment of actual infinity *only*, as well as for the necessity of developing a detailed theory of the infinite (transfinite). And there is no question that (*a*) remained a central tenet in classical mathematics after Cantor, *nor* that it was adopted by the axiomatizers of set theory. Indeed in axiomatic set theory it takes a somewhat clearer form. What is axiomatized in effect is the notion of completed and fixed range or extension. The fundamental primitive is *set*—sometimes the notion of *class* is added—it being assumed that *all* mathematical objects are, or are representable as, such objects; and these sets (and classes) are treated extensionally as objects existing *completed*, and independently of and separated from, any intensions which may help us to define or recognize them. Infinite extensions are explicitly included via the adoption of an axiom of infinity, and it then follows that *all* infinities appearing in the system are of this completed form. Indeed, the only way to 'produce' or 'exhibit' an infinity here is to exhibit a set, and any notion of variation must be explained in terms of membership of some fixed completed extension. Effectively, then, (*a*) is embraced by set-theoretic reductionism.

The absorption of principle (*a*) by axiomatic set theory is thus reasonably clear. Axiomatic reductionism was also connected to principle (*b*) (Cantorian 'finitism'), for it takes extensional sets as single objects, as logical individuals. Thus, in a strong sense, reductionism regards all mathematical infinities as finite. Moreover, the *logic* of axiomatic set theory treats the infinite objects just like finite objects, at least as regards attribution of predicates and admission to ranges of quantification. But while this was what one might call the logical core of Cantorian finitism (finite and infinite sets are objects of the same kind), there was a heuristic side too, for it provides a partial answer to the question: given actual infinities, how should they be handled theoretically? This was of especial importance in Cantor's theory of number: infinite sets should have associated numbers just as finite sets do, and these numbers should be of the same kind. But, while Zermelo's and Fraenkel's reductionism clearly embraces 'logical' finitism, it was not at all designed to embrace the results of Cantor's numerical finitism. Thus, to a certain extent one can say that it does not take Cantor's finitism sufficiently into account. The same is not true of von Neumann's reductionism. What changed crucially with von Neumann was the decision about which objects treated in the pre-axiomatic system are *essential* mathematical objects. For von Neumann these included all the Cantorian numbers, and his attempt to regain the fruits of Cantor's numerical finitism (by reconstructing the Cantorian theory of number axiomatically) led to a quite different axiom system. The situation here is made more complicated by the fact that von Neumann did not just *add* to Zermelo's reductionism, but made substantial changes to its logical core as well. Thus, tracing the development of principle (*b*) involves a detailed examination of Zermelo's reductionism and von Neumann's modifications. These are thus central topics of discussion in Chapters 7 and 8.

The main umbrella topic of part 2, though, is the limitation of size idea. Cantor's principle (*c*) originally acted as a regulatory or shaping principle giving (admittedly rather minimal) information about the overall shape of the set-theoretic universe. However, in Chapter 4 we saw that with the advent of the paradoxes, this principle and the doctrine of the Absolute from which it derives suggested more precise regulatory principles. The doctrine of the Absolute indicates in advance *why* the Cantorian system *ought not* to fall foul of the kind of contradictions that were later discovered. The theory of absolute collections which this generated then became a source of ideas as to *how* paradoxes can be avoided. It is, of course, interesting to ask quite generally how far principle (*c*) and its derivatives are observed by the axiomatic systems which superseded Cantor. But since (c) gave rise to explicit suggestions for avoiding the paradoxes, and since the standard axiom systems *do* avoid them, it is of particular interest to ask what influence, if any, these suggestions had on the shaping of the axiom systems. In Chapter 4 I explained that the 'protective' core of Cantor's later ideas is the notion of limitation of size, an idea developed to some extent by Cantor himself, and then later by Jourdain and Russell. Ideally, limitation of size would yield a precise version of (c) which puts exact bounds on 'logical' finitism. That is, it would specify in terms of size precise conditions for the formation of those infinities which can legitimately and safely be treated as single objects. The central issue, and what I want to focus on in Part 2, is whether and how far axiomatic set theory can be said to pursue this line in avoiding the traditional paradoxes.

This question has both a foundational and a heuristic element. The major foundational question is whether limitation of size provides a plausible explanation of why the sets of the standard axiom systems do not include the traditional paradoxical sets. This is of interest not only because of the historical connection with Cantor, but also because many standard expositions of set theory adopt limitation of size (or something similar) as just such an explanation. This element of limitation of size is seriously examined in Chapter 5. I suggest that, although the limitation of size argument embodies an important statement about the set-theoretic universe, it is by no means a complete, plausible explanation of the axioms. But this raises another question, this time about the heuristic efficacy of limitation of size. The argument plausibly captures the essential nature of certain of the central axioms. So the question arises: did 'limitation of size' considerations at least guide the selection of these? This question is pursued in Chapters 7 and 8.

5

THE LIMITATION OF SIZE ARGUMENT

One way of giving a (weak) justification for the standard axioms of set theory is to put forward plausible arguments which explain why these axioms are chosen and how systems based on them improve on 'naïve' (or pre-axiomatic) set theory. One such argument is what I call the 'limitation of size argument'.

The argument is in two parts. The first tries to show that what is wrong with the troublesome, or potentially troublesome, sets or collections is that they are in some sense 'too big'. The second is to the effect that the basic axioms of set 'construction' (the axioms of union, of separation or replacement, and the power-set axiom) do not allow the construction of excessively big sets. Hence, the axioms are justified at least to the extent that if the systems based on them do go wrong, they will not go wrong for the same reason that naïve set theory does. (The argument could be called 'heuristic' because it purports to explain what guides the selection of axioms which are to take the place of pre-axiomatic disorder.) This argument has been employed, or hinted at, by many distinguished commentators. For example, Bernays (comparing set theory with the theory of types) writes

However, in axiomatic set theory the guiding idea for the avoidance of the paradoxes is not that of the vicious circle principle but of 'limitation of size' . . . (Bernays [1946], p. 77)

Weyl, describing the main ideas behind axiomatic set theory, states

Such classes as are too 'big' are here excluded from admission to the club of the decent sets, and in this way the disaster of the antinomies is avoided. (Weyl [1949], p. 232)

(Cf. Weyl [1949], p. 234 and Weyl [1946], p. 11.) And Quine: 'Zermelo's protection against the paradoxes consists essentially in eschewing too big classes.' (Quine [1969], p. 278.) (Cf. also Quine [1969], p. 284 and p. 200 below.) The argument has been most explicitly stated by Levy (though in this form it goes back to Fraenkel):

In ZF (the Zermelo–Fraenkel system) the guiding principle in writing down the axioms is the doctrine of limitation of size, i.e. we do not admit very comprehensive sets in order to avoid the antinomies. (Fraenkel, Bar-Hillel, and Levy [1973], p. 135)

Levy spells out very clearly the heuristic nature of the argument. In reconstructing set theory, he says, we choose certain instances of the axiom of comprehension to act as axioms. And

Our guiding principle, for the system *ZF*, will be to admit only those instances of the axiom schema of comprehension which assert the existence of sets which are not too 'big' compared to sets which we already have. We shall call this principle *the limitation of size doctrine*. (Fraenkel, Bar-Hillel, and Levy [1973], p. 32)

Indeed, Levy sees the exercise as very closely related to the modified comprehension principle of Russell's proposal (§4.3):

According to this point of view the axiom schema of comprehension is only in need of some tinkering to avoid the antinomies: the guide on how to do it will be the *doctrine of limitation of size*. (Levy [1979], p. 7)

(See also Levy [1979], p. 18.) Thus the axioms are weakly justified, or at least partially so: they cannot 'create' sets of the kind which have traditionally led to contradiction.

It seems to me that this argument does not and indeed *cannot* work. There are no good reasons for accepting that all sets are small, and on one of the usual characterizations of excessive size there are good reasons for accepting that some sets are excessively big. But before this, we have to look at the argument in more detail.

5.1. Fraenkel's argument criticized

In §4.2 we saw that Russell and Jourdain considered a simple modified comprehension principle (*MCP*) based on a notion of size. Now certainly axiomatic set theory is not based on a *single* such principle. However, the 'limitation of size argument' says that axiomatic set theory does, in effect, implement this proposal either partially or fully. According to the argument, it is not necessarily the case that all and only small collections be sets as *MCP* demands; however, it is the case that all sets created by the axioms are small. Thus Wang, after mentioning Russell's limitation of size theory, states:

Zermelo's system (1908) may be taken as a theory of limitation of size. Thus, a set term or propositional function can define a set only if its extension is not 'too big'. (Wang [1962], p. 385)

And Gödel [1944] says something similar in his famous essay on Russell. As Gödel describes it, Russell's limitation of size theory makes

. . . the existence of a class or concept depend on the extension of the propositional function (requiring that it be not too big), . . .

And

Axiomatic set theory as later developed by Zermelo and others can be considered as an elaboration of this idea as far as classes are concerned. (Gödel [1944], p. 132)

Quine actually goes further, implying that *MCP* is *fully* embodied in axiomatic set theory, at least in the system *ZF* which strengthens *Z* by adding the axiom of replacement:

We saw Zermelo's protection against paradoxes in his avoidance of too big classes. For the converse assurance, that classes fail to exist *only* if they would be bigger than all that exist, there was only partial provision in Zermelo's schema of *Aussonderung*. Full provision was added rather by Fraenkel and Skolem in the axiom schema of replacement. (Quine [1969], p. 284)

The claim that Zermelo's system embodies (or partially embodies) Russell's proposal goes back to Fraenkel's expositions of Zermelo's set theory in the 1920s and 1930s. Fraenkel's discussion is important for several reasons, not least because he was the first staunch defender of axiomatic set theory against the claims of intuitionism and constructivism. More important in the present context is the fact that Fraenkel's 'limitation of size argument' is the forerunner of more recent defences and expositions of set theory (by Gödel, Wang, and Shoenfield in particular) founded on the so-called iterative concept of set. Moreover, Fraenkel's arguments were taken over by Levy and used explicitly by him not only in his marvellous textbook [1979], but also earlier in the second edition of Fraenkel and Bar-Hillel [1958] (Fraenkel, Bar-Hillel, and Levy [1973]), undoubtedly the most important expository book on axiomatic set theory. I shall thus consider Fraenkel's argument in some detail.

Fraenkel tried to carry through both parts of the limitation of size argument. The notion of size adopted by Jourdain and Russell (and later von Neumann) directly involves infinity and focuses on the *cardinal size* of collections. However, Fraenkel's notion of bigness does not specifically involve infinity or cardinal size but rather the idea of 'comprehensiveness' or 'extension'. Fraenkel says in effect that a collection is 'too big' if it includes too much, or is of 'unbounded comprehension' or of 'unlimited extent'. This is associated with a view that, intuitively speaking, sets should be 'bounded', or of 'restricted extent'.

Fraenkel states the limitation of size hypothesis (*LSH*) in his [1927] (p. 115) as follows: '. . . contradictory sets arise from the formation of all-too inclusive sets.' (The phrase 'all-too inclusive' is also used in Fraenkel [1923] (not in the first edition [1919]), although Fraenkel does not present a limitation of size argument here.) Likewise in his [1924] he refers to the 'all-too comprehensive sets of the antinomies of Russell and Burali-Forti' (pp. 99-100). And in Fraenkel and Bar-Hillel [1958], p. 97, he writes:

When we considered the antinomies of the logical type, e.g. those of Burali-Forti and of Russell, we perceived as a common characteristic trait the un-limited extent of the sets concerned. . .

Having stated *LSH*, Fraenkel then claims that the sets created by the 'constructional' axioms of Zermelo's system are all of bounded extent.

One can distinguish two arguments which Fraenkel puts forward for this claim. Both are based on the 'iterative' nature of the constructional axioms in Zermelo's system, that is to say, on the fact that each instance of a constructional axiom is of the form $\exists x \forall z [z \in x \leftrightarrow \phi(z, a)]$ where a is a parameter which must be a previously 'constructed' set. The result of applying an axiom gives rise to a set *b* which can then itself figure as parameter in an axiom. Fraenkel's first argument plays on the presence of the parameter sets. In Zermelo's system we start from a small number of initial parameter sets—just the one given by the axiom of infinity will do—which are presumably *assumed* by Fraenkel to be of restricted extent. We then proceed to form new sets by the iteration of the 'constructional' axioms. Now, says Fraenkel ([1927], p. 116), these axioms cannot create the troublesome big collections because they '. . . allow us to form as further sets only those which closely depend according to both content and extent [*Umfang*] on already known sets, and never completely freely.' It is clear that this must apply to those axioms like separation which postulate sets *less* extensive than given sets. But the important part of Fraenkel's argument is that the same applies to the *expansive* axioms as well:

It does not matter whether the extent of the new set is expanded compared to the old (as for example happens by using the power-set axiom) or whether it is cut down (as happens with the axioms of *Aussonderung* and choice), *the scope of the new set is never boundless*. (Fraenkel [1927], p. 116; the italics are mine)

The same claim is made in Fraenkel's [1928*b*]:

. . . the formation of the sets afflicted with the logical antinomies, that is, the all-too comprehensive sets, . . . is excluded by our axioms. For, starting from one or several given sets, the axioms permit only the formation of sets which are either more restricted than given sets (the axioms of separation and choice) or which are more 'comprehensive' ['*umfassender*'] (through the axioms of pairing, union and power-set) but only relatively so, and both quantitatively and qualitatively only in a sharply circumscribed sense. (Fraenkel [1928*b*], p. 322)

(Cf. Fraenkel and Bar-Hillel [1958], p. 95.) The argument is therefore that the extension of the 'constructed' *x* is somehow limited (it is not stressed how) by its dependence on the set *a* which itself is taken to be of limited extent. Thus, at each stage the constructed sets are small or limited with respect to those already constructed. So while we do construct bigger and more extensive sets, the extension is never absolutely or dangerously unlimited. Thus:

Axioms of such a conditional character are appropriate for the goal of excluding antinomies, for the sets guaranteed by these axioms have an extension which refers to the extension of the sets introduced previously and not the absolute comprehensiveness characteristic of the sets which appear in the antinomies. (Fraenkel and Bar-Hillel [1958], pp. 33–34)

(Cf. Fraenkel and Bar-Hillel [1958], p. 95, and Fraenkel [1928*b*], pp. 330-1. For Fraenkel's particular defence of the union set axiom, see his [1927], p. 71, echoed by Levy in Fraenkel, Bar-Hillel, and Levy [1973], pp. 33-4.) Note that this is the form of limitation of size argument favoured by Levy:

The [limitation of size] doctrine says that we should use the axiom schema of comprehension only in order to obtain new sets which are not too "large" compared to the sets whose existence is assumed in the construction. (Levy [1979], p. 7)

And concerning the crucial axioms of power-set, union and replacement:

From the point of view of the limitation of size doctrine the justification of these axioms is rather obvious; the sets whose existence is claimed by these axioms are not much bigger than the sets whose existence is assumed by the axioms. (Levy [1979], p. 18)

The second Fraenkel argument again relies on the relative nature of the axioms, but this time stresses the ability to iterate 'constructions' at any stage and thus to create *new* sets. What Fraenkel says is that whenever we have a collection given by one of the axioms we can apply the axioms to this collection and obtain a different, indeed more extensive, collection. Hence the given collection cannot have been 'exhaustive', or 'unbounded'; consequently, the axioms cannot give rise to the collections responsible for the antinomies. While the first argument stresses the presence of a parameter, this argument depends on the fact that the collection b resulting from the application of an axiom can *itself* figure as the parameter base of a new construction. Fraenkel particularly stresses this approach in his defence of the power set axiom. He claims that, while the power-set of an infinite set might be of 'unsurveyable extent *in itself*', nevertheless it is of limited [*umgrenzt*] extent when compared with the universe of all existents, the 'outside world' as Fraenkel calls it. (See Fraenkel [1927], pp. 124-5, or [1928*b*], pp. 330-1.) In other words, however extensive a power set may appear, particularly in view of its impredicative nature, it still cannot be *over*-extensive because iteration of the constructional axioms shows how small it must be relative to the universe.

Gödel's explanation of why the axioms of set theory avoid the traditional contradictions is strikingly similar to Fraenkel's second argument. Speaking of the iterative concept of set Gödel writes:

This concept of set, however, according to which a set is something obtainable from the integers (or some other well-defined objects) by iterated application of the operation 'set of', not something obtained by dividing the totality of all existing things into two categories, has never led to any antinomy whatsoever; that is, the completely 'naïve' and uncritical working with this concept of set has so far proved completely self-consistent. (Gödel [1964], pp. 262-3)

(Cf. Fraenkel [1927], p. 116. This way of looking at sets mimics the historical development of the mathematical *use* of sets.) And he goes on to argue in a footnote that neither the universal class, nor any other unbounded class can be reached in this way:

It follows at once from this explanation of the term 'set' that a set of all sets or other sets of a similar extension cannot exist, since every set obtained in this way immediately gives rise to further applications of the operation 'set of' and, therefore, to the existence of larger sets. (Gödel [1964], p. 263, n.15)

(Cf. Wang [1977], p. 311.) And in his [1944] Gödel remarks that 'the most characteristic feature' of the limitation of size theory consists 'in the non-existence of the universal class' (p. 132). There are also similarities between Fraenkel's argument and other expositions of the iterative concept of set. I discuss some of these in Chapter 6. My purpose here is to show what is unsatisfactory about Fraenkel's arguments.

The central difficulty is in specifying what is meant by 'boundedness' or, conversely, 'unboundedness' or 'over-comprehensiveness'. One mistake Fraenkel makes, particularly with respect to the second form of his argument, is in closely tying boundedness to diagonalization and extendability. Fraenkel claims that all Zermelo sets are extendable, and this is the basis of the further claim (cf. Gödel's remark just quoted) that sets are 'small with respect to the universe', i.e. 'bounded' and therefore safe. The claim that all Zermelo sets are extendable relies on Zermelo's method of showing that there will always be sets lying *outside* any given set. Let a be a set; we can form a subset $u = \{x : x \in a \wedge x \notin x\}$ of a which is such that $u \in a \rightarrow (u \in u \leftrightarrow u \notin u)$. Hence $u \notin a$; consequently a is 'extendable'. This analysis is fortified by the observation that neither Russell's collection nor the universal collection are extendable in this way; we cannot diagonalize out of them by extracting subsets. Thus neither the troublesome Russell collection nor the universal collection can be present as sets in the Zermelo system. Fraenkel sometimes relies on a physical, even spatial, metaphor to express this contrast. Sets, he says, have around them 'a closed wall which separates them from the "outside world"'. On the other hand, to constitute Russell's collection, say, 'elements have to be taken from outside every wall, no matter how inclusive.' (See his [1927], pp. 123-4 or [1928b], p. 330.) (Cf. Gödel's comment on the universal set quoted above. This analysis goes back to Zermelo [1908b], p. 203.)

Fraenkel's argument stated in this form has certainly gained in clarity. But it does not work, largely because the notion of extendability is not strong enough to bear the weight put on it.

First, the argument is misleading because the analysis of extendability does not provide a uniform account of how Zermelo's systems avoids *all* the known contradictions. Specifically it does not, as it stands, cover the Burali-Forti case. (In one of his books Fraenkel, no doubt unintentionally, hinted that it *does*. For, after pointing out that Russell's paradox fails because 'each set m has at least one subset which is not an element of m' he adds: '. . . in this way, the possibility that a set could be constituted by the completely limitless assignment of elements is avoided from the beginning.' (Fraenkel ?1927], pp. 117-8.) The non-specific nature of this statement together with the description a few lines earlier of the set of all ordinal numbers as 'contradictory by virtue of its *limitless* extent' suggests a common solution to all the contradictions.) If Fraenkel's analysis is to be quite general, one must be able to show that all the known contradictory collections share a common non-extendability property.

But it is not at all clear that they do. Certainly they do not share the property of closure under the extraction of subsets on which Fraenkel bases his argument. This property is actually rather specialized; in particular, the collection *Ord* of all von Neumann ordinal numbers does not possess it, being rather meagre with regard to the subsets which belong to it. (Only segment subsets of *Ord* also belong to *Ord*.) So at least one contradictory collection fails to have the putative distinguishing property. (Note that if the axiom of foundation is adopted then the *only* collection with this property is the universe itself.) To cover *Ord* as well 'non-extendability' would have to be defined in a more liberal way. For example, one could adopt the definition '*x* is non-extendable iff *x* is closed under subset extraction or cardinally equivalent to such a collection' and simply decree that *Ord* is cardinally equivalent to the universe. But to put such emphasis on the importance of the ordinals was certainly against the spirit of Fraenkel's approach to set theory. (See Chapter 8.)

More generally, Fraenkel's discussion and particularly his 'wall' metaphor, suggests perhaps that closure under diagonalization (of which closure under extraction of subsets, or non-extendability, is one aspect) may be at the root of the contradictions. But again this is not of much help. Of course, all the guilty collections *are* closed under diagonalization in various ways. But they do not seem to share a *common* closure property. Moreover, it cannot be merely such closure which causes them to go wrong. For example, *Ord* is closed under ordinal sequences, in the sense that no (set) sequence of ordinals can exhaust *Ord*. In other words, if (crudely) we think of *Ord* as the 'limit' of the sequence of ordinals, the cofinality of *Ord* is *Ord* itself. But there are ordinals (i.e. legitimate sets) which are closed in a similar sense. For example, as we saw in §2.3(*a*), if f is any enumeration of countable ordinals (i.e. $f \in \omega_1^\omega$) then $\exists \lambda \in \omega_1 \forall n \in \omega$ $[\lambda > f(n)]$, i.e. the cofinality of ω_1 is ω_1.

However, there is a much more serious difficulty with the reliance on extendability. Suppose we accept that all contradictory collections are non-extendable; the point of the Fraenkel approach is then to use extendability as a guarantee of safety. But it seems that one must already use a safety assumption in arguing for the thesis that all Zermelo sets are extendable. Thus the point of the Fraenkel argument is quite lost. Suppose X is a Zermelo set. To show that X is extendable one has to be able to show that X is not exhaustive. In other words one must be able to diagonalize out of X; for this one needs to apply the axiom of separation to obtain the sub-collection $Y = \{x : x \in X \land x \notin x\}$. (One could use the power-set axiom to obtain a strictly bigger collection. But the crucial step, showing that it *is* strictly bigger, relies on just the application of the axiom of separation mentioned here.) Certainly one must depend on the assumption that X is a set. But one needs more than this; one needs to assume that X is a legitimate consistent set, and equivalently, that the axiom or axioms giving rise to X are consistent. For surely we do not want to establish the extendability of X by *default*, by arguing from a contradiction. But now we are assuming that X (in effect the Zermelo system) is *safe*, and this is what we are supposed to be giving a plausible argument for. This circularity undermining Fraenkel's argument is really what lies behind Skolem's remark on the limitation of size idea:

It is sometimes asserted that to put everything in order all that is required is to prohibit the formation of sets which are too 'big', sets like the set of all sets, of all cardinal or ordinal numbers and so on. However, it is easy to see that such a prohibition is powerless without a more precise specification of set-theoretic methods. For example, it would no longer be possible to introduce infinite sets in a non-circular way. If one wishes, say, to form the set of all finite whole numbers, then because of the prohibition laid down this is only possible if one already knows that these numbers do not constitute *all* ordinal or cardinal numbers, i.e. if one knows in advance that there are infinite sets. (Skolem [1929], p. 214)

Thus, because the intuitive idea of boundedness is too closely tied to that of extendability, it seems that one cannot successfully argue, as Fraenkel and Gödel wish, that no Zermelo set is large with respect to the universe. Worse still for the second limitation of size argument, one *can* plausibly argue that this central thesis is false because there are grounds for accepting that the extent of an infinite power-set *is* indissolubly linked to the *unlimited* extent of the universe. This I will argue below.

What of Fraenkel's first argument (used by Levy) that all Zermelo sets are small with respect to the parameter sets used in their construction? Is this sound? In the case of some of the axioms there are perhaps some grounds for accepting this, particularly if one treats 'relative smallness' cardinally. For example, Levy remarks:

If we are given a set a and a collection of sets which has no more members than a it seems to be within the scope of our guiding principle to admit that collection as a new set. (Fraenkel, Bar-Hillel, and Levy [1973], p. 50)

Certainly a sub-collection X of a set a has 'no more' members than a; thus, the separation axiom is legitimate. If further, as Levy assumes, X is taken to have 'no more' members than a when there is a functional relation taking a onto x, then Levy's principle immediately justifies the replacement axiom. The union set axiom is rather harder to justify along these lines, for it is certainly expansive, the union of a frequently having more members than a. Fraenkel's and Levy's justifications of union are much too weak (see Fraenkel [1927], p. 71, or Fraenkel [1928b], p. 277, Fraenkel and Bar-Hillel [1958], p. 35, and Fraenkel, Bar-Hillel, and Levy [1973], pp. 33–4), but they can be strengthened. Perhaps the best method would again be to rely on cardinality arguments. For if one knows the cardinality of a collection X and the cardinality of each of its members one can quite easily assign a cardinal bound to the union of X. However, the situation is quite different when we consider the power-set axiom. In the first place, there are no good grounds for assuming that the power-set $P(a)$ of an infinite set a is small, either with respect to the universe or with respect to the parameter a. Certainly the burden of proof lies with the proponents of the limitation of size argument, and neither Fraenkel nor Levy give such proof. (As we have seen, Fraenkel even admits that an infinite power-set may have 'unsurveyable extent'.) Note that the appeal to cardinality considerations would not help here, for in the light of Cohen's work there is no way of estimating the cardinality of any *infinite* power-set, not even the smallest, $P(\omega)$. But not only is there an absence of positive argument in favour of the smallness of power-sets, it seems to me that there are good reasons for taking the contrary view, namely that infinite power-sets are extremely large.

The importance of any $P(a)$ in classical mathematics is precisely that it represents the set of all extensions of properties defined on a, an extension belonging to $P(a)$ if it is sanctioned by the axiom of separation. The nature of the ϕ appearing in the axiom of separation is quite unrestricted. Indeed, if we take a second-order view of properties the number of possible ϕ is vast. For example the ϕ can involve sets from anywhere in the universe as parameters. Therefore, since the universe is taken to be of 'unrestricted extent' the number of *possible* extensions of properties (i.e. members of $P(a)$) is also quite unrestricted. Of course, when a is a finite set we can avoid this difficulty by proving that the number of *different* extensions is finite. In other words, although there is a vast number of possible *properties* defined on a, all but a finite number of them coincide with one another extensionally. Thus, in this case $P(a)$ is of restricted extent. But whenever a is infinite the number of different subsets cannot be enumerated; indeed we have no knowledge at all of how many of the enormous variety of possible properties with domain a will be extensionally equivalent. In other words, taking this second-order view of properties, it seems that the extent of the power set $P(a)$ is still indissolubly linked with the (unrestricted) extent of the universe.

We can reach this conclusion by a rather different route without taking a second-order view of the properties. It is known that the term $P(x)$ is an absolute term. To say that a term t is absolute is to say, informally, that to ascertain its value it is enough to ascertain it in any transitive class or set containing all the objects to which the formulas making up t refer. (Each set theorist seems to have his own definition of absoluteness; for one clear account, see Bell and Machover [1977], pp. 502-8 (see also Gödel [1940], p. 42, or Drake [1974], pp. 84-8).) Thus, if t is absolute, in the ascent of the cumulative hierarchy some stage will be reached beyond which the addition of sets will not affect the value of t. Conversely, if t is not absolute then at no stage in the ascent of the cumulative hierarchy will it be possible to assess the true value of t. This applies to $P(x)$. To be sure that one has 'captured' all subsets of x one has to take into account the whole universe; if one considers only a part of the universe one might get a false picture. As Cohen [1966], p. 94, puts it: 'The relation $y = P(x)$ is not absolute since y may be the set of all subsets of x in the transitive class B yet not be the true power set.' And thus, to quote another source (Bell and Machover [1977], p. 509): '. . . the size of the power set $P(u)$ of a given set u is proportional not only to the size of u but also to the "richness" of the entire universe . . .'. Not only does this appear to refute the claim that all sets are small with respect to the universe, it runs directly counter to the Fraenkel–Levy claim that ZF admits only sets 'which are not too "big" compared to sets we already have' (Levy; see p. 199). One is led to suspect that there is simply *no link* between the size (comprehensiveness) of an infinite a and the size of $P(a)$. This, of course, is underlined by the independence of the generalized continuum hypothesis (GCH) from ZF and subsequent independence results; there is no link between the cardinal size of a and that of $P(a)$ (apart from certain minimal limiting information; see §§2.3(b) and 2.3 (c)).

The subject of *cardinal* size, which Levy focuses on, raises again the question of whether ordinals can be used as an absolute measure of set smallness. According to Cantor, legitimate sets must be countable by ordinals. This is completely captured in ZFC by the so-called enumeration theorem $\forall x \exists \alpha [x \sim \alpha]$. Can one not argue, therefore, that if being equivalent to an ordinal is an indication of smallness then all Z or ZF sets, including power-sets, are small, even though in the case of infinite power-sets we cannot pinpoint any specific ordinals which count them? This is quite correct, but it does not get us anywhere. In the first place, there is no intrinsic reason why ordinals *must* be small. For instance, in von Neumann's system *Ord* itself is an ordinal, indeed the *greatest* ordinal; but according to the limitation of size argument *Ord* (being a proper class) should be *large*. Of course, what we are interested in are *set* ordinals. But how can we guarantee that these are small? If they are simply decreed to be small the argument for the smallness of all sets becomes otiose; indeed, the enumeration theorem would say no more *vis-à-vis* smallness than does the trivial 'x is small if $\exists y [x \sim y]$'. If the set ordinals are not just decreed to be small, their smallness has to be argued for. Now one might successfully argue that all constructive ordinals, or even all countable ordinals, are small. But there seems to be no argument for the smallness of uncountable ordinals which would not involve the assumption that power-sets are small. And, of course, such an assumption would render circular any subsequent argument for the smallness of power-sets. Hence, while the involvement of the ordinals appears to shift the central problem, it does not do so in any helpful way. As Russell remarked in the context of Jourdain's ordinal-based limitation of size theory (see §§4.2 and 4.3):

A great difficulty of this [Jourdain's] theory is that it does not tell us how far up the ordinals it is legitimate to go. It might happen that ω was already illegitimate: in that case all proper classes [i.e. sets!] would be finite . . . Or it might happen that ω^2 was illegitimate, or ω^ω or ω_1 or any other ordinal having no immediate predecessor. We need further axioms before we can tell where the series begins to be illegitimate. (Russell [1906*a*], p. 153)

5.2. The explanatory role of limitation of size

This discussion shows that the limitation of size argument *by itself* does not do anything to make the classical power-set axiom more plausible or more intelligible. Stated crudely, from this argument alone, we have no positive reason to assume that even only *one* application of the power-set axiom to an infinite set will not exhaust the whole universe. There is no reasonable *internal* criterion of 'extent' or 'size', so the limitation of size conception has to use a notion of *relative* size. And there is no good reason to suppose that a power set $P(a)$ always remains small with respect to any measure not involving $P(x)$. Without a special treatment of power-sets, or a recognition that they fall *outside* the scope of the argument, blithe acceptance of the limitation of size argument simply obscures the nature of the power-set axiom. (Its nature is even further obscured in some presentations of the iterative conception of set closely related to Fraenkel's arguments.)

The power-set axiom just is a mystery. Axiomatic set theory was constructed as much to capture the structure of Cantor's transfinite number-scale as it was to capture 'classical' mathematics. Yet we are still no closer to knowing how these domains fit together. Indeed, given the effort and ingenuity expended *since* Cantor's time, one has to say that the mystery is *deeper* today than it ever was. This mystery clearly has something to do with why the limitation of size argument does not work, and in this respect the two following passages from Cohen and Scott are interesting:

A point of view which the author feels may eventually come to be accepted is that CH [the continuum hypothesis] is *obviously* false. . . $\aleph_1$ is the set of countable ordinals and this is merely a special and the simplest way of generating a higher cardinal. The set C [the continuum] is, in contrast, generated by a totally new and more powerful principle, namely the Power Set Axiom. It is unreasonable to expect that any description of a larger cardinal which attempts to build up that cardinal from ideas deriving from the Replacement Axiom can ever reach C. Thus C is greater that $\aleph_n$, $\aleph_\omega$, $\aleph_\alpha$ where $\alpha = \aleph_\omega$ etc. This point of view regards C as an incredibly rich set given to us by one bold new axiom, which can never be approached by any piecemeal process of construction. Perhaps later generations will see the problem more clearly and express themselves more eloquently. (Cohen [1966], p. 151)

Scott, commenting on the proliferation of models of *ZF* after the discovery of Cohen's method of forcing, remarks:

I still feel that it ought to be possible to have strong axioms which would generate these types of models as submodels of the universe, but where the universe can be thought of as something absolute. Perhaps we should be pushed in the end to say that all sets are *countable* (and that the continuum is not even a set!) when at last all cardinals are absolutely destroyed. (Scott [1977], p. xiv)

Thus, the continuum evades all our attempts to characterize it by size (Cohen), so maybe we should start with this transcendence as a datum (Scott).

But this negative note by no means marks the end of our interest in limitation of size. For one thing, the fact that it is so widely appealed to prompts one to ask: if the argument is not a plausible justification of the axioms, can we at least extract from it a reasonable explanatory core? Reference to Cantor's principle (c) is important here. This gave information about the limitations of mathematization. This does not at all mean that it gives precise information about what are mathematical objects or that it provides a *justification* for taking certain things as mathematical objects. Rather, given a certain adequate list of objects, it puts a restriction on what else can be subject to mathematization. Does the limitation of size argument do something similar if we look behind its presentation as a justification of the axioms?

I think it does. In comparison with Cantor's ideas about what is included in the set-theoretic universe, axiomatic set theory is very precise. And we know that hitherto the axioms of ZF have been quite sufficient for ordinary mathematical practice. We can interpret the limitation of size argument as explaining why we cannot get beyond these to the large inconsistent sets. The discussion here also connects with the heuristic role of limitation of size. Let us take this first. The circularities which vitiate the limitation of size arguments derived from *hidden* assumptions that certain key sets (most importantly, at least one infinite set, and all power-sets) are legitimate or *small*. If these assumptions are *openly* adopted, or if one can provide independent arguments for them, then the idea of limiting the size of sets can guide the search for a substitute for comprehension, that is, for an axiom which 'fills out' the universe by allowing arbitrary properties to be converted into sets in some way. Since the axioms of infinity, power-set, and union are the most difficult to justify on grounds of size, let us *suppose* that these permit only small (legitimate) sets. Then, adopting a comprehensiveness view of size, the axiom of separation can be regarded as capturing the principle: given any predicate ϕ, the collection $\{x : \phi(x)\}$ is a set if it is of restricted extent. This can be seen as follows. The power-set and union-set axioms assert that the extensions of two particular predicates ($\exists y[y \in a \land x \in y]$ and $\forall y[y \in x \to y \in a]$) always constitute sets for any particular set parameter a. The axiom of separation then allows the extension of infinitely many more predicates to be converted into sets, namely all predicates whose extensions are bounded by a set. But the point is now this: because these extensions are all bounded by sets and sets created by the power-set and union axioms are *assumed* to be of restricted extent, it follows that sets created by the axiom of separation *must* be of restricted extent also. Therefore, iteration cannot lead out of the class of small sets. Various other comprehension principles could be justified if one changes one's view of size. For example, if one adopts Levy's cardinality approach and assumes that countable sets and all power-sets are small then the axiom of replacement becomes a plausible and intelligible substitute for comprehension. (As remarked above, on cardinality grounds one might regard union as already justified by limitation of size.)

This restricted, less dramatic, use of limitation of size strongly reflects its historical importance, for I suggest in Chapters 7 and 8 that the main heuristic role of the idea of 'limiting the size of sets' was as a general heuristic guide in the search for comprehension principles *once* certain key axioms (e.g. infinity, power-set) were adopted. The interesting differences between the way the various pioneers (Zermelo, Fraenkel, and von Neumann) used this idea are brought about by their quite different views of size. Note that von Neumann explicitly adopted the *assumption* that the axioms of infinity, union and power-set create only small sets (see §8.3).

The sensible explanatory core of the limitation of size argument is already contained in this heuristic and historical core. While some instances of comprehension are adopted in any case (partly on grounds of mathematical necessity) and their output is assumed to be small, *full* comprehension itself is replaced by a *restricted* comprehension based on size. It is now guaranteed that we can never come out of the class of small collections, for we can always apply the axioms to prove that there is a small set *beyond* any given one. In other words, the argument says, we never begin to reach collections of the order, or the *size*, of the universe or the paradoxical classes. What this shows is that the assumption of the legitimacy and smallness of power-sets allows the terms 'small' and 'large' to be invoked as useful metaphors in drawing a type distinction between sets on the one hand and the universe and collections of the same order on the other. Since (as can be shown by applying the axioms) none of the contradictory collections can be included in or is equivalent to a set, then we know that they cannot be sets—they are indeed too large, that is, outside the scope of the axioms. The axioms therefore capture the original intuition that the contradictory collections are too big, based on the idea that they go wrong because they include too much. The metaphor thus clarifies the 'shape' of the set-theoretic universe, what lies inside it and what outside.

This, I suggest, is the most appropriate way to interpret Fraenkel's limitation of size arguments, particularly in the light of his statement that infinite power-sets are of 'unsurveyable extent in themselves', yet 'of limited or circumscribed [*umgrenzt*] extent with respect to the outside world' (Fraenkel [1927], p. 124). While a great deal of weight does rest on the assumed smallness of power-sets here (similarly on the axiom of infinity) this does not after all render the arguments vacuous, provided that we do not look to them as a plausible explanation of these particular axioms. The limitation of size argument *is* a useful metaphor. It does to *some* degree explain why the traditional paradoxes are avoided. Moreover, there is now a clear parallel between the role played by the limitation of size argument in this conditional form and the role played by Cantor's doctrine of the Absolute with respect to the transfinite.

Limitation of size

(i) Type distinction between small and large.

(ii) The small sets are adequate for mathematics; certain key sets necessary for mathematics are assumed to be small.

(iii) The overly large collections are excluded from the domain of mathematics, as can be shown by the 'piecemeal' or iterative nature of the axioms.

Cantor's Absolute

(i)′ Type distinction between finite/transfinite and the Absolute.

(ii)′ The finite/transfinite exhausts the realm of mathematical objects; certain key mathematical objects are assumed to lie within the transfinite. This assumption was at first implicit and then made openly in Cantor's 'axiom of the transfinite'. (See §4.1, p. 175. Recall that Jourdain also had to assume that the continuum is a 'consistent' aggregate (p. 184).)

(iii)′ The Absolute is excluded—the iteration of normal mathematical operations to the transfinite can never reach it.

Note that, thus stated, neither Cantor's doctrine of the Absolute nor the limitation of size argument yield much information about the nature (absolute size) of the objects which lie inside the universe.

5.3. The power-set axiom

The above criticism and reworking of the limitation of size argument serves to isolate the power-set axiom since it cannot be made plausible on grounds of size alone. The difficulties outlined above show that it is not much good trying to justify this axiom by arguing that its extent is small or limited. Various other metaphors for 'small' have been used (for example, bounded, completed, surveyable), but, as I show in Chapter 6 (§6.3) these do not help either. There is no sense in which the range of a power-set is intuitively graspable in such a way. Any attempt to construe it so runs into the barrier of impredicativity. But if such *direct* means of legitimizing power-sets fail, do we have any *indirect* means? In particular, is there any clear sense in which classical power-sets are legitimate when looked at from the framework of Cantorian principles? This is important because, as we have seen, the limitation of size argument, which is weak on the power-set axiom, itself fits well with Cantorian principles.

Much of Cantorian set theory was developed on the basis of analogy with the finite; that is to say (by principle (*b*)) our guide as to what happens in the transfinite realm is what happens in the finite, at least so far as 'fundamental' mathematical operations are concerned. Cantor himself had no clearly formulated power-set principle. But the development of set theory *after* Cantor, under the pressure of set-theoretic reductionism, showed clearly that power-set must be a 'fundamental' Cantorian principle. For, if one is dealing *only* with sets, use of a power-set principle is the only way we have of dealing with properties of sets and relations over sets. If a set *a* is taken as a legitimate object, then arguably we must permit properties and relations over it to be legitimate objects also, for without these there can be no serious investigation of *a*. But reductionism says that these must be extensional objects, i.e. sets. In the case of properties this forces us to consider subsets of *a*. Thus far, we do not know how many or which properties exist, and there is apparently as yet no pressure to consider them as collected together into a (power-) set. But Zermelo's work on the well-ordering problem showed that this *must* be the case when relations are considered. In his conception an ordering relation on *a* is a collection of subsets of *a* ordered by inclusion (see §7.3). If this collection is to be an object, it must be a set, and the only way to show that it is a set is to separate it off from the collection of all subsets of *a*, which means that this latter collection must itself be a *set*. Since Zermelo, the reductionist treatment of relations has changed somewhat, but the point remains: use of power-sets is essential. It is essential, that is, if one regards relations as essential. But for Cantor and thus for Cantorian set theory, given a set *a*, treatment of its ordering relations is not just essential but lies at the most fundamental mathematical level, for it is through orderings on *a* that we approach its *numerability*. Thus, through axiomatic reductionism it became clear that the power-set principle must be a fundamental principle of Cantorian set theory; without it, the mathematical analysis of sets cannot proceed. Furthermore, if the legitimacy of finite sets is assumed, then the power-set principle is certainly safe in the finite realm. Its fundamental nature and its finite legitimacy together therefore indicate, according to Cantorian 'finitism', that the principle should be adopted in the transfinite realm as well.

Of course, the mathematical *necessity* of applying power-set to infinite sets does not say anything at all about the *legitimacy* of so doing. Moreover, nothing so far said indicates that we must be dealing with the full impredicative power-set. Is there any evidence that this latter is legitimate in the Cantorian transfinite realm? In a certain sense, there is. The reason here is quite simple, namely the crucial involvement of the impredicative power-set axiom in the set theoretic construction of the continuum. (This is, in effect, one of Fraenkel's justifications for assuming the power-set principle.)

Actually the position of the power-set principle is somewhat stronger than this rather bald last statement suggests. After all, there are other ways of building the continuum. But to repeat some of the discussion of §1.2, before the set-theoretization of analysis, the precise structure of the linear continuum was unclear—sufficiently unclear that it was a matter of doubt whether the Bolzano–Weierstrass property could be assumed or not. Of course, geometrical intuition was a strong guide, but not sufficiently strong (see Dedekind [1872], p. 1). Cantor's and Dedekind's application of set-theoretic methods succeeded in giving a model of the continuum that was reasonably clear, or at least reasonably successful, in such a way that previously strong geometrical intuitions are not violated. Now the power-set principle is not explicitly applied in the original work of Cantor and Dedekind on the continuum. But one might say that it is implicitly, and indeed 'all but' applied in the following sense. According to Dedekind (see Chapter 1, p. 29), the main purpose of the 'arithmetization' of the continuum was to define real numbers in such a way that the Bolzano–Weierstrass theorem is provable. His definition (or Cantor's) achieves this. But the proof uses exactly the kind of impredicative subset formation which allows one to prove that infinite power-sets are uncountable, that is to say 'classical' (impredicative). Thus assumptions equivalent to the power-set principle were crucially involved in this first clear, and apparently consistent and intelligible, 'arithmetic' definition of the linear continuum.

This does not prove the legitimacy of the power-set principle. For the argument is not: we have a perfectly clear intuitive picture of the continuum, and the power-set principle enables us to capture this set-theoretically. Rather, the argument is: the power-set principle (or principles which imply it) was revealed in our attempts to make our intuitive picture of the continuum analytically clearer; in so far as these attempts are successful, then the power-set principle gains some confirmatory support. At least the development of the set-theoretic continuum shows that one application of the power-set principle to an infinite set has productive and intelligible, although strong, consequences.

6

THE COMPLETABILITY OF SETS

6.1. The iterative conception

Fraenkel's limitation of size arguments have some important descendants amongst modern defences of the Zermelo-Fraenkel axioms. In this section I want to concentrate on two of these defences, given by Shoenfield and Wang, and on some of the issues they raise.

The second form of Fraenkel's argument, which relies heavily on the notion of the extendability of sets, can be paraphrased in the following way. Sets are built up in successive layers or stages by successive application of the axioms; there can be no *final* stage, because the iterative nature of the axioms ensures that any given stage can be transcended; because of this, neither the universe nor any other similarly extensive collection can be reached, or *completed*, within the system of sets. The crucial distinction in this paraphrase appears to be between the *accessible* or *completable* sets or stages and the *inaccessible*, *incompletable* collections, a class which is supposed to include all the traditional contradictory collections. 'Completable', 'accessible', 'incompletable', and 'inaccessible' are here acting as substitutes for 'bounded' (or 'small') and 'unbounded' (or 'overly large'). Using these terms, one can also give a rendering of the first Fraenkel argument. According to this, the contradictory collections are regarded as incompletable; the axioms, on the contrary, generate completable collections, given, that is, the availability of the 'completed' parameter set *a*. But note that any attempt to justify particular axioms along these lines has to explain in what way the sets generated by the axioms *are* completed, just as Fraenkel's argument ought to justify why the generated collections are small.

Shoenfield gives a form of this paraphrase in his [1967] and [1977] (pp. 238–40 and 322–7 respectively). (Another important and closely related account is in Boolos [1971] (see pp. 221-3 below).) His procedure (see [1977], p. 322) is to describe and justify the von Neumann cumulative hierarchy as the natural universe of sets and to argue that all the *ZF* axioms are *true* in this hierarchy. The description of the hierarchy rests heavily on the notion of *stage* and of *precedence* among stages (Shoenfield [1977], p. 323), and this is what makes Shoenfield's concept of set specifically an iterative one. Sets are said to be formed at certain stages and these then give rise to later stages, etc. This crucial notion is stated thus:

Sets are formed in *stages*. For each stage S, there are certain stages which are before S. At each stage S, each collection consisting of sets formed at stages before S is formed into a set. There are no sets other than the sets which are formed at the stages. (Shoenfield [1977], p. 323)

(Note that Shoenfield here makes the stages *cumulative*.) Since the cumulative hierarchy is well-founded, one would expect the axiom of foundation to be built into Shoenfield's stage view. And indeed he claims that his hierarchy and his sets *are* well-founded:

When we are forming a set z by choosing its members, we do not yet have the object z, and hence cannot use it as a member of z. The same reasoning shows that certain other sets cannot be members of z. For example, suppose that $z \in y$. Then we cannot form y until we have formed z. Hence y is not available as an object when z is formed, and therefore cannot be a member of z. (Shoenfield [1977], p. 323)

This statement is particularly important, since Shoenfield explicitly extends the relation of precedence to the sets and makes it a relation of precedence of actual *formation*. Note that properly speaking this indicates an intrusion of constructivist notions, for it appears to rule out impredicatively defined sets. (I take this up later.) Moreover, it is just this 'priority in formation' which is appealed to in Shoenfield's explanation of how the contradictions are avoided. Take his explanation of why Russell's contradiction fails:

A closer examination of the [i.e. Russell's] paradox shows that it does not really contradict the intuitive notion of set. According to this notion, a set A is formed by gathering together certain objects to form a single object, which is the set A. Thus before the set A is formed, we must have available all of the objects which are to be members of A. It follows that the set A is not one of the possible members of A; so the Russell paradox disappears. (Shoenfield [1967], p. 238)

Similar explanations could be given for why the other contradictions fail.

Having adopted the notions of stage and precedence, Shoenfield proceeds to the crucial question:

When can a collection of sets be formed into a set? For each set x in the collection, let S_x be the stage at which x is formed. Then we can form a set of this collection iff there is a stage S which follows all the S_x. However, such a stage may fail to exist. For example *every* stage may be an S_x. Thus we want an answer to the following question: given a collection of stages, under what conditions is there a stage which follows every stage in the collection? Since we wish to allow a set to be as arbitrary as possible, we agree that there shall be such a stage whenever possible *i.e. whenever we can visualise a situation in which all the stages in the collection are completed*. (Shoenfield [1967], p. 239)

(The italics here are mine.) Something similar appears in Shoenfield's [1977] treatment:

. . . there is a stage after all the stages in the collection of stages S provided that we can imagine a situation in which all of the stages in S have been completed. In the case in which S contains all stages, we cannot imagine such a situation, since we can always imagine a further stage. (Shoenfield [1977], pp. 323–4)

The introduction of the notions of *completion* and *completability* is the second intrusion of constructivist terminology, or terminology suggesting our ability to carry out an operation. Its purpose here, or so I suggest, is to endow the axioms with the inherent soundness and reasonableness of constructivist notions. The problem is that, although Shoenfield uses constructivist terms (formation, completable, etc.), the 'operations' he describes are not constructive ones. It transpires that what Shoenfield tells us are completed collections are just *assumed* to be completable; their completability is not established by appeal to plausible argument. Indeed, one suspects that if pressed Shoenfield would have to base his notion of completability on the notion of *ZF* sets. I shall come back to these criticisms later. First, let us look at how Shoenfield justifies some of *ZF* axioms.

How do we know which collections of stages are completable and which are not? In general we do not know, but we can, apparently, take some completions for granted:

Specifically, there are three cases in which our vague answer leads us to conclude that there is a stage after each member of the collection of stages S. The first is the case in which S consists of a single stage. The second is that in which S consists of an infinite sequence $S_0, S_1, \ldots$ of stages. The third case is that in which we have a set x and a stage S_y for each y in x, and S consists of the stages S_y for y in x. (Shoenfield [1977], p. 324)

In the first two cases, says Shoenfield, '. . . it is clear that we can imagine a situation in which all of the stages in S have been completed'. But is it clear? The first case might be, but the second is not. Some constructivists would argue that *no* infinite collection can be completed. (This assumption that a denumerably infinite S can be completed matches the assumption in the limitation of size argument that a denumerably infinite set is *small*.) In the third case, Shoenfield provides an argument:

Suppose that as each stage S is completed, we take each y in x which is formed at S and complete the stage S_y. When we reach the stage at which x is formed, we will have formed each y in x and hence completed each stage S_y in S. (Shoenfield [1977], p. 324)

A slightly different argument is given in his [1967]:

Suppose that we have a set A and that we have assigned a stage S_a to each element a of A. Since we can visualise the collection A as a single object (viz. the set A) we can also visualise the collection of stages S_a as a single object; so we can visualise a situation in which all these stages are completed. Hence there is to be a stage which follows all the stages S_a. This result is called the *principle of cofinality*. (Shoenfield [1967], p. 239)

Shoenfield makes much of this principle because it is used crucially to justify the axiom of replacement ([1977], p. 326). Let x be a set and $F(y)$ a unary operation on the $y \in x$:

. . . let S_y be a stage at which $F(y)$ is formed. Then there is a stage S after all the stages S_y for $y \in z$. At stage S, we can form the desired set [i.e. $\{F(y) : y \in x\}$]. (Shoenfield [1977], p. 326)

This is justified by appeal to the 'principle of cofinality'. The argument is meant to suggest that since the 'operation' of 'forming' the set x is 'completed' (for this is what it means to call x a set) then so can the 'operation' of 'forming' the collection $\{F(y) : y \in x\}$ be 'completed'; hence the latter is legitimately a set. But appeal to 'formation' and 'completion' here is illicit. Let us assume with Shoenfield that x is completed, that is that the collection of operations of forming x can actually be carried out. Even so, it is not automatically the case that any equivalent (i.e. cardinally, or even ordinally, equivalent) sequence of operations can be carried out. In the case of replacement, there is a mediate operation F, and there are no restrictions at all on the F. F may well involve arbitrary parameters and unbounded quantification, both features which would lead us to doubt whether any of S_y, $F(y)$ and $\{F(y) : y \in x\}$ could ever be 'completed'. More specifically, F may involve reference to as yet unformed sets or stages; not only does this impredicativity imply that Shoenfield is not taking 'completability' seriously in a constructivist sense, it also seems to violate his own strictures on the 'formation' of sets. Shoenfield insists that there is a precedence of formation among sets, that in forming a set z 'we do not yet have the object z'. But, surely, if this hierarchy of set formation is taken seriously, not only would a set z be unavailable as a *member* of z (as well-foundedness insists) but it should also be forbidden to *refer* to z in z's formation. After all, if z is not 'available' how can we refer to it? Yet this kind of reference to the 'future development' of the hierarchy is quite permissible in the arbitrary 'operation' F. In the last analysis, it would seem that the axiom of replacement rests on the axiom that any *set* of arbitrary operations can be 'completed', an axiom in as much need of justification as, and no more obvious than, replacement itself.

There is an interesting contrast here with the Fraenkel-Levy argument, for replacement is one of the axioms which can easily be justified on grounds of size, at least if size is treated cardinally. In this respect, Shoenfield seems to be worse off than Fraenkel and Levy. How does he fare with the power-set axiom, the bane of the limitation of size argument? At first sight Shoenfield has no difficulty:

. . . suppose x is formed at S. Since every member of x is formed before S, every subset of x is formed at S. Thus the set of all subsets of x can be formed at any stage after S. (Shoenfield [1977], p. 326)

(Cf. also [1967], p. 240.) That there always will be a stage after S is the first of Shoenfield's 'clear cases' mentioned above, an appeal in effect to the fundamental

successor operation of primitive arithmetic. Let us take this as read. But all
Shoenfield has justified is the axiom 'the collection of those subsets of x proved
to exist is a set' and this collection may be quite small, as small as x itself. To
be sure that this collection is the classical power-set in all its full-brown, impredicative majesty one needs to appeal to the classical axiom of separation. This
Shoenfield does, as one would expect. Shoenfield can, of course, derive this
from the axiom of replacement, but his justification of this is weak as I have
pointed out. Actually, he justifies separation separately, so let us see whether
this justification is any more satisfactory. Suppose every member of a collection
Y belongs to a given set x, i.e. suppose for some predicate ϕ, $Y = \{y : y \in x \wedge$
$\phi(y)\}$. By assumption, x is formed as some stage S: 'Then every member of x
is formed before S, and hence so is every member of the collection. Hence the
collection can be formed into a set at stage S.' (Shoenfield [1977], p. 325.)
There are two problems here. First, we are asked to accept that Y can be completed or 'formed' (at stage S) without regard to the nature of the predicate ϕ
involved in its specification. At this point, one could repeat the objections
raised against the use of arbitrary F in the replacement axiom. Secondly, if
the permitted ϕ are quite unrestricted what justification do we have for thinking
that S *could* ever have been completed? One could argue that, if the notion of
completion is taken seriously, when x is infinite S could never be completed
and mark the completion of all the subsets of x. Of course, the completion of
S is *assumed* by Shoenfield's argument. But what does this assumption amount
to? Namely that it is possible to complete the collection of stages $\{S_y : y \in P(x)\}$,
and this amounts to assuming that the classical power collection $P(x)$ can be
completed!

Shoenfield's difficulty over the power-set axiom is one which faces the usual
attempt to justify the cumulative hierarchy as the natural universe of sets.
The argument usually asserts that the shift upwards from level R_α in the hierarchy to level $R_{\alpha+1}$ is justified because it is just a natural (desirable and necessary) shift upwards in *type*, $R_{\alpha+1}$ representing the collection of all (extensions
of) properties over R_α which exist as sets. Fair enough, but this does not justify
(as it is usually taken to) adoption of the full classical power-set axiom, for it
leaves open the question of *which* extensions of properties of R_α can legitimately exist as sets. Usually it is *assumed without argument* that these will be
all the subsets of R_α guaranteed by the classical separation axiom, thus yielding
the full power-set of R_α at level $\alpha + 1$. Scott's axiomatization of set theory via
the cumulative stage or type approach (Scott [1974]) is interesting in this
respect. For Scott, unlike Shoenfield, openly adopts the axiom of separation
before he proceeds to develop his stage axioms. Thus, separation is left out
of the justification of axioms on purely type-theoretic grounds. This procedure,
it seems to me, is quite correct; there is a limit to what the appeal to type
theory can do. Scott's stage or type axioms imply that there is a 'power' set,
but it is the *prior* adoption of the classical separation axiom which allows one

to show that the 'power'-sets are *the* (classical) power-sets. As for justifying separation, Scott ([1974], p. 208) appeals directly to realism: 'There is no reason to place any restriction on how the property $\Phi(x)$ [allowed to appear in the axiom schema] is formulated: We believe in the existence of arbitrary subsets.'

Scott's system presents an interesting comparison with Shoenfield's. The latter's system is clumsy because he operated with two primitives, stage and set, and with a rather hazy notion of precedence among stages. Scott apparently sets out to axiomatize the notion of *level* (or stage) directly, but he assumes that the levels are themselves *sets*, so what he really axiomatizes is the notion of set. Moreover, the relation of precedence among levels or stages is defined simply to be set membership: V precedes V' if $V \in V'$. Scott's type or stage axioms are particularly simple: to wit, *an accumulation axiom* which states that a new stage simply accumulates all subsets and members of previous levels, and an *axiom of restriction* which states that every set is included in a level (Scott [1974], p. 209). (This amounts to Shoenfield's assumption that sets can only be formed at some stage.) From these axioms (with separation and extensionality) Scott proves that the stages or levels are well-ordered, and that the axiom of foundation must hold. The union-set axiom is derived immediately. As for power-set, Scott proves that a has a power-set just in case $V[a \in V]$ ([1974], p. 212). This means, since by restriction there is a V' with $a \subseteq V'$, that (by accumulation) a has a power set just in case there is a V following V'. This, of course, is very close to Shoenfield's argument that if a is formed at S, its power-set is formed at any stage after S. Scott's prior adoption of unrestricted separation guarantees that these power-sets are the genuine classical power-sets. One might add that Scott derives neither the axiom of infinity nor replacement from his stage axioms; rather he adds a further principle, a reflection principle for stages, in order to develop the full ZF system (Scott [1974], p. 213).

Thus it would appear that appeal to 'natural' stage or completion principles can justify very little of the full ZF system. The circularities in Shoenfield's arguments for power-set and replacement are introduced because he broadens the notion of completability so much that both arbitrary power collections $P(x)$ and arbitrary replacement collections $\{F(y) : y \in x\}$ are *assumed* to be completable. But this is surely what he is supposed to be arguing *for*. Moreover, this vitiates his explanation of how the contradictions are to be avoided. For this appears to rest exactly on his interpreting the notion of set formation in a quasi-predicative way.

Wang's attempt to justify the iterative concept of set and the ZF axioms is much vaguer than Shoenfield's, but falls into similar traps. Wang's approach differs from Shoenfield's in several respects, notably in its emphasis on the ordinals for indexing the stages of the iterative hierarchy. But in many respects it is similar. For instance, Wang insists ([1974], p. 181) that 'a set is a collection of *previously given* objects', and he too seems to appeal to quasi-constructivist notions. He does not use 'completability', but rather the notion of 'ideal overview', which one might describe as the ability of an ideal agent to 'collect together' or 'comprehend' a collection of objects. Wang's idea is that sets are 'multitudes' which are 'intuitive ranges of variability'. According to his explanation:

The natural way of getting such intuitive ranges is by the use of intuitive concepts (defining properties). An intuitive concept, unlike an abstract concept . . ., enables us to overview (or look through or run through or collect together), in an *idealised* sense, all the objects in the multitude which make up the extension of the concept, in such a way that there are no surprises as to the objects which fall under the concept. Hence, each intuitive concept determines an intuitive range of variability and hence a set. (Wang [1974], p. 182)

Wang recognizes that 'overviewing' or 'running through' or 'collecting together' cannot be tied closely to human capacity to construct or comprehend. For, in attempting to argue for *ZF* one wants to legitimize actually infinite collections. He tries to get round this by assuming an 'infinite intuition' which is supposed to do the 'running through' or 'collecting together':

The overviewing of an infinite range of objects presupposes an infinite intuition which is an idealization. Strictly speaking, we can only run through finite ranges (and perhaps ones of rather limited size only). (Wang [1974], p. 182)

Thus, the constructing capacities of the human agent are replaced by the constructing capacities of an ideal agent. Appeal to this notion recalls both Shoenfield's 'completability' (if we interpret it as 'completability by an ideal agent') and Fraenkel's 'bounded comprehension'. Moreover, the aim of the argument is much the same: the *ZF* sets are all 'comprehensible' by the ideal intuition, while the universe, and presumably the other paradoxical collections, cannot possibly be comprehended, even by the ideal intuition; they are not intuitive ranges of variability. (See Wang [1977], §1, and especially the last paragraph on p. 313.)

But Wang's argument suffers from the same fallacies as Shoenfield's, for the justification of the power-set axiom is built into Wang's interpretation of what can be 'comprehended' by his 'ideal intuition'. For example, Wang states:

. . . not only are the infinitely many integers taken as given, but we also take as given the process of selecting integers from this unity of all integers, and therewith all possible ways of leaving integers out in the process. So we get a new intuitive idealisation (viz. the set of all sets of integers). (Wang [1974], p. 182)

And in iterating this 'process', 'we do not', says Wang, 'concern ourselves over how a set is defined, e.g. whether by an impredicative definition' (p. 183). In his [1977] Wang expresses some slight doubt about the power-set axiom, but nevertheless he sticks to his view that there is a recognizable difference, in effect a difference in comprehension or extent, between, say, $P(\omega)$ and the universe:

In applying the idea of having an overview of a range, we are helped by a contrast between the power-set of ω and the totality V of all sets obtainable by the iterative concept. Admittedly the intuition of the former calls for a strong idealisation as is seen from the fact that we have no constructive consistency proof of the assumption of its existence. But clearly we do not have even a similarly weak intuition of the range of V. (Wang [1977], p. 313)

But again it is this 'contrast' which ought to be argued for and not simply unproblematically assumed.

These presentations of the iterative conception of set go wrong for much the same reason that the limitation of size argument (in its original form) goes wrong. Namely, because they attempt to explain the power-set axiom by reference to concepts which do not at all capture, and indeed contradict, its nature. For this type of explanatory attempt to succeed there would have to be a demonstrable connection between the 'size' (extent or comprehensiveness), cardinality, or 'completability' of a and the 'size', cardinality, or 'completability' of $P(a)$. But the impredicativity or absoluteness of $P(x)$ appears to rule out such a connection, which is why 'limitation of size' arguments always collide with the absoluteness or impredicativity of $P(x)$. Gödel's work on the consistency of the GCH perhaps confirms this view. In order to construct a model of ZF in which GCH holds, i.e. a model in which the cardinal size of infinite $P(a)$ is always tied closely to that of a, Gödel had to 'absolutize' the power-set axiom. In effect, in defining his constructible hierarchy, he replaced $P(x)$ by a new *absolute* term $D(x)$ consisting of the 'constructible' subsets of x: 'For a given set u, $D(u)$ is to contain only those subsets of u whose existence can be ascertained by, so to speak, examining only u itself, without scanning the whole universe.' (Bell and Machover [1977], p. 509.) This is why one can establish a connection between the size of u and the size of $D(u)$. $D(u)$ is what one might call a predicative operation, in the sense that for a subset y of u to be in $D(u)$ it must be defined by a predicate whose quantifiers and parameters are restricted to u. If one is to justify the *classical* power-set operation one must accept its blatantly impredicative nature; this means giving up any limitation of size argument for it, and also, I would contend, requires an appeal to realism in some form. This issue is touched on again in §6.2(*b*), as are some other issues connected with completability and impredicativity.

We saw that in working out his stage theory Scott is quite clear about the exceptional (and also the realist) nature of the impredicative separation axiom, the backbone of the classical power-set principle. Boolos [1971] has also given a presentation of an iterative stage theory which, though related to Shoenfield's theory, is in this respect much clearer. Like Shoenfield, Boolos begins with a kind of 'constructivist' intuition of set which tells us that a set should not be a member of itself. As before, the intuition is based on a notion of successive— or 'earlier' and 'later'—formation. Thus

. . . when one is told that a set is a collection into a whole of definite elements of our thought, one thinks: Here are some things. Now we bind them up into a whole. *Now* we have a set. We don't suppose that what we come up with after combining some elements into a whole could have been one of the very things we combined. . . . (Boolos [1971], p. 220)

As with Shoenfield this is the intuitive starting point for a theory of stages, and again as with Shoenfield, the motivation is clearly the idea that *we* are actually forming sets (through time) by putting elements together. But while Boolos develops from this an intuitive picture of a rank hierarchy, he does not make Shoenfield's mistake of allowing this intuitive picture to creep into the formal justification of the axioms of Z, thus creating tension between the 'constructivist' motivation and the *realist* outcome. Indeed Boolos, after giving the intuitive picture, is careful to lay it aside and present instead a formal theory of stages based only on certain key properties which the intuitive stages have, specifically that they are well-ordered. To this are added various principles concerning the 'formation' of sets at stages (see Boolos [1971], p. 223). As a simple consequence it follows that no set can belong to itself; thus the paradoxes fall away. The axioms contain two predicates 's E t' and 'x F t' which are to be informally understood as meaning 'stage s is earlier than stage t' and 'set x is formed at stage t'. Thus, in understanding the formal stage theory there is more than a backward glance to the intuitive explanatory model. But here we need not pretend that 'earlier' means any more than the idea of linear order, nor that 'formed at t' means anything more than 'belongs to (or is correlated with) t but no earlier stage'.

Boolos's presentation, where the original intuitive motivation is set aside, suggests a comparison with Cantor's theory of number where the early reliance on the intuitive notion of counting was later dispensed with in favour of the existence of well-orderings. The later 'formal' theory develops and relies on theoretical presuppositions *revealed* by the idea of counting though without necessarily appealing to this intuitive notion. Indeed, it seems that the presence of such sets as $P(\omega)$ make appeal to counting in any strict sense impossible, and similarly with Boolos's model. The presence of the very strong *specification (or collection) axioms* $(\exists y)(\forall x)$ $[x \in y \leftrightarrow (\psi \wedge (\exists t)$ $[t$ E $s \wedge x F t]]$ governing the 'formation' of sets at any given stage s makes impossible any appeal to formation in the intuitive sense, that is the putting together of something *new* from something given. For the ψ here are quite general, and indeed this freedom is necessary if (as required) the classical separation, and therefore classical power-set, axioms are to be derived. But, for a given s, many formulae ψ will contain reference to sets formed at stages *later* than s, and even reference to y itself. Thus, 'formation at s' *cannot possibly* be interpreted like 'formation' in the intuitively presented sense of successive building of new things from given. Indeed, 'formation' here can only really refer to the structural hierarchical organization of a *given* universe of objects (see §6.2(*b*)). Thus, in both cases it seems as if the original intuitions are considerably weakened.

Boolos's explanation of the specification schema is that 'every *possible collection*' (Boolos [1977], p. 223) should appear as a set at any given stage *s*. In view of the quite general (and therefore impredicative) nature of the ψ which can appear, we can take this as a realist assertion that 'as many sets as possible exist'. One slight source of confusion, and one place where Boolos's approach threatens to repeat the mistake of Wang and Shoenfield, is that Boolos uses the same term 'all possible collections' in his account of the intuitive hierarchy. If we are really *forming* sets in stages, than it may be that at stage *s* we do form all *possible* sets of objects formed at previous stages. But here 'all possible' will not be very many, certainly not many compared with the 'all possible' of the later 'formal' (impredicative) stage theory (see §6.2(*b*)). Here again we have to be a little careful.

Nevertheless, Boolos's picture is much clearer than those of Shoenfield and Wang. What it captures, with its impredicative specification axioms, is something like that captured by the limitation of size argument when reframed with the power-set axiom treated separately. Thus, it is not a justification of the power-set axiom, but rather a useful explanation of what the universe looks like when (or even when) impredicative principles are included.

6.2. Completability and Kant's first antinomy

Russell was the first to suggest that 'completing the incompletable' was the source of the set-theoretic contradictions, and the first to tie this idea to limitation of size:

> We saw, in the first part of this paper, that there are a number of processes, of which the generation of ordinals is one, which seem essentially incapable of terminating, although each process is such that the class of all terms generated by it (or a function of this class) ought to be the last term generated by that process. Thus it is natural to suppose that the terms generated by such a process do not form a class. And, if so, it seems also natural to suppose that any aggregate embracing all the terms generated by one of these processes cannot form a class. Consequently there will be (so to speak) a certain limit of size which no class can reach; and any supposed class which reaches or surpasses this limit is an improper class, i.e. is a non-entity. The existence of self-reproductive processes of this kind seem to make the notion of a totality of all entities an impossible one; ... (Russell [1906*a*], p. 152)

The connection pointed out by Russell persists in the modern limitation of size argument or the iterative theory at least when recast in the following form: the contradictions arise from assuming completed essentially incompletable collections, collections which *cannot* be completed by the Z or ZF axioms. The suggested involvement of the notion of *completion* in the set-theoretic contradictions has caused some writers to see a connection between these contradictions and certain of Kant's Antinomies of Pure Reason. For example, Martin in his [1955] notes that Russell's paradox is 'concerned with the concept of totality' and 'totality' in this context is clearly used in the Kantian sense of completed whole. Martin goes on:

I consider all classes and form the concept of their totality, namely the class of all classes. But then I regard this totality as an element in the formation of a new class. A concept of totality is again formed and this new totality is again used as an element. This conflict between concluding and beginning anew, between forming a totality and using this totality as a new element, is the actual ground of the [set-theoretic] antinomy. It is this conflict that gives the connection with the Kantian antinomies. Kant saw quite clearly that the antinomies rest on this antithesis between making a conclusion and going beyond the conclusion. In principle this has already been seen by Archytas, when he wanted to go to the end of the world and then stretch out his arm . . . (Martin [1955], p. 55)

It is quite possible that Fraenkel also saw a connection of this kind between the Cantorian and Kantian antinomies, and a resemblance to the limitation of size argument, for he noted:

There is a certain kinship between these Cantorian paradoxes and the antinomies which Kant set out in his Kritik der reinen Vernunft (e.g. those which arise if we treat nature as a closed whole). (Fraenkel [1928b], pp. 212–13)

(The same statement is made in the first and second editions of Fraenkel's book [1919], p. 133, and [1923], p. 155; see also Fraenkel [1927], p. 24.) Indeed we might recall both Fraenkel's use of *spatial* metaphor in drawing a distinction between sets and contradictory collections *and* that part of Kant's first Antinomy concerns the universe of spatially situated objects. Moreover, Fraenkel's comment echoes a remark of Hessenberg's:

The paradox of the set *W* of all ordinal numbers recalls those antinomies which, according to Kant, arise if we consider Nature as a closed whole. (Hessenberg [1906], p. 633)

(Cf. Hessenberg [1906], p. 706.) And Martin, who, as we have seen, explicitly states the connecton, attributes his recognition of it to Hessenberg and Zermelo, Fraenkel's set-theoretic ancestors. In his [1955] Martin remarks:

The connection [of the Cantorian antinomies] with the Kantian doctrine of antinomies has often been noticed and I myself am greatly indebted to Ernst Zermelo for drawing attention to it in his lectures. (Martin [1955], p. 54)

(However the only reference Martin gives is to the passage from Hessenberg [1906] quoted above. In his [1968] he attributes the connection *directly* to Hessenberg (see p. 306) though he mentions that Zermelo emphasized it. Zermelo mentions the Kantian Antinomies in two places, [1930], p. 47 (see p. 239 below), and [1932], p. 377 where he asserts a 'formal analogy' between them and the set theoretic antinomies.)

Now it is undoubtedly the case that Kant is very careful to distinguish between completable and incompletable collections. And he certainly suggests that his Antinomies are all caused by assuming completed what cannot be

completed. However, the connection between the Cantorian and Kantian antinomies is only superficial, or so I shall argue here. Nevertheless what Kant has
to say is of great interest to the philosopher of mathematics. And in the light
of this I shall return (at the end of this section) to the question of whether
there is any sense at all in using the notion of completability in a discussion
of the set-theoretic contradictions. Curiously Bennett in his [1974] states
abruptly, and specifically *contra* Martin, that there is *no* similarity between
the two sets of antinomies. But Bennett does not argue his point (he does
not even mention completability). This is wrong, partly because there *are*
apparent similarities which ought to be discussed, and partly because Kant's
'Solution' of the Antinomies involves him in a form of constructivism which
is pertinent to the philosophy of mathematics. Bennett's reason for dismissing
the analogy is that, first, he does not accept Kant's attribution of all four of
his antinomies to a common cause, and secondly, because in Martin's 'one
account of a logical [or set-theoretic] "antinomy" he gets it wrong' (Bennett
[1974], pp. 115-6). But this is too superficial. Certainly, in his explanation
Martin suggests that he is dealing with Russell's paradox, but what he then
sketches is the paradox of the universal set. But to dismiss for this reason the
analogy Martin is trying to promote is too easy. The question of whether there
is any connection among Kant's four Antinomies is a difficult one which (for
lack of competence, space, and time!) I shall not go into here. Most commentators split the Antinomies into two groups, commonly called the mathematical
(the first and second) and the dynamical (third and fourth). I shall concern myself
exclusively with the first which itself embraces two so-called cosmological Antinomies. This is the Antinomy which above all concerns the completion of collections and sequences and might therefore, of any of them, be expected to resemble
the set-theoretic antinomies. Moreover it is this Antinomy which is evoked by
Fraenkel's and Hessenberg's reference to treating 'Nature as a closed whole'.

(a) Contradiction or sleight of hand?

One way of analysing the set-theoretic contradictions is by attributing them to
naive mathematical realism. The comprehension principle bids us treat any
collection as an individual object of the same kind as the elements which go
to make it up, and thus as an object which must either belong or not belong
to any object, etc. If this naive realist position is adopted, then contradiction
ensues as a matter of logical course. One need not as a consequence of this
abandon set-theoretic realism, since systems free from the known contradictions
can be constructed which embrace the principle of bivalence, and blithely
accept completed infinites and impredicative definitions, all or any of which
might be taken as indicative of realism. But the unchallengeable nature of
the logical step from comprehension to contradiction must force even a mathematical realist to abandon his naivety. Keeping naive realism in mind let us
now turn to Kant's first Antinomy.

The Antinomy concerns two pairs of mutually contradictory statements, namely 'the world has a beginning in time' (Thesis), 'the world has no beginning or is infinite as regards time' (Antithesis); and 'the world is limited in space' (Thesis), 'the world has no limits in space . . . or is infinite as regards space' (Antithesis). (Kant [1787], p. 396.) Kant gives proofs for all four propositions, claiming validity for them (see Kant [1783], p. 103). (I shall come to this shortly.) Thus, we appear to have landed in contradiction. Moreover, Kant clearly attributes the contradictions to naive realism concerning the physical world. Consider any of the four propositions in question, say the spatial Antithesis 'the world has no limits in space' or synonymously for Kant 'the world occupies an infinite amount of space'. This latter statement has the same form as the statement 'my pen occupies an infinite amount of space'. Both attribute a certain spatial property to an *object*, as do their finite counterparts. Indeed, it is a crucial premise of the Antinomy that the world *can* be treated as an ordinary physical object of the same kind as my pen, that is, as an object which, by the law of excluded middle, either has or has not ordinary spatial and temporal properties. (Note the involvement of excluded middle in Kant's proofs. The proofs are often cast as *reductios*; we start from $S(a)$, derive a contradiction and conclude $\sim S(a)$, the assumption being that a, the world, must possess either S or $\sim S$). It is just this premise that the world is an object in the ordinary sense that Kant regards as the source of the trouble which the Antinomies present.

Consider what the ascription of spatial properties means for Kant. For Kant, to ascribe to the putative object a a spatial property S is to say something about a possible perceptual experience, to make, as it were, an observational claim. But according to Kant there can be *no* possible experience corresponding to $S(a)$, where a is taken to denote the world. Kant calls a physical object as experienced or perceived an 'appearance'. The spatial world is thus the 'totality of all possible appearances' (see Kant [1787], p. 393). So what Kant denies is that there is any possible 'appearance' (i.e. by definition, any physical object) corresponding to the 'idea' of the totality of all possible appearances. In other words Kant is saying that the world cannot naively be treated as an object of the same kind as those which go to make it up. That *this* is the lesson we must draw from the Antinomy is suggested by Kant in the following passage:

All these questions, i.e. as to whether the thesis or antithesis of each antimony is correct, refer to an object which can be found nowhere save in our thoughts, namely the absolutely unconditioned totality of the synthesis of appearances. If from our own concepts we are unable to assert and determine anything certain, we must not throw the blame upon the object as concealing itself from us since such an object is nowhere to be met with outside our idea, it is not possible for it to be given. The cause of failure we must seek in our idea itself. For so long as we obstinately persist in assuming that there is an actual object corresponding to the idea, the problem as thus viewed allows of no solution. (Kant [1787], p. 434; cf. ibid., p. 393)

The scenario sketched so far seems to show an analogy between the two sets of antinomies: an over-indulgent realism leads to contradiction both in set theory and in cosmology. One can extend the analogy. For Kant, there is no harm in 'aggregating' or 'collecting together' as such (see his [1770], p. 51); so one might, for example, want to use 'the world' or 'the universe' as convenient rhetorical devices, as terms referring to mere collections. What is wrong is to assume that they refer to 'totalities', objects of the same kind as the elements which constitute them. The same might equally well be said of the universe of sets. So, the analogy goes, in both cases to avoid serious trouble one has to restrain what Weyl calls 'that theoretical desire . . . alive in us which urges toward totality.' (Weyl [1949], p. 66)

But to see an analogy here is, I think, very misleading. The logical situation in Kant's case is quite different from that in the set-theoretic case. As I pointed out, 'naive realism' in set theory cannot be defended; the principle of comprehension in its naive form has to be abandoned. But 'naive realism' with regard to the physical universe is not obviously false in anything like the same way. In the first place, Kant's proofs of theses and antitheses are vague, even obscure, and certainly not clear-cut enough to be declared unequivocally (let alone self-evidently) valid. (Arguments about their validity still rage; for example, see Whitrow [1978], Popper [1978], Bell [1979], and Craig [1979].) Their obscurity forces commentators to rewrite them, and many have done so in such a way that they either turn out to be invalid or to involve false premises. This uncertainty over Kant's proofs leaves open the possibility of defending naive realism with respect to the universe. One could, quite coherently, take up the following position. 'There *is* a world and, applying excluded middle, it either began or not, is either spatially bounded or not. Doubtless we will never discover the *truth* of the matter. The best we can do is to develop, test and gather solid confirmation for, a physical theory which "settles" it one way or the other. Kant's Antinomies stir up a fuss about nothing (no pun intended); there is a *fact* of the matter, regardless of whether it will ever be revealed to us or not. Consequently, one side of his Thesis/Antithesis dichotomy must be false. If we examine his proofs we will find at least one to be at fault, either invalid or using false (or debatable) premises.' (Cf. Strawson [1966], pp. 199-201.) This 'common sense' approach motivates many of the discussions of Kant's Antinomies. (Russell's [1914] is a good example; see especially pp. 159-60.) Whatever one might think of it as a piece of philosophy, it is certainly not obviously false. However, its set-theoretic counterpart, 'There *is* a universe of sets, it is either the biggest set or not *etc*.', certainly is.

But there is another more serious sense in which the Kantian contradictions are dissimilar from the set-theoretic contradictions. Kant argues that the Antinomies demonstrate that pre-Kantian metaphysics is untenable, that the difficulties they cause can only be overcome by adopting the Kantian view of the world. This argument is already undermined by the lack of suasive force of

Kant's proofs. But Strawson suggests another reason why Kant's argument is not convincing, namely that the doctrines which Kant advances as his solution to the cosmological problem are already to some degree responsible for creating the problem in the first place. This is an important clue, and I want to follow it up here, and specifically to suggest a stronger version of it. This is: Kant's *proofs* are already predicated on a particular view of the physical world and what we are permitted to say about it which is inconsistent with the starting point of the proof, namely, that it is meaningful to ascribe spatial and temporal properties to the world as a whole. Kant's world view is shaped in particular by what appears to be a strong empiricist or phenomenalist theory of meaning and a restrictive theory of infinity.

To put it shortly: Kant conceives of the physical world as the world of appearances; this is quite different from the naive view of the world as independent from us, and from which the argument starts; nevertheless Kant frequently manipulates his proofs by covert appeal to his own special conception of the world. If my suggestion is correct, then the respective analyses of the set-theoretic contradictions and Kant's first Antinomy will be as follows. Set theory: comprehension principle (or naive realism) plus logic yields contradictions. Kant: naive realism plus principles inconsistent with naive realism plus logic yields contradictions. Stated crudely, it would seem that Kant injects the contradictions into his proofs, and what he claims to be a fundamental and inevitable problem is, in fact, an artificial one.

Let us see what evidence there is of this in Kant's proofs. Consider Kant's proof of the temporal Antithesis, 'the world has no beginning'. Assume that the world has a beginning in time. Then the world is preceded by empty time. But nothing can come to be in empty time 'because no part of such a time possesses, as compared with any other, a distinguishing condition of existence rather than non-existence.' (Kant [1787], p. 397.) From this Kant concludes immediately that the world had no beginning and thus has an infinite past. Presumably what he means to say is: 'The proposition that the world began is absurd; therefore the world had no beginning and thus an infinite past.' What can we make of this proof? The crucial passage is that concerning 'distinguishing conditions of existence'. I suppose what Kant is referring to here is *evidence* of existence. In other words he is saying one of two things. Either 'there can be no evidence that there is a beginning of the universe in time at all', or 'given t, there can be no conclusive evidence that the universe began t years ago and not, say, $t + 1$ years ago'. But why should Kant conclude from either of these that it is absurd to say that the world began? The reason, I suggest, is Kant's empiricism or phenomenalism, the demand that assertions about the physical world are only meaningful if they can be given a direct interpretation in terms of sense experience. This doctrine (which Strawson [1966], p. 16, calls Kant's principle of significance) is summed up in the following passage:

All concepts, and with them all principles, even such as are possible *apriori*, relate to empirical intuitions, that is, to the data for a possible experience. Apart from this relation they have no objective validity, . . . (Kant [1787], p. 259)

Of course, it is not clear what would constitute 'data for a possible experience'. The (naive) realist might take a line similar to that indicated above (p. 227): 'I have a general theory *T* which, with various initial conditions, tells me that the universe began *t* years ago. *T* has empirical consequences and is well confirmed. Hence I do have solid evidence that the universe began *t*, not *t* + 1, years ago.' Why cannot Kant accept this? The only reason I can suggest is that he puts much more rigid constraints on 'data for a possible experience'. Indeed he holds that 'concepts' as well as 'principles' (i.e. theories or laws) must relate to 'empirical intuitions', i.e. to *direct* experiences. I take this to mean that before it makes sense to assert that *a* has a property *S* we have to be able to specify what direct experiences would lead us to assent to or dissent from *S(a)*. This is just what Kant himself says in the following passage:

We demand in every concept . . . the possibility of giving it an object to which it may be applied. In the absence of such object, it has no meaning and is completely lacking in content . . . Now the object cannot be given to a concept otherwise than in intuition . . . (Kant [1787], p. 259)

that is, in direct apprehension. This even extends to 'bare concepts', concepts apparently without any sensible connections, e.g. the concepts of pure mathematics:

We therefore demand that a bare concept be *made sensible*, that is, that an object corresponding to it be presented in intuition. Otherwise the concept would, as we say, be without *sense*, that is, without meaning. (Kant [1787], p. 260)

Correspondingly, a term *a* purporting to refer to a (physical) object presumably cannot be used unless we can find a physical concept *S* (that is, a concept involving space or time) which can meaningfully be applied to *a*, and this, as we have seen, would mean that *S* must be 'given to *a* in intuition'. One must therefore specify what possible perception would correspond to the assertion *S(a)*.

This much stronger reading of 'data for a possible experience' clearly rules out the realist response outlined above. It also explains why Kant would claim that there is no evidence in this strong sense that the world began, no evidence that one time and not another marks the first event. For example, it would be absurd to ask what we would observe if we witnessed the beginning of the world. This would presuppose that we were somehow *outside* space and time, and if we were, observation by definition would be impossible. Similarly, if we were travelling back in time what possible experience could we specify which would mark a particular event as being the first?

So far, this is coherent. Where Kant's proof (as I have reconstructed it) goes wrong is in claiming that from this inability to specify perceptual evidence for the beginning of the universe it follows that the universe is infinitely old. Surely all that follows from it, using Kant's criterion of meaning, is that we cannot sensibly operate with the 'concept', or property, of being that unique time which marks the first event. Because nothing would count as evidence that a discretely (say yearly) spaced sequence of past events has a first member, it is absurd to say that the world began. The 'concept' of being the unique first event, or the unique time marking the first event, cannot possibly have any *recognizable* object which can fall under it, i.e. which can be 'given to it in intuition'. By Kant's criterion, then, the concept is meaningless. It does *not* follow, therefore, that it is *false* to assume the world began, but rather that it is *meaningless to ask* whether the world began. Thus, Kant's procedure is illicit. He starts from the assumption that the question *does* make sense, from the truth of 'the world either began or did not' (a realist assumption), and then rejects the first disjunct by implicit appeal to principles inconsistent with realism, i.e. inconsistent with the initial assumption. In effect, Kant proves too much.

Interestingly, in his exposition of his 'solution' to the antinomy Kant makes explicit the anti-realism which I claim is implicit in the above proof. Suppose the world is

. . . limited on the one hand by empty time and on the other by empty space. Since, however, as appearance, it [the world] cannot in itself be limited in either manner—appearance not being a thing in itself—*these limits of the world would have to be given in a possible experience, that is to say, we should require to have perception of limitation by absolutely empty time or space. But such an experience, as completely empty of content, is impossible.* Consequently, an absolute limit of the world is impossible empirically, and therefore also also absolutely. (Kant [1787], p. 457; the italics are mine.)

Kant makes his perception-dependent theory of meaning quite clear. But it can in fact be derived from his doctrines of space and time (cf. Strawson [1966], pp. 193–6). Suppose Kant's quoted remarks about relating 'concepts' to empirical intuitions are taken to apply only to what we might for convenience call *secondary properties*, properties which involve the perceptual mechanism of the observer. Indeed one might consider it natural to demand that ascription of a secondary property to a putative object a is meaningless unless there is the possibility of verifying or refuting this ascription through direct perception. For if ϕ is a secondary property, to assert $\phi(a)$ is actually to make an observational claim. According to this, if one regards the property of being red as secondary, it would only make sense to assert 'a is red' if one can specify what perceptual experience would count as conclusively verifying (or refuting) it. For example, it would make sense to assert 'Pluto is red', since we can imagine ourselves travelling to the outer limits of the solar system to look at Pluto, but it would not make sense to assert 'the universe as *a whole* is red' since we cannot possibly envisage being in a position to view the universe as a whole. As it stands, there is nothing particularly radical in this, for as well as secondary properties one might allow a range of *primary* properties quite unconnected with our perceptual mechanism whose meaningful use is not subject to perceptual constraints. So while one could not meaningfully describe the universe as a whole as red one might well meaningfully ascribe some primary properties to it, despite the fact that the universe as a whole is not observable. But in Kant's hands even this much weaker doctrine is extremely radical, since for Kant *all physical properties* are necessarily secondary: all physical properties involve space and time and these are indissolubly bound up with our perceptual and intellectual mechanisms. (See in particular B.42 of Kant [1787].) This is precisely why Kant calls physical objects 'appearances' and why the spatial world, the universe of all spatially situated objects, is the world of all 'appearances'. Note that it is just this which leads Kant to conclude that the universe as a whole cannot itself be regarded as a physical object, since it cannot make sense to ascribe *any* physical properties to it.

Consider now Kant's proof of the temporal Thesis (the universe must have had a beginning) which goes as follows. Assume to the contrary that there was no first event. Then if b is some event signifying the present, say the latest tick of my clock, then b is the last event in an infinite sequence S of past events. But says Kant:

. . . the infinity of a series consists in the fact that it can never be completed through successive synthesis. It thus follows that it is impossible for an infinite world-series to have passed away. . . (Kant [1787], p. 396)

Hence the assumption that there is no first event must be wrong and the Thesis is established. What are we to make of this proof? Let us assume (as we must if the argument is to stand any chance of success) that the events making up S are discretely spaced, i.e. separated by a finite, constant interval of time. The crucial point is that concerning 'completion by successive synthesis'. Why should we conclude from a premise about the 'completion' of a series through 'successive synthesis', a premise which presumably concerns *human* capacities, that the universe is finitely old? Surely, this is only possible if there is a connection between the universe and its past and our constructing capacities. But according to Kant's special doctrines there *is* a connection. According to Kant whatever is not given in intuition, by direct perception, must be constructed or synthesized by the human intellect out of direct perception. Given this, Kant's argument would go as follows. We obviously have no direct experience of the universe's past. But we *are* operating with the concept of 'the world' or, in this case, 'the world in past time'; hence it must have been 'synthesized' by us. (Note that this step is taken quite openly in Kant's proof of the spatial Thesis. After supposing that there is an infinite world he *immediately assumes* that we must have synthesized it ([1787], pp. 397-8).) Let us relieve Kant of the burden of arguing that *every* world event has been synthesized and assume only that some representative sequence S of discrete (say yearly spaced) events has been synthesized. But, according to Kant, it is impossible for us to synthesize or 'construct' a completed infinite sequence. Hence, since by hypothesis we *have* constructed 'the world in past time', or its representative *S*, then this world, or *S*, must be finite. Hence, the world began.

This argument seems valid. But consider more closely its key steps: (*a*) the assumption that we can only operate with 'concepts' given in direct intuition or such as are 'constructed' by the human intellect, and (*b*) that it is impossible for the human intellect to construct a completed infinity. One could argue that the doctrine (*a*) stems from the strong empiricism outlined above. This empiricism demands that there are no physical objects except those for which we can specify conditions of observation, i.e. if *a* refers to a putative object we must specify a physical predicate *P* and possible observational conditions under which one would assent to or dissent from *P*(*a*). (Strawson's discussion of 'transcendental idealism' is interesting on this; see his [1966], especially pp. 193-6.) These conditions would either involve a direct 'intuition' or, according to Kant, the construction of a sequence of steps which would eventually put us in the position to receive a direct perception. This already imposes some restriction on what can be allowed as physical objects. But the restrictions are further tightened by demanding that 'construction' is tied to human capacities: it must be stepwise, effected in discrete time, etc. From this it follows that the sequence of steps in a completed construction cannot be infinite, i.e. doctrine (*b*). There is thus a clear sense in which Kant might be called a constructivist. In any case, it is quite clear that (*a*) and (*b*) are inconsistent with realism, and therefore with the premise (the world either began or did not) which the argument starts out from. Indeed, in his 'solution' to the Antinomy Kant makes use of precisely the doctrines (*a*) and (*b*) to argue that it makes no sense to ask whether the world began or not. His position is neatly stated as follows: the only way we can achieve the concept of the universe's past is to construct it; if we try to construct the past we proceed in the construction of a series (a 'world series' as Kant calls it); there is no criterion we can apply which will allow us to conclude that this sequence has come, or will come, to an end; moreover, we cannot complete an infinite sequence, so we have no right to refer to 'the infinite past'; the most we can conclude is that the process of 'constructing the past' is indefinite, or unending.

Thus, I can say nothing regarding the whole object of experience, the world of sense; I must limit my assertions to the rule which determines how experience, in conformity with its object is to be obtained and further extended. (Kant [1787], pp. 457-8)

(Strawson is also very helpful on Kant's 'solution'; see his [1966], pp. 194-6.) Again it seems that we can only make sense of Kant's proof by interpreting it in the light of doctrines which render the proofs unnecessary. Similar conclusions could be drawn, I suggest, from a careful analysis of the proofs of the spatial Thesis and Antithesis. The 'problem' of Kant's first Antinomy is artificial; it does not arise naturally from naive physics or metaphysics (as the set-theoretic paradoxes arise naturally from naive mathematics); rather it is created by Kantian sleight of hand.

(b) Completability and constructivity

I have argued that the logical structure of the two sets of antinomies is quite different. What then of completability and 'completing the incompletable'? Certainly Kant suggests that it is a mistake to assume 'world series' to be completed and thus to treat the series as objects. But even if illicit completion of world series is taken to be the cause of the Antinomy (and to argue this one first has to show that there *is* a genuine contradiction involved) 'completability' is such a strong notion for Kant that it can have no relation to any sense of 'completability' appropriate to the limitation of size argument which was the starting point for our consideration of Kant. For Kant 'completion' must be 'constructive completion', and his 'construction' (as I have pointed out) is closely tied to the capacities of the human intellect. Synthesis, the process of bringing separate things (e.g. appearances, or intuitions) together into a whole, must take place in time (Kant [1770], p. 47), presumably with a lower bound on the interval between successive steps. Hence, no synthesis can be completed unless the process is a *finite one*. A 'whole' will not emerge by synthesis '. . . unless it should be possible to carry out the respective processes in a finite assignable period of time' (Kant [1770], p. 48). Thus infinite synthesis is impossible. As Kant ([1787], p. 398) remarks in his proof of the spatial Thesis: 'An infinite aggregate of actual things cannot therefore be viewed as a given whole, nor consequently as simultaneously given.' 'Incompletable', therefore, must be co-extensive with 'infinite'. And sure enough, this is actually how Kant ([1787], p. 401) chooses to *define* infinity: 'The true transcendental concept of infinitude is this, that the successive synthesis of units required for the enumeration of a quantum can never be completed.' Certainly any concept of 'incompletable' which rules out the completion of an infinite set would be anathema to the classical proponents of the limitation of size argument.

There is more than a passing resemblance between Kant's doctrines outlined here and the doctrines of intuitionism. Certainly, Kant's constructivist treatment of 'aggregates', 'wholes', and 'manifolds' hints at the intuitionists' treatment of sets. Of course, Kant's construction principles are rather vague. But they do emphasize construction *via* mental or intellectual processes performable in time, and they rule out the construction of completed infinities, features which are characteristic of the intuitionists treatment of sets. Not only this, but Kant's anti-realist theory of meaning is quite similar to that which Dummett has argued is the philosophical basis of intuitionism (see Dummett [1975] and [1977]). Of course, Kant's doctrines deal with the 'construction' and treatment of physical objects, and not directly with mathematical objects. But I think Kant's principles both of construction and of meaning do extend to mathematics. For example, in the following passage Kant states his constructivist approach to mathematics

... mathematical knowledge is the knowledge gained from the *construction* of concepts. To *construct* a concept means to exhibit *a priori* the intuition which corresponds to the concept. For the construction of a concept we therefore need a *non-empirical* intuition. The latter must, as intuition, be a *single* object, ... (Kant [1787], p. 577)

That Kant's empiricist principles of meaning apply to mathematics should be clear from his statement (quoted above) that even 'bare concepts' must have *empirical* meaning.

There are, of course, differences between Kant's theory and intuitionism. One major difference is that Kant, unlike the intuitionists, was not a logical revisionist. Moreover, intuitionism is much less hampered by its refusal to accept completed infinite sets as single objects than Kant appears to be. Intuitionism develops an extensive mathematics of processes. Thus, what may appear to be reference to a completed infinite totality (say quantification over the natural numbers) is in fact treated as reference to a finitely specifiable process of generation. The mathematics can then be made more sophisticated by 'nesting' the processes, i.e. with processes depending on continuing processes, and so on. And while this introduces technical complications, it allows intuitionism relative freedom, particularly with regard to the continuum. (Cf. van Dalen's remarks; Fraenkel, Bar-Hillel and Levy [1973], p. 254.) Kant does not have this freedom and his constructivism is correspondingly more rigid.

Kant it seems is a constructivist, an anti-realist with a strong notion of completability. But, it may be asked, even though Kant's notion of completability is too strong to be literally the same as any advanced by the limitation of size argument, could it not be qualitatively similar? No, because classical set theorists cannot successfully claim that *impredicatively defined* sets are completable.

Let me tackle this issue in a different way. Can we start from a rigid constructivism like Kant's and, by weakening the conditions which a construction must satisfy, arrive at a classical set theory? One way of liberalizing Kant's theory would be to develop, as the intuitionists have done, a mathematical theory of processes. Another liberalizing move would be to drop the insistence on the human intellectual capacity for completing processes. This would then permit the formation of some (countably) infinite sets.

Dropping the insistence on *human* capacity for completing processes would amount, in mathematical effect, to dropping the intuitionists' insistence that for infinitely proceeding processes the output of the process cannot be divorced from the process itself (see Dummett [1977], p. 56). According to the intuitionists, a species (which is perhaps the cloest intuitionistic analogue of a set) is given by specifying a 'well-defined' domain D and a definite property ϕ. The species is then $S = \{a \in D: \phi(a)\}$. But the terms used here do not have their usual classical sense. A well-defined domain D is a domain of constructions already carried out or specified. Indeed one must be able to enumerate the 'members' of D in such a way that the enumeration carries each natural number

n onto a specification of a construction, which is denoted by a_n. A property ϕ is definite for D if it is known for each element a_n of D what is to count as a proof of $\phi(a_n)$; specifying S then consists in specifying a constructive map taking each a_n onto a proof of $\phi(a_n)$, if there is one, or onto 0 if there is no such proof (or rather no proof *so far*). If it can be proved that from some n_0 on no $\phi(a_n)$, $n > n_0$, can be proved, then the construction of S will terminate after a finite number of steps. If this cannot be proved, the construction will not terminate. If 'construction' is interpreted in the light of *human* ability to construct through time, this means that the species can never be completely finished; i.e. the *extension* will never be completely formed. Thus, no infinite extension will ever be completely formed, and, regarding a set as an extension which can be divorced from its process of generation, no infinite sets can be formed.

One way of weakening this is by allowing that once a process has been specified in such a way that each individual step could be performed by a human agent then one can *assume* that the process is completed even though the human agent cannot himself complete it. This would then mean that for effective processes the outcome can be divorced from its process of generation and then treated as a single completed object. If such a procedure is applied to the natural numbers and then to the construction of the real numbers, the natural numbers can immediately be considered as a set, and the resulting (denumerable) continuum would be something like that conceived by Borel (see §2.3(*a*)).

The claim that infinities generated by effective processes can be considered as completed may be taken as postulating the existence of a being or agent who *can*, unlike the human agent, actually complete the process. (Such an agent would be something like Kant's imagined agent who can synthesize an infinite manifold (see Kant [1770], p. 49).) The mathematics thus generated is still in a broad sense constructive since, although we have now moved away from *human* intellectual capacity, it still depends on the ability of some (real or imagined) being to construct.

It may be thought that once we begin to tread this path there is no limit to what can be allowed as broadly 'constructive', for we can simply postulate a more powerful agent who can complete more and more processes. For example, the next step would be to relax the condition which insists that processes develop stepwise through discrete time. Thus, we might postulate an agent who can run through any number of steps in finite time, thus allowing more and more infinite extensions to be 'synthesized'. Indeed, classical mathematics is often represented as the mathematics of an ideal agent with unlimited construction powers. (Dummett seems to represent Platonism this way in his [1977], pp. 59–60, thus, as he sees it, reducing it to absurdity; and recall Wang's 'ideal agent' discussed in §6.1.) However, it seems to me that there is a limit to what can be allowed as constructive even if we do allow greater and greater powers to a constructing agent, and this limit can also be taken as a limit on how far Kant's principles can be weakened. The limit, I contend, comes with the distinction between predicative and impredicative processes. No matter what powers of surveillance, or what ability to run through infinite collections in a finite time, are ascribed to a postulated constructing agent it seems to me that the constructing agent can never complete what we might call an impredicative process, i.e. can never construct a set (or number) via an impredicative definition. Since, therefore, classical set theory depends very much on such 'completion' the gulf between constructive principles, however broadly conceived, and classical mathematics is too wide to be bridged.

The key point here is that a process clearly cannot be carried out if any one step in the process depends on the outcome, and this applies to *all* constructions, whatever powers of surveillance or of 'running through in a finite time' are granted the constructing agent. This is connected with what Gödel calls Russell's 'vicious circle principle in its first form', namely '. . . no totality can contain members definable only in terms of this totality. . . .' (Gödel [1944], p. 133.) As Gödel says ([1944], p. 136) '. . . the construction of a thing can certainly not be based on a totality of things to which the thing to be constructed itself belongs.' If by 'definition' one understands 'specification for construction', then clearly this vicious circle principle would epitomize constructivity in its very broadest form.

Earlier I criticized Shoenfield's use of 'completable' and Wang's use of 'formation' because these terms were characterized so broadly as to make the arguments based on them circular. My point now is to emphasize how misleading and wholly inappropriate the use of constructivist terminology is here. Gödel for example, claims that the 'operation "set of" ' is actually closely related to Kant's concept of synthesis (see Gödel [1964], p. 272, n. 40). This simply cannot be the case. Similar criticism can be levelled at Wang. Parsons, in his excellent [1977], has criticized the iterative theorists, particularly Wang, for misuse of constructivist imagery. Parsons sees Wang's use of 'overviewing' or 'intuition' as an appeal to something like Kant's idea of *de re* intuition. But, as Parsons points out, this idea clearly has no sense as soon as one considers the collection of all sub-collections of the integers:

The divorce from *sensible* intuition involved in treating this totality as 'intuitable' seems complete . . . I no longer understand Wang's talk of 'intuitively running through' where it is applied to the set of all sets of integers. In the above I have perhaps connected intuition more closely with the senses (more abstractly Kant's "sensibility") than Wang would find acceptable. But even quite abstract marks of sensibility, such as the structure of time, are lost in this case. (Parsons [1977], p. 343)

Parsons is certainly right. However, I am quite prepared to *grant* Wang's agent intuitions which are not Kantian intuitions, for example sweeping powers of surveillance, access to 'super time', etc. My point is that even granted these powers, sets like $P(\omega)$ will never be *constructed*.

These points are not trivial. It seems to me that the use of constructivist terminology in these arguments for set theory is actually dangerous because it serves to obscure the realist nature of the theory. In short, it obscures the difference between the definition of an object in the sense of a unique description of an object *assumed already to exist*, and constructive definition, i.e. the specification of a process whereby a *new* object is to be formed. As both Gödel and, before him, Poincaré pointed out, one need not worry about impredicative definitions if the business of mathematics is conceived as the description of, and the establishment of truths about, already existing objects, 'objects that exist independently of our constructions' (Gödel [1944], p. 136). For instance, the term 'the least upper bound of the bounded point-set A' is perfectly harmless as a *description* of a particular real number provided that one is assumed to be describing existing objects. So is the description 'the set of objects in a satisfying the property ϕ' where a is a set and ϕ involves, say, the parameter $P(a)$ or quantification over the whole set-theoretic universe. But if mathematics is concerned with constructing *new* objects, the vicious circle principle has to come into play. One *cannot* construct objects using specifications which refer to the object itself or to sets to which it belongs, or which use unbounded quantification. It may be that the least upper bound for A can be, or has already been, constructed by a quite different specification. But it cannot be *constructed* by an impredicative definition. This confusion occurs in both Wang's and Shoenfield's explanation of the axioms of set theory.

Consider Wang's explanation. Wang is quite prepared to assume that his 'ideal agent' can 'form' sets *via* predicates which contain unbounded universal quantifiers. This means that to check whether a given object belongs to the set or not the agent must be able to 'run through' or 'survey' the whole universe. Very well, let us grant him that power of surveillance. It must mean, however, that since the quantification *is* unrestricted ('all' literally means 'all') the collections which are going to be sets *already exist as sets*; they are not being formed by the agent. In other words, Wang must be assuming implicitly that the complete universe of sets already exists. This strong realist assumption is obscured by the constructivist terminology.

There is a similar obscurantism in Shoenfield's treatment of hierarchy. Shoenfield wants to justify the von Neumann cumulative hierarchy as the natural universe of set theory. Therefore, among other things, he must show that sets are well-founded. But to argue this he appeals to a version of the vicious circle principle, namely that the members of a set must be formed before the set itself is formed. Thus, according to Shoenfield there should be a natural precedence relation between sets, namely precedence of formation. Now it is true that in the von Neumann hierarchy the relation of priority (the rank relation) follows that of membership and *seems* very close to Shoenfield's precedence of formation. But there is a crucial difference. Priority in the cumulative hierarchy is only priority in *de facto* constitution. Since unbounded quantification is used in the characterization of sets the assumption must be again that *all* sets pre-exist. Given this the axioms *ZF* (including *AF)* allow us to show, or to discover, that the universe can be arranged in cumulative stages with each set assigned a unique rank such that $a \in b \rightarrow \text{rank}(a) < \text{rank}(b)$. If one is defining a hierarchy in which there is a *genuine* precedence of formation among sets then one would have to accept that in forming sets at stage α there can be no reference to sets of levels $\geqslant \alpha$. And this hierarchy would be nothing like as rich in sets as the cumulative hierarchy. Since Shoenfield ends up with the *ZF* axioms and the cumulative hierarchy one can only conclude that the insistence on 'precedence of formation' has been quietly dropped in favour of *de facto* (indeed *post facto*) constitution. Shoenfield's basic assumption, like Wang's, is a realist one. But his constructivist terminology obscures it.

Finally, to return to our starting point, Martin's analogy between the Kantian and Cantorian antinomies suffers from a similar confusion over realism and constructivism. In describing the paradox of the universal set, Martin starts from the assumption that all sets exist, and then indicates that a contradiction can be obtained by proceeding to the 'formation' of a new set. But this violates the original assumption. There is no question of 'forming' a *new* set using V (the universe of all sets); it has been *assumed* that all sets already exist. Martin's reference to *formation* only makes sense if one adopts a constructivist approach to sets. And if this is adopted one could not begin with a premise like 'consider all sets as given'. There is thus a confusion between realism and constructivism. The same objection applies, I think, to Gödel's remarks on the non-existence of the universal set (p. 202). The existence of the universal set would *not* 'immediately give rise to further applications of the operation "set of" ' as Gödel claims. The paradox arises simply because comprehension, which naively is supposed to be the central axiom for describing sets, describes 'sets' which have contradictory properties. The paradoxes have nothing to do, classically, with set *formation*.

Something of this confusion is reflected in Zermelo's striking remarks on the Kantian Antinomies in his [1930]. There he presents a (second-order) view of the cumulative hierarchy $\bigcup_{\alpha \in Ord} R_\alpha$ where, allowing that the α can reach inaccessible size, one obtains set models of the *ZF* axioms. He writes:

Scientific reactionaries and anti-mathematicians have so often eagerly and lovingly appealed to the 'ultrafinite antinomies' in their struggle against set theory. But these are only apparent 'contradictions' and depend solely on a confusion between set theory itself, which is not categorically determined by its axioms, and particular models of these axioms. What appears in one model as an 'ultrafinite non- or super-set' is in the succeeding model already a perfectly good, valid 'set' with a cardinal number and ordinal type, and is itself a foundation-stone for the construction of a new domain. To the unlimited series of Cantor ordinal numbers there corresponds a likewise unlimited double series of essentially different set-theoretic models in each of which the whole classical theory is expressed. The two opposite tendencies of the thinking spirit, the idea of creative *advancement* and that of collective *completion* [*Abschluss*], which also lies behind the Kantian 'antinomies', are symbolically represented and reconciled in the transfinite number series based on well-ordering. This series in its unrestricted progress reaches no true completion; but it does possess relative stopping points, namely those 'limit numbers' which separate the higher from the lower models. (Zermelo [1930], p. 47)

(By the non-categoricity of the axioms Zermelo is certainly not referring to the Skolem paradox and the problem of first-order axiomatization. Rather he means just what is expressed in this passage, namely, that if a range of large cardinals exist, then perhaps there will be *many* non-isomorphic models of *ZF*.) In his reference to the Kantian antinomies, Zermelo hints at the idea that after the universe of set theory, or the sequence of ordinals, has been 'completed', one proceeds again to form more ordinals and *new* universes or models. But again, properly speaking, what Zermelo is considering has nothing to do with *completion*, certainly not Kantian completion. If a collection R is the whole universe of sets, then it is 'ultra-finite' or 'paradoxical' and cannot be a set. And it either contains or does not contain a first inaccessible ordinal θ and thus a set R_θ. But R cannot be both the whole universe *and* an R_θ: if it is one it is not the other. The theoretical idea that Zermelo is using here is that there may well be sets which can act as models for (some of) the axioms, specifically sets R_θ which are models of all the *ZF* axioms. But this by itself is no more remarkable than that R_ω is a model of *ZF* minus the axiom of infinity, or $R_{\omega+\omega}$ a model of *Z*. None of this means that any of R_ω or $R_{\omega+\omega}$ or R_θ once *was* the whole universe, and that we then proceeded to 'build' more extensive universes based on these as 'foundation stones'. Indeed, even to speak of R_ω, $R_{\omega+\omega}$, R_θ we need to use axioms which these sets do *not* satisfy (infinity, replacement, inaccessible ordinals, respectively) outside the ones they *do* satisfy, and the presence of these further axioms says that R_ω, $R_{\omega+\omega}$, R_θ cannot be true 'ultra-finite sets'. What we are witnessing here is not the successive 'unfolding' of various universes of set theory, but rather the results of a genuine unfolding of theories about *the* set-theoretic universe. This is perhaps how we might best interpret Zermelo's concluding remark:

And so the set theoretic 'antinomies' properly understood, instead of leading to a contradiction and mutilation of mathematical science, lead rather to an unsurveyable unfolding and enrichment of that science. (Zermelo [1930], p. 47)

7
THE ZERMELO SYSTEM

We have now seen how the limitation of size idea emerged from basic Cantorian ideas connected specially with the Cantorian Absolute and also the status of this idea with respect to modern axiomatic set theory. It is now time to consider in more detail how limitation of size fitted in with the development of the two major axiomatizations of the theory, Zermelo's axiomatization, which is considered in this chapter, and von Neumann's axiomatization, which is considered in the next.

7.1. Zermelo's separation axiom as a limitation of size principle

In Chapter 4 I explained how the Jourdain-Russell 'limitation of size' substitute for comprehension could not work because it lacked additional specific principles of set existence with which to fill out the universe, for example principles which would provide enough sets to develop an adequate theory of number. Zermelo's system does not suffer from this defect. It does introduce a substitute for Russellian comprehension, as Zermelo himself pointed out:

By giving us a large measure of freedom in defining new sets Axiom III [i.e. *Aussonderungsaxiom* or Separation Axiom] in a sense furnishes a substitute for the general definition of set that was . . . rejected as untenable. (Zermelo [1908*b*], p. 202)

And this, as we shall see, is a limitation of size principle. But this size principle is not and cannot be a *full* substitute for comprehension; as Zermelo says, it is a substitute only 'in a sense'. It does convert arbitrary properties into sets (though, of course, a set need not be the *whole* extension of the property which gives rise to it). But the qualification is necessary because in order to apply separation one needs *set* parameters and hence a fund of sets. Zermelo, unlike Jourdain, *had* such a fund of sets, namely the sets provided by his axioms of power-set, union, infinity, etc. I shall discuss the important question of why he chose these axioms in subsequent sections; for the time being, let us concentrate on separation.

Separation is a natural limitation of size principle since it says that the extension of a predicate can be a set if that extension is *restricted*. Or rather, it captures this idea provided that we assume *a priori* that inclusion in a set (specifically in any set given by Zermelo's power-set and infinity axioms) is sufficient evidence of restriction. This fits exactly with the strengths and weaknesses of the limitation of size argument and with what I called the heuristic use of limitation of size. In other words Zermelo's system assumes that the sets given by the power-set and infinity axioms are themselves legitimate (small or restricted). So if the limitation of size hypothesis (that contradictory sets are too big) is right, then there is apparently no way in which separation (on its own or together with the other axioms) could give rise to any of the over-comprehensive or unrestricted paradoxical sets. If power-sets etc. are small, so are the sets given by the partial substitute for comprehension. Indeed, it says precisely that *smallness* (restrictedness) is a sufficient condition for set-hood, as we can see if we state separation in the form: given ϕ, $\exists b \forall x \, [\phi(x) \to x \in b] \to \exists a \forall x \, [x \in a \leftrightarrow \phi(x)]$. Thus, Zermelo states:

. . . sets may never be *independently defined* by means of this axiom but must always be *separated* as subsets from sets already given; thus contradictory notions such as 'the set of all sets' or 'the set of all ordinal numbers', and with them the 'ultrafinite paradoxes', to use Hessenberg's expression . . . are excluded. (Zermelo [1908*b*], p. 202)

Zermelo's reference to Hessenberg's term 'ultrafinite paradoxes' is a fairly clear indication that he did accept the limitation of size hypothesis. For Hessenberg remarks ([1906], p. 627): 'It can be shown of the paradoxical sets which we investigate here that they are of greater power than any aleph.' And such sets are just what he calls 'ultrafinite' meaning that they fall outside (are too big for) transfinite set theory. Although this characterization of 'too big' is really a cardinal characterization, something which Zermelo himself did not explicitly rely on, by using the term 'ultrafinite' Zermelo is clearly saying that the separation axiom cannot approach the *very big* sets which cause the paradoxes. In this sense, separation acts as a limitation of size principle. But since he is saying that this is the case even when separation is conjoined with the power-set and infinity axioms, then he must be *assuming* also, as I said above, that these axioms do not create sets which are too big. That this *is* an assumption was first explicitly recognized by von Neumann:

. . . Zermelo and Fraenkel do not define exactly when the extension of a set is 'too big', but only establish lower bounds on the idea through some postulates . . . (von Neumann [1929], p. 494)

In a footnote he continues: 'For instance: if a set is "not too big" neither is its power set or union set.'

There are also historical reasons for seeing the separation axiom as a limitation of size principle. For what it does is to make more precise Cantor's [1882] 'definition' of set discussed in §1.4 according to which sets are 'defined' out of domains which are themselves limited. When associated with Cantor's theory of the Absolute, this is clearly a principle of limited comprehension. Cantor was not the first to reject the use of the universal domain as a means of defining sets or collections. For example, De Morgan had (before Cantor) argued against the use of it on practical grounds, arguing that in the all-inclusive domain intuitively clear extensions do not necessarily have intuitively clear complements, whereas if we choose a suitably restricted domain they do. (I suppose an example would be that in the universal domain the class of all men does not have an intuitively clear complement, whereas in the domain of all humans it does.) This pragmatic line of thought suggests that classes, at least those which we make any use of, are of restricted extent.[1] But more importantly, mathematics it seems (at least as long as it operated with implicit principles which generated sufficiently large domains) had no need to rely on a universal domain. Certainly, as Cantor pointed out in his [1883b] (p. 587, n. 1), his own studies up to this time and (he could have added) those of the other mathematicians who had begun to use set theory in analysis dealt only with rather restricted sets. His [1882] characterization of set might be seen as an attempt to capture this *de facto* limitation.

Of course, using a universal domain is logically simpler, and in the end it avoids having to detail more specific set existence principles. And Dedekind, Frege, and Russell all relied on this logical simplicity. But when this use of a universal domain had clearly broken down (one could call it a logical oversimplification) it was not surprising that those interested in producing an axiomatization of *mathematical* set theory (as opposed to set-theoretical logic or logicistic mathematics) should want to return to and clarify the original limited operation with sets. This is, in effect, Cantor's first 'way out' discussed on p. 47 and perhaps what Zermelo means when he says we must

. . . start from set theory as it is historically given, to seek out the principles required for establishing the foundations of this mathematical discipline. (Zermelo [1908b], p. 200)

As Scott has put it:

It must be understood from the start that Russell's paradox is *not* to be regarded as a disaster. It and the related paradoxes show that the naive notion of all-inclusive collections is untenable. That *is* an interesting result, no doubt about it. But note that our original intuition of set is based on the idea of having collections of already fixed objects. The suggestion of considering all-inclusive collections only came in later by way of formal simplification of language. The suggestion proved to be unfortunate, and so we must return to the primary intuitions. (Scott [1974], p. 207)

With the insight that the paradoxical collections are all 'very big', a 'limitation of size' or 'restricted comprehension' approach to the paradoxes was perfectly natural. With hindsight, it seems quite natural, too, that a restricted comprehension approach like Zermelo's should embody or build on Cantor's attempt at a limited comprehension characterization, the first expression of the mathematical concept of set.

---

[1] See, for example, De Morgan [1858]. It is worth noting that Schröder also objected to the universal class (specifically to Boole's use of it), though this time for logical reasons, in effect because it blurs type distinctions (see Bochenski [1970], pp. 391–3). Bochenski also points to a similar objection to the universal class in Aristotle. Interestingly Bochenski ([1970], p. 393) quotes Frege *against* Schröder's position.

Perhaps the best illustration that restricting domains was one of the 'primary intuitions' to which mathematicians wanted to return is the textbook published by the Youngs in 1906. The Youngs' comments are particularly interesting, since they had close contacts with Göttingen at this time, contacts which arose in the first place through Mrs. Young who, as Grace Chisholm, took a doctorate under Felix Klein. She recalls (see Grattan-Guinness [1972]) that many important issues concerning the foundations of set theory were discussed in Hilbert's seminar, and Zermelo would undoubtedly have taken part in these discussions. Thus the Youngs' views, expressed in their [1906], may well reflect those of the Göttingen circle around Hilbert on how to clarify the set concept, for the position they present gives a view of the notion of set somewhere between Cantor's 1882 concept and that lying behind the central axiom (separation) of Zermelo's system. In effect the Youngs argue for restricted comprehension without specifying exactly, as Zermelo did, how the restriction is achieved.

After giving Cantor's 1882 definition of set the Youngs [1906], p. 146, comment: 'Here Cantor emphasises the fact that in any logical and more especially mathematical thinking, we must confine our ideas to some particular field . . .' They then underline (though they do not refer explicitly to the paradoxes) the danger associated with arbitrary 'conceptual spheres' or 'fundamental regions' as they call them: namely that there is a temptation to allow the fundamental region itself to be a set. Certainly this would be logically convenient and many of the natural 'fundamental regions' *are* treated as sets in other contexts. But, say the Youngs:

. . . it should be postulated that the fundamental region itself is not to be regarded as a set *unless there is another fundamental region containing the first fundamental region together with other elements.* (Young and Young [1906], p. 146; the italics are mine)

This statement encapsulates not only the demand that 'conceptual spheres' be *restricted*, but also the demand that sets be extendable. This latter is further emphasized:

For most practical purposes this distinction is immaterial, since such an extended fundamental region can be found, but it is perhaps a proper law of thought that *we cannot regard a lot of objects in their totality unless we can get beyond them.* (Young and Young [1906], p. 146; the italics are mine)

This is already very close to the axiom of separation. For it suggests that invariably the fundamental regions we use in practice *are* sets: we can treat the objects of these regions as a 'totality' since we can 'get beyond them', say by embedding them in further regions. The gap between the Youngs' demands and the axiom of separation (and thus between Cantor and Zermelo) is bridged in two steps. The first and most important step is to demand that fundamental regions themselves be mathematical objects or subject to mathematical investigation. This crystallizes into a condition the impression stated above that invariably fundamental regions are embedded in further fundamental regions and thus themselves become the object of mathematical investigation. The second step is the reductionist step taken by Zermelo, which decrees that the only mathematical objects, in particular the only extensions which are mathematical objects, are sets. (Interestingly, in their [1913], Schoenflies and Hahn present the axiom of separation as a natural formulation of the demand that sets can only be formed inside domains which are themselves known to be mathematical objects.) This extends or strengthens a tendency already strongly present in Cantor.

Thus Zermelo's key idea goes back to principles deeply rooted in Cantor's work: sets in practice are limited or restricted in extent, and absolute collections cannot be reached by the use of our set principles. However, the central place of the separation axiom riases two questions: How is the axiom to be formulated? And how did Zermelo decide which axioms to use with it?

7.2. Zermelo's reductionism

It is not hard to discover the source of Zermelo's fund of sets. Stated shortly, what Zermelo did was to analyse his own proof of the well-ordering theorem and take the central principles used there as set axioms. But this is only the beginning of the problem. For example, why did Zermelo prove the well-ordering theorem in the form 'every set can be well-ordered' and not in the Cantorian form 'every set can be counted by an ordinal'? Why did Zermelo shun the Cantor–Jourdain proof of the aleph theorem, and with it the ordinal-based limitation of size theory? To answer these questions we need some idea of the heuristic principles which guided Zermelo in constructing and analysing his proof.

The most important component was Zermelo's set-theoretic reductionism. This is built around the doctrine that sets are the only primitive entities we can or need use in mathematics. In other words, a system adequate for the construction of classical mathematics need only have sets as its primitive objects. Or again, all mathematical problems, whether foundational like the contradictions or technical like the well-ordering problem, have to be tackled at the level of sets and solved in *purely* set-theoretic terms. Such a heuristic view may now appear banal. But this was not the case at the turn of the century. For instance, there was no fully successful reductionist treatment either of ordinal and cardinal number or of the notion of a function, the two most fundamental notions of mathematics. Neither was there any reductionist treatment of relations or the theory of ordered sets. And recall that Cantor's and Jourdain's proof of the aleph theorem, a possible alternative to Zermelo's, involved the *totality* of all ordinals, demonstrably *not* an ordinary set, in an essential way. In the face of all this Zermelo's reductionism was undoubtedly a bold step. Not surprisingly, it had a profound effect on the shape of his system, particularly *via* its influence on his treatment of the contradictions and the well-ordering problem. But I begin with Zermelo's reductionist theory of number.

(a) Zermelo's reductionist treatment of number

Zermelo's approach is most evident in the following statement:

. . . for me, every theorem stated about finite numbers is nothing other than a theorem about *finite sets*. (Zermelo [1909a], p. 185)

And then:

If one wishes to base arithmetic on the theory of natural numbers as the *finite enumerals*, then we must deal above all with the definition of *finite sets*. For the enumeral is according to its nature a property of a set, and every statement about finite enumerals can always be expressed as a statement about finite sets. (Zermelo [1909b], p. 8)

Nevertheless it is important to distinguish two quite different approaches to the problem of attempting to reconstruct a theory of number inside a pure theory of sets. One can either attempt the definition of numbers as set objects and hope to prove set-theoretic versions of the basic 'axioms' governing them. Or one can eschew number objects altogether and attempt to build the theory solely on the basis of relations of numerical equivalence. Both courses are reductionist, the latter being more economical than the former in the sense that it appears to demand the development of less theoretical machinery. This is essentially the course Zermelo chose. In this respect, Zermelo's work was close to that of Hessenberg. They had similar views, particularly on number, the paradoxes, and ordering, and they were familiar with each other's work. According to Hessenberg, Zermelo read the proofs of his [1906], and in the introduction to that paper he thanks Zermelo '. . . above all for the communication of some as yet unpublished investigations and the permission to make use of them' (p. 483). These 'unpublished investigations' were presumably early versions of some or all of Zermelo's later [1908a], [1908b], [1909a], and [1909b] papers. In these papers, Zermelo also refers to Hessenberg. However, despite this interchange of works it is not clear whether either was a direct influence of the other, or whether they arrived at similar views independently.

The basic idea of the Zermelo (and Hessenberg) treatment of number is somewhat similar to that of Frege. As with Frege and Cantor the relations of equinumerosity and isomorphism between sets are taken to be fundamental. The idea is then, as far as possible, to reduce statements involving numerical terms to statements involving only equinumerosity and isomorphism (see Frege [1884] or Parsons [1965]). Statements which cannot be so reduced are regarded as devoid of meaning. This doctrine is neatly encapsulated in Zermelo's firm rejection of Jourdain's theory (Zermelo [1908a], p. 193). Recall that this attempts to dissolve the Burali-Forti contradiction by admitting the set of ordinals but then denying that it has a type:

But now since in Cantor's theory 'order-types' and 'cardinal numbers' are nothing but convenient *means of expression* for the comparison of sets with respect to similarity or equivalence of their parts, I cannot extract any intelligible meaning from the proposition that a well-ordered set possess no type or cardinal number . . . (Zermelo [1908a], p. 193; the italics are Zermelo's)

In so far as it describes Cantor's theory, Zermelo's statement is quite wrong. But as a statement of his own intent it is revealing. And compare Hessenberg discussing various attempted solutions of Burali-Forti's paradox:

Second attempt [i.e. Jourdain's] : W [the set of ordinal numbers] is well-ordered but has no order-type. I can make no sense of this statement. The statement that a well-ordered set M possesses an order-type μ asserts that there is a set M' which is similar to M. Since now in all cases M is similar to itself, each well-ordered set defines an order-type. (Hessenberg [1906], p. 631)

(See also p. 550, quoted below.)

What effect does this doctrine have? Consider first which statements will have a meaning according to this doctrine. Statements of the form $card(x) = card(y)$ or $ord(x, r) > ord(y, s)$ are interpreted straightforwardly as 'x and y are equinumerous', 's is a well-ordering of y, r a well-ordering of x and (y, s) is isomorphic to a proper initial segment of (x, r)', etc. According to this, to assert '$card(x) = card(y)$' is to assert *just* 'x and y are equinumerous'. As Hessenberg notes ([1906], p. 536): '. . . of two equivalent sets we say that they are of equal power. Of two similar sets we say that they possess the same order-type.' Introduction of the term 'power' or 'number' is just a convenience. In constructing a pure theory of sets, therefore, informal statements of the form $card(x) = card(y)$, or $ord(x, r) > ord(y, s)$ have a *formal* rendering provided that we can formally render equinumerosity and isomorphism. In other words, we go back to before Cantor's [1878] and deny the step of asserting that powers are objects. So far this fits with the early part of the Frege programme. But how are other statements involving number terms to be interpreted in the pure theory of sets? For example, how do we interpret ordinary statements of informal set theory like '$card(a) = \aleph_1$, or '$card(b) = n$' or even '$card(a) = b$' where b stands for a set? It was statements of this kind, especially the latter, which convinced Frege that it was necessary for $card(a)$ to be intrepreted as referring to some object. Zermelo's solution was in general simply to shun the terms $card(x)$, $ord(x, r)$, and the special number signs ω, ω^ω, $\aleph_1$, etc. According to this position, the informal statement $card(a) = \aleph_1$, simply has no direct counterpart in the formal system. Of course, it may have an *indirect* counterpart. For example, it may be that some statements of the form $card(a) = \alpha$ have an informal equivalent which *can* be rendered in the formal system. Indeed the Zermelo position is that all *meaningful* numerical statements *will* have formal counterparts, direct or indirect. The theoretical principle behind this is that nothing is lost by doing without number terms or number objects: all meaningful numerical statements can be cast in terms of equinumerosity and isomorphism. We find this view in Hessenberg, for example. He seems to imply that the numbers are introduced for our convenience, as a linguistic device for rewriting statements involving only equinumerosity or isomorphism. For instance, he remarks:

An ordinal number μ is regarded as given when a set M is known which appears with it. The statement that a set N is well-ordered according to the number or type μ then asserts nothing more than that N is similar to M. (Hessenberg [1906], p. 550)

In other words, if we use a sign μ it is because we wish to mark a special (ordered) set M. So with respect to a pure set theory, translating statements involving μ presents no difficulty: they can be rendered using just isomorphism and (a formal counterpart to) M.

The view that number objects are theoretically redundant was put most strikingly somewhat later by Fraenkel. For example, in his [1922a], noting the failure of some attempts to define numbers as sets he goes on:

This in itself does not constitute a blow to the foundations of set theory, since strictly speaking for this theory the concept of cardinal number (like that of order-type and ordinal number) is not necessary. For, there are no judgements about cardinal numbers in set theory other than those which state that either two cardinal numbers are equal or that one is smaller than the other: both kinds of judgements can in known ways be reduced to judgements about the equivalence or non-equivalence of sets. (Fraenkel [1922a], p. 154)

This sums up Zermelo's attitude too, at least in practice. Consequently, in working out the basic theory of cardinality and ordinality in the Zermelo system both Zermelo ([1908b]) and Fraenkel ([1925] and [1926]) carefully avoided any suggestion of number objects and number symbols. (See for example, Zermelo's statement and proof of the Cantor and König theorems ([1908b], pp. 211-4).) Of course, Zermelo and Fraenkel originally had no set-theoretic definition of numbers available, so one could argue that their attitude was forced on them. But this is not convincing. Their original attitude is that such a theory was *unnecessary* (cf. §2.3(b)). Nowhere do they state that their position is a provisional one; nowhere do they suggest the development of a set-theoretic theory of number as a goal to work towards. Indeed, even after von Neumann had shown how to develop such a theory Fraenkel still insisted that it was unnecessary (see e.g. [1927], pp. 139-40). Fraenkel relaxed his position later: see, for example, [1932a], p. 1. Zermelo seems to have had doubts about it much earlier. He developed a set-theoretic definition of ordinal number very much like von Neumann's by 1915 at the latest (see §8.1). This, of course, does not indicate conclusively that he thought that this notion of number should be incorporated into the axiom system, or that he had a method of doing so. Nevertheless, by 1930 Zermelo was using the so-called Zermelo–Fraenkel system (with the axiom of replacement) where the von Neumann theory is available, and was making explicit use of cardinal and ordinal numbers. But anyway, whenever these changes of mind took place, it was von Neumann who broke the heuristic pattern that Zermelo and Hessenberg had laid out.

There are in any case various problems with the Zermelo approach. First, there is the problem of inconvenience. Fraenkel, for instance, pointed to the 'intolerable complications' caused by doing without numbers. For example, it is inconvenient not to be able to render simple and important informal statements like $2^{\aleph_1} = \aleph_2$ in the formal system in any direct way. Fraenkel [1922a] tried to get round the problem by introducing an axiomatic theory of *number signs* to stand alongside Zermelo's system of sets. But in doing so he strongly violated Zermelo's reductionism since he effectively treats the signs themselves as primitive objects, allowing them to be elements of sets. Thus we can quantify over them, and so on. He did not pursue the theory. Indeed, as I have mentioned, in his later work of the 1920s and 1930s he stuck rigidly to the Zermelo-Hessenberg idea of doing without numbers and number terms altogether.

In a formal, or semi-formal, axiomatic exercise inconvenience is not necessarily a serious problem. A far more serious problem is that the theory of real numbers appears to depend on the presence of natural number objects. For, following Cantor and Dedekind, rational numbers are defined as ordered pairs of natural numbers and real numbers as certain sets or sequences of rational numbers. As I have pointed out, in his [1909a] Zermelo regards number theory as *nothing but* the theory of finite sets, finitude being defined in purely set-theoretic terms. Here he seems to eschew natural number objects, as we can see in the following passage on the induction principle:

If a proposition is proved on the one hand for every set containing a single element and on the other hand it holds for any finite set whenever it holds for this set minus one of its elements, then the proposition holds for all finite sets. This is what is called reasoning from n to $n + 1$. (Zermelo [1909a], p. 188)

This strong position is somewhat relaxed in his [1908b]. After proving the existence of a set Z_0 containing $\emptyset$, $\{\emptyset\}$, $\{\{\emptyset\}\}$, etc., Zermelo remarks (p. 205) that Z_0 'may be called the *number sequence*, because its elements can take the place of the numerals.' Perhaps he *had* recognized the need for natural number representatives in order to develop a theory of real numbers. But the relaxation is not all that great. For example he makes no attempt to define natural numbers, and he certainly did not go on to develop any number theory based on Z_0. All we have is the very weak statement that Z_0 'can take the place of the numerals'.

More important than these problems, however, is the effect of the Zermelo-Hessenberg–Fraenkel position as a philosophical or heuristic attitude. Reluctance to take numbers objectually meant that, although Zermelo embraced one aspect of Cantorian 'finitism' (the treatment of (some) infinite collections as single objects), he could not embrace the other main aspect which gives rise to the counting theory of infinite size. According to this, size is measured by counting sets off against a fixed stock of ordinal numbers, and the Cantorian well-ordering problem is that of showing that all sets can be so 'counted'. But number objects are *essential* for this approach, because they must be elements in the correspondence which does the counting. Thus a Cantorian reductionist would be forced to take the *first* reductionist course outlined at the beginning of this section. The search for a set-theoretic definition, and an extensive theory, of number would be high on his list of priorities. Not only this, but he would have to look for principles of set existence which guarantee that there are always enough ordinal numbers to count any set. We can sum all this up by saying that a Cantorian reductionist would look to build a system in which it is possible to prove not only 'every set can be well-ordered' but more particularly $\forall x \exists \alpha [x \sim \alpha]$. Zermelo was a reductionist of the *second* kind, regarding the theory of number-objects as unnecessary, and therefore also the theorem 'every set can be well-ordered' as a perfectly adequate solution to the well-ordering problem. This already explains why he developed his axiomatic theory around the (pure) well-ordering theorem rather than one of its number theoretic versions.[2]

(b) *Zermelo's reductionist treatment of the contradictions*

Zermelo's reductionist attitude to numbers already had a strong constraining influence on his approach to the contradictions. For instance, Jourdain wanted to use the ordinal numbers structurally in his criterion of what is a genuine set. Since Zermelo intended to proceed as if there are no number objects, this line of action was denied him. But simple reductionism also had a direct effect on Zermelo's analysis of the contradictions. Suppose one decides that a certain collection X cannot be a set. Then reductionism (sets are the only mathematical objects) dictates that X cannot be regarded as a mathematical object. Consequently, according to reductionism, contradictory collections cannot figure in mathematical proofs. This was precisely Zermelo's position. Recall that Cantor and Jourdain used the collection *Ord* in their proof that every cardinal is an aleph. From Zermelo's point of view this is illegitimate, as may be seen from his [1932] comment on Cantor's 1899 proof:

But even then the doubt would remain that the proof operates with 'inconsistent' multiplicities, indeed possibly contradictory concepts, and hence is already logically impermissible. Doubts of this kind determined the editor [i.e. himself] some years later to base his own proof of the well-ordering theorem purely on the axiom of choice *without* application of inconsistent multiplicities. (Zermelo [1932], p. 451)

This was particularly important. For, as von Neumann showed, to make the proof using *Ord* work axiomatically in addition to an axiom of choice one needs not only a theory of proper classes (non-sets) but also some version of the axiom of replacement. However, Zermelo, according to this account, chose to work solely with the axiom of choice.

---

[2] It is interesting to note that the positions of Bourbaki and other modern structuralists are strikingly reminiscent of the Zermelo–Hessenberg–Fraenkel position. For example, one might consider category theory as concerned not so much with objects but rather with isomorphisms and classes of isomorphisms.

But there is more to Zermelo's reductionism than this. The simple reductionist position is that the 'contradictory' sets are inadmissible; they must be expunged from the theory along with the comprehension principle. But Zermelo adopted a much stronger position which is hinted at in the passage quoted above. This is that the contradictory collections are 'contradictory in themselves', or, to put it more coherently, are *solely* responsible for the contradictions. This greatly strengthens set-theoretic reductionism since it focuses attention on the comprehension principle alone, and away from other possible sources of trouble, in particular away from other more specialized principles of set creation variously involved in the contradictions.

Consider Zermelo's treatment of the Burali-Forti contradiction. According to Zermelo's number-theoretic reductionism, the Burali-Forti contradiction must be about well-ordered sets and not about independent number objects. Zermelo concludes immediately from this that *Ord* cannot be a set, and that comprehension fails:

Since now, on the other hand, 'order-type of a well-ordered set' is certainly a logically admissible *notion*, it follows further—as already appears in a much simpler way from Russell's antinomy, to be sure—that it is not permissible to treat the extension of every arbitrary notion as a set and therefore that the customary definition of set is too wide. (Zermelo [1908a], p. 195)

Or, as he states specifically with regard to Jourdain's treatment of the contradiction (p. 193), '. . . this attempt to resolve the antinomy while retaining *W* [the set of all ordinal numbers] seems to amount to a mere word game.' The implication is that *W* cannot be retained. But of course the Burali-Forti contradiction involves much more than just the set *W*. It also involves crucial principles in the theory of well-ordered sets, in particular the extendability principle: 'every well-ordered set can be extended to an ordinally greater well-ordered set'. Various of these principles had been challenged by critics, some of whom wished to preserve the set *W*. As we have seen, Jourdain challenged the assignment of ordinal numbers to all well-ordered sets, and Schoenflies and Bernstein challenged the extendability principle itself (see Schoenflies [1905] and Bernstein [1905a]). But Zermelo dismisses all these challenges to the theory of well-ordered sets:

On the other hand, the 'Burali-Forti antinomy', recently again so much the subject of discussion, concerning the 'set *W* of all Cantor ordinals' seems to have instilled in some [critics] an all too pervasive scepticism toward the theory of well-ordering. And yet, even the elementary form that Russell gave to the set-theoretic antinomies could have persuaded them that the solution of these difficulties is not to be sought in the surrender of well-ordering but only in a suitable restriction of the notion of set. (Zermelo [1908a], p. 191)

'Suitable restriction' already means 'nothing which interferes with the theory of well-ordering'. The principles involved in the theory of well-ordered sets are hereby protected and absolved from any responsibility for the Burali-Forti contradiction. But this is an extremely strong heuristic doctrine because Zermelo makes it very clear that 'principles involved in the theory of well-ordering' include *all those principles which he had previously used in his 1904 proof of the well-ordering theorem*. Thus he continues the passage quoted above:

Already in my 1904 proof, having such reservations in mind, I avoided not only all notions that were in any way dubious but also the use of ordinals in general; I clearly restricted myself to principles and devices that have not yet by themselves given rise to any antinomy. (Zermelo [1908a], p. 192)

The greatly strengthened form of reductionism appealed to by Zermelo can thus be simply stated: the contradictions are due to a defect in the notion of set, not to the application of any of the 'basic' operations on sets. The principle is also unequivocally stated by Hessenberg:

It is not the operation with the concepts *W* [the set of all ordinals] and *D* [the set of all 'things'] which is the source of the contradictions *but rather the concepts themselves are untenable*. (Hessenberg [1906], p. 634)

For Zermelo, as for Hessenberg, the 'basic' operations are those crucial to his 1904 proof, namely impredicative subset formation, the power-set operation, and the union-set operation. These are also precisely the operations which according to various commentators are implicated in the contradictions. And they (together with the axiom of choice) are precisely the operations which *for this reason* were most under attack from critics of Zermelo's 1904 proof. Baire, Borel, and Lebesgue attacked the power-set operation, Poincaré attacked impredicative subset formation, and Schoenflies and Bernstein attacked the union-set operation by attacking the extendability principle. Zermelo's strengthened reductionism in effect assumes the consistency of these operations when arbitrarily applied. That is, it assumes that they can be safely applied to *all* sets, and that, whatever sets we eventually decide do exist, among them must be the sets described by these procedures. Thus, not only does Zermelo's strong reductionism force the abandonment of the comprehension principle, it puts rigid constraints on how comprehension is to be modified. It assumes that comprehension *can* be applied to the properties mentioned in the power-set and union-set operations respectively:

$$[x \subseteq a] \quad \text{and} \quad \exists z[z \in a \wedge x \in z]$$

for any given parameter set *a*. And it assumes that, in 'selecting' subsets from a given set *a*, impredicative procedures are legitimate. Note how close we are already to the 'reconstructed' or modified limitation of size argument of §5.2.

One may question whether the consistency of the operations mentioned is merely *assumed*. Does not Zermelo argue for it? There are two possible lines of argument suggested in the two passages from his [1908a] recently quoted. The first is suggested by the frequent reference to the apparently elementary nature of Russell's paradox, and the second by reference to 'principles and devices that have *not yet by themselves given rise to any antinomy'*. The two lines are in fact woven supportively together.

The first argument would go as follows. Forgetting about impredicativity (perhaps by an appeal to realism), Russell's paradox involves none of the threatened operations. It involves only the Russell set and the law of excluded middle applied to set membership. Since Zermelo did appeal to realism (in defending the axiom of choice as well as impredicative definitions), he would presumably not countenance rejection of excluded middle here. Indeed let us accept that excluded middle applied to membership is basic to the concept of set: any independently existing set must either belong or not belong to any other set. Given this we are forced to conclude that Russell's contradiction must strike at the notion of set itself. This argument (or something like it) is what I take to lie behind Zermelo's attempt to direct attention away from the 'complicated' Burali-Forti contradiction:

And yet, even the elementary form that Russell gave to the set-theoretic anti-nomies could have persuaded [us] that the solution of these difficulties is not to be sought in the surrender of well-ordering but only in a suitable restriction of the notion of set. (Zermelo [1908*a*], p. 191)

This argument has some plausibility. Russell's contradiction does have a more elementary form than the other contradictions, and, given excluded middle for membership, simple reductionism (the only objects are sets) allows us to conclude that Russell's 'set' is no object. Zermelo assumes that because the set concept (based on comprehension) is at fault in Russell's paradox it must be at fault in the Burali-Forti paradox also. Of course, it may be that some of the 'basic' operations are also defective, but Zermelo's assumption that they are not is a natural one. In effect, the argument uses a principle of minimum mutilation: we have found one source of difficulty; let us leave it at that. In particular, it helps to answer the question about where the set-theoretic axioms come from: since the theory of well-ordering and the proof of the well-ordering theorem are protected by the simplicity argument, we can use any of the principles involved as axioms if we so choose. And this is exactly what Zermelo *did* choose.

While this first argument throws a protective blanket around the specific set principles challenged by Zermelo's various critics, the second would help to support them directly. It goes as follows. Any challenge to the operations mentioned above is also a challenge to large and unexceptionable parts of classical mathematics. Baire's attack on the power-set axiom (see Baire's letter to Hadamard in Baire, Borel, Hadamard, and Lebesgue [1905], pp. 151-2) is an attack on the non-denumerable in mathematics, and thus is an attack on the continuum, on the use of the set of all natural number-valued functions, etc. (see Hessenberg [1906], p. 639). An attack on the union-set axiom is an attack on the successor operation, on Cantor's second number class, and even on the theory of finite sets (that is to say, Zermelian number theory). (For the attacks see Bernstein [1905a] and Schoenflies [1905]. For the consequences see Hessenberg [1906], pp. 632-3.) An attack on impredicative definitions and procedures is an attack also on classical methods in analysis (see Zermelo [1908a], pp. 190-1). But no contradictions have ever appeared in any of these threatened fields. The operations under attack appear to be harmless. Consequently, if application of the operations to a putative set gives rise to a contradiction, then the putative set must be to blame and not the operation itself.

This type of argument is actually used by Hessenberg ([1906], p. 633) and Zermelo ([1908a], p. 195) in defending the union-set operation. And Zermelo uses it to reply to Poincaré's challenge to impredicative definitions, stiffening this reply with an explicit appeal to realism (see [1908a], p. 191). The argument is similar to that for the power-set axiom outlined in §5.3. Of course, it does not directly appeal to Cantorian finitism or to the need to mathematize properties as the argument there does. But it does implicitly say that not only are the operations mentioned safe in the realm of finite sets, but their application in the production of those infinite sets crucial to classical mathematics is also apparently safe. By itself this argument does not justify arbitrary use of these operations. But it does *support* them, and if we apply Cantorian finitism (as Zermelo did, for example in the case of the axiom of choice) and the demand to be able to treat properties and relations set theoretically, arbitrary application is quite natural.

7.3 Reductionism and well-ordering

(b) Zermelo's 1904 proof

As I have hinted above, Zermelo's 1904 proof of the well-ordering theorem played a crucial role in his treatment of the paradoxes. It was criticism of this proof which prodded Zermelo into a full-scale axiomatization of his theory, which caused him to make explicit his reductionism and to defend the set existence principles on which the proof is based. His reply is contained in the two papers of 1908. Before we consider these we have to consider the 1904 proof itself.

The basis of the 'naive' Cantor-Hardy-Jourdain proof of the aleph theorem is the idea of successive and dependent choices. The intuitive idea (if it were possible to carry it through) is that one starts with the set M, and keeps on selecting elements and placing them in sequential order until M is exhausted. But Zermelo eschewed any notion of 'successive selection', as he makes clear in his [1932]. Criticizing Cantor's proof that 'any definite multiplicity V which has no aleph as its cardinal must contain a subset V' equivalent to the whole ordinal sequence' (see §4.1), he writes:

Here temporal intuition is applied to a process which surpasses all intuition, and a fictitious being posited which it is assumed could make *successive* arbitrary choices, and thereby define a subset V' of V which, by the conditions imposed, is plainly *not* definable. Only by applying the 'axiom of choice' which postulates the possibility of a *simultaneous* choice and which Cantor unconsciously and instinctively uses everywhere, but nowhere formulates explicitly, could V' be defined as a subset of V. (Zermelo [1932], p. 451)

Zermelo's assumption and formulation of the axiom of choice in his 1904 is an explicitly realist one. It assumes, for any arbitrary set M, that there *exists* a *simultaneous* 'choice' of an element from each non-empty subset of M. Using Zermelo's notation and terminology, let **M** be the set of all subsets of M; **M** is then said to be 'covered' by a covering γ' if γ assigns to each (non-empty) $M' \in \mathbf{M}$ an element $m' \in M'$. Then

The present proof rests upon the assumption that coverings γ actually do exist, hence upon the principle that even for an infinite totality of sets there are always mappings that associate with every [non-empty] set one of its elements . . . (Zermelo [1904], p. 141)

Thus, the axiom is a pure existence assumption; all reference to construction, even by an ideal 'super being', is eschewed. Note also that as well as the axiom of choice, Zermelo is assuming that for arbitrary M the power set **M** must also exist. (The dependence of the choice postulate on the power-set postulate is strongly emphasized by Hessenberg in his [1906], p. 638.)

The proof proceeds as follows. Let γ be a given (arbitrary but fixed) covering of **M**. Then Zermelo defines a γ-set as a set $M_\gamma \subseteq M$, which is well-ordered and such that if $a \in M_\gamma$ and if $A = \{x : x \in M_\gamma$ and $x < a$ in the well-ordering of $M_\gamma\}$, then a is the distinguished element of $M - A$ according to the covering γ. Considered intuitively, say from the Cantor–Jourdain point of view, the γ-sets are those well-ordered parts of M which are 'built up' from $\emptyset$ by tacking on the distinguished elements of the remainders. Zermelo proves that there are γ-sets, e.g. $\{m_1\}$ where m_1 is the distinguished element of M itself, $\{m_1, m_2\}$ where m_2 is the distinguished element of $M - \{m_1\}$, and so on. He then defines L_γ as the union of all the γ-sets, i.e. $L_\gamma = \{x : x \in M \land \exists M_\gamma \, [x \in M_\gamma]\}$. The proof is completed by showing that L_γ is itself a γ-set, and hence well ordered, and that $L_\gamma = M$.

There are various important points here. First, the definition of L_γ is impredicative: it is itself a γ-set and yet is characterized in terms of the set of *all* γ-sets. This is precisely the kind of definition to which Poincaré objected and on which be blamed the contradictions. Second, the proof that L_γ is a γ-set and equal to M itself is important. The argument is in two parts:

(i)   showing that for any two non-equal γ-sets one is an initial segment of the other

(this shows that L_γ is a γ-set, indeed the greatest such);

(ii)   showing that $M - L_\gamma = \emptyset$.

The proof of (ii) is as follows. Assume that $M - L_\gamma \neq \emptyset$; then $M - L_\gamma$ has a distinguished element, say m'. It is not difficult to show that $L_\gamma \cup \{m'\}$ itself forms a γ-set where the ordering among members of L_γ is retained and m' is assumed to follow all elements of L_γ. But L_γ is the greatest γ-set; hence contradiction. Therefore $M - L_\gamma = \emptyset$. (For the whole proof, see Zermelo [1904]. This proof is reproduced in various guises in Kamke [1950], pp. 110-5, Kelley [1955], p. 35, Halmos [1960], pp. 68-9 and Drake [1974], p. 62, exercise 6.)

Step (ii) is precisely the step that worried Bernstein and Schoenflies. The argument for (ii) uses the extendability principle for well-ordered sets, 'every well-ordered set can be extended to an ordinally greater one' (or every ordinal β has a successor $\beta + 1$), and moreover mimics the use of this principle in the Burali-Forti argument. (In both cases it is used to derive a contradiction.) Thus if L_γ was already the set of all ordinals or some similar set the whole proof would collapse, not just the assumption $L_\gamma \neq M$. So:

The possibility that for a definite set, for example the continuum, the set L_γ of γ-elements could be similar to W [the set of all ordinals] is not disproved. The conclusion that $M = L_\gamma$ is, moreover, only permissible *if* $L_\gamma \neq W$ . . . (Bernstein [1905a], p. 193)

And Schoenflies ([1905]) declared that the extendability principle '. . . leads to a contradiction in the theory of well-ordered sets: it therefore cannot be considered as the basis of a mathematical proof.' The criticism is clear: the proof takes no precautions against the intrusion of the Burali-Forti contradiction. Cantor and Jourdain, remember, *had* taken such precautions. Indeed, in their proof the 'contradiction' emerges as an argument by *reductio*, just as it does in the modern textbooks.

Zermelo's reply in his [1908a] is cutting and dismissive (as are Hessenberg's remarks in his [1906]). But the point being made is a good one. Whatever Zermelo's intention, there is no attempt to exclude the possibility that $L_\gamma = W$ and no suggestion that any danger of contradiction could threaten. Of course, Zermelo in his [1908a] (p. 198), referring to critics who 'base their objections upon the Burali-Forti antinomy', declared that this antinomy is '. . . in fact . . . without significance from my point of view, since the principles I employed *exclude* the existence of a set W [of all ordinals].' Even if it referred to the *second* proof and the accompanying axiom system the correctness of this remark is not obvious. But as a reply to criticism of the 1904 proof it is unfair. Zermelo does *not* repudiate the comprehension principle in his [1904], he does *not* say that the existence principles on which the proof is based are the *only* set existence principles, and he does not divorce the proof of the theorem from the Cantorian assumptions about well-ordering and ordinals.

This last point is especially important. In the first place, Zermelo assumes that 'every set can be well-ordered' is equivalent to the Cantorian 'every cardinality is an aleph' ([1904], p. 141). Secondly, despite his later claim in his [1908a] (p. 192), he does *appear* to use the ordinals and the 'naive' Cantorian theory of well-ordering in his definition of γ-sets. His proof that there are γ-sets involves 'rank ordering', i.e. the notion of ordering according to an ordinal (rank) index. And the definition of a γ-set is 'any well-ordered M_γ ...' without specifying how 'well-ordered set' is to be defined. How do we know that *this* can be reduced to Zermelo's principles?

In recasting his proof for his [1908a] Zermelo employed his reductionism to get rid of any association with ordinals and (as we have seen) to defend his main set existence principles against the accusation that they are implicated in the contradictions. But the major change involved in the shift (and what makes the second proof immeasurably more complicated) was Zermelo's reductionist treatment of well-ordering. It is to this that we must now turn.

(b) Inclusion orderings and the 1908 proof

The trick underlying Zermelo's second proof of the well-ordering theorem is to use only inclusion orderings. The effect of this was to emphasize the importance of the power-set axiom, as we shall see. Zermelo's proof is best understood through the earlier work of Hessenberg ([1906], pp. 674–85) and the later work of Kuratowski who took up the cause of reductionism vigorously in his [1921] and [1922].

First one should remember that the modern method of reducing relations to sets via the definition of the ordered pair was not discovered until 1921 by Kuratowski. We shall see that this was a direct consequence of the Hessenberg-Zermelo–Kuratowski reduction. In his [1906], Hessenberg first pointed out a simple way of approaching the reduction of ordering to purely set-theoretic principles. If M is a totally-ordered set ordered by some relation $<$, then the collection of all remainders (or segments) determined by $<$ form an inclusion chain of subsets of M. For a remainder under $<$ is a subset of M determined by an arbitrary given element a of M, namely $R_a = \{x : x \in M \land x \geqslant a\}$. And given any two remainders R_a and R_b we have $R_a \subseteq R_b$ or $R_b \subseteq R_a$. What is more, for segments the ordering $\subset$ on the R_a mimics the $<$ ordering on M, and for remainders $\subset$ inverts the ordering. So if M is well ordered by $<$, the R_a are well-ordered by *reverse* inclusion and the two orderings are *isomorphic*. This simple observation opened the way to a reduction of ordering to Zermelo set theory. For instead of considering an abstract ordering $<$ on M one can consider a certain subset **M** of $P(M)$ ordered by the 'pure' relation '$\subset$', or its inverse. This is not the end of the matter. For one might want to consider (as Zermelo does in his [1904]) *all* total- or well-orderings on a given set. Before this can be done, it is necessary to specify necessary and sufficient conditions for an inclusion chain **M**, with $\mathbf{M} \subseteq P(M)$ and $\mathbf{U}\mathbf{M} = M$, to be the set of all remainders (or segments) of an abstract (well- or total-) ordering $<$ on M. Given this, the theory of ordering is reducible to two basic principles (power-set, and 'arbitrary' subset definition) and the relation '$\subset$' which itself is reducible to the '$\in$' relation. Kuratowski summed this up in his [1921]:

Thus, the theory of classes establishing order [i.e. of inclusion chains] can be regarded as equivalent to the classical theory of ordered sets based on the intuitive notion of order (Cantor). At the same time, one can deduce it from the general theory of (non-ordered) sets without the necessity of introducing any prior, supplementary notion: the idea of order is then given in the fundamental terms of Zermelo's system of axioms, namely in terms of set and element. The importance of this method is manifest. (Kuratowski [1921], p. 162-3)

(This method of reducing ordering relations to sets is explained briefly in Fraenkel [1927], pp. 135-6, and Fraenkel and Bar-Hillel [1958], pp. 127-32.)

Let us begin by considering necessary and sufficient conditions for an M to be a set of remainders of an ordering on a given set M. Such conditions were first achieved by Hessenberg. Consider the following:

(A) M uniquely represents the set of all remainders of a total ordering on a set M

if and only if

(1) $\mathbf{M} \subseteq P(M)$
(2) if $X,\ Y \in \mathbf{M}$ then either $X \subseteq Y$ or $Y \subseteq X$
(3) $\forall x,\ y \in M$, if $x \neq y$ then there is an $X \in \mathbf{M}$ which contains one of $x,\ y$ but not the other.
(4) if $\mathbf{M}' \subseteq \mathbf{M}$ then $\mathbf{U}\,\mathbf{M}' \in \mathbf{M}$
(5) if $\mathbf{M}' \subseteq \mathbf{M}$ then $\bigcap \mathbf{M}' \in \mathbf{M}$
(6) $\emptyset \in \mathbf{M}$ and $M \in \mathbf{M}$.

This result is Hessenberg's, as processed successively in Hartogs [1915], p. 440, theorem 7, and appendix, pp. 442-3, Kuratowski [1921], pp. 161-2, Fraenkel [1926], pp. 135-5, and Fraenkel [1927], pp. 135-6. Hessenberg's original result was stated for segments, not for remainders (a minor point), and was considerably more complicated. It shows that a system $\mathbf{M}$ satisfying conditions (1), (2), and (3), and such that $\mathbf{U}\,\mathbf{M} = M$ is *complete* (i.e. will contain all segments of the derived ordering) if and only if

(4') if $\mathbf{M}'$ is a segment of $\mathbf{M}$ in the subset ordering on the latter then $\mathbf{U}\,\mathbf{M}' \in \mathbf{M}$
(5') if $\mathbf{M}'$ is again a segment, $\mathbf{N} = \mathbf{M} - \mathbf{M}'$ and $N = \bigcap \{A : A \in \mathbf{M} \cap \mathbf{N}\}$, then $N \in \mathbf{M}$.

(See Hessenberg [1906], pp. 681-2.) Hartog's, stating the result for remainders, greatly simplified it by getting rid of Hessenberg's (4') and the complicated (5'), and replacing them by his own

(4'') If $\mathbf{M}' \subseteq \mathbf{M}$ then $\mathbf{U}\,\mathbf{M}' \in \mathbf{M}$

(Hartogs [1915], p. 433.) Kuratowski states the result using conditions (1)-(5), pointing out with a counter-example that Hartogs had in fact oversimplified. Fraenkel added condition (6). (See his [1927], p. 136, n. 1. Kuratowski's omission of $\emptyset$ ($M \in \mathbf{M}$ follows from (3) and (4)) had highly important and fruitful consequences; see below.)

Conditions (1)-(6) are natural conditions on $\mathbf{M}$. If $<$ is a total ordering on M then the set $\mathbf{M}$ of all remainders under $<$ satisfies (1)-(6). Conversely, if $\mathbf{M}$ satisfies (1)-(6) then $\mathbf{M}$ is the set of all remainders of M under the relation $<$ defined by

(*) $\forall x, y \in M[x < y \text{ iff } \exists X \in \mathbf{M}[x \notin X \wedge y \in X]]$.

(To state the theorem for segments not remainders replace (*) by

(**) $\forall x, y \in M[x < y \text{ iff } \exists X \in \mathbf{M}[x \in X \wedge y \notin X]]$.)

Conditions (1) and (2) establish that $\mathbf{M}$ is an inclusion chain of subsets of M, while (3)-(6) are essentially *maximality* conditions. (Maximality was first singled out as being important by Kuratowski; I shall show below that it is precisely to guarantee the maximality of an inclusion chain that one needs the axiom of choice.) Given these or similar conditions it is not difficult to show (using the power-set and separation axioms) that if M is a Zermelo set there is also a Zermelo set representing all the orderings on M (see Fraenkel [1926], pp. 136-7). (A) covers total orderings. For well-orderings there is the closely related

(B) $\mathbf{M}$ uniquely represents the set of all remainders of a well-ordering of M

if and only if

(1)-(6) and, in addition, $\mathbf{M}$ is well-ordered by inclusion.

(See Hessenberg [1906], p. 680.)

In the light of theorem (B) (or Hessenberg's more complicated version of it), to prove the well-ordering theorem it is now sufficient to show that for any set M there is a well-ordered inclusion chain $\mathbf{M}$ satisfying conditions (1)-(6). This is just what Zermelo's well-ordering proof of 1908 *does* show, albeit clumsily. Zermelo's central theorem is rather obscure:

THEOREM. If with every non-empty subset of a set M an element of that subset can be associated by some law as 'distinguished element', then $\mathfrak{U}(M)$, the set of all subsets of M, possesses one and only one subset $\mathbf{M}$ such that to every arbitrary subset P of M there always corresponds one and only one element P_0 of $\mathbf{M}$ that includes P as a subset and contains an element of P as its distinguished element. The set M is well-ordered by $\mathbf{M}$. (Zermelo [1908a], p. 184)

It transpires that the $\mathbf{M}$ of Zermelo's theorem *is* well-ordered and satisfies (1)-(6), and thus, by (B) represents a well-ordering on M. Zermelo's Theorem specifies a rather strange-looking condition on $\mathbf{M}$, namely that there is a map $f: P(M) \to \mathbf{M}$ such that $\forall X \in P(M)\ [X \subseteq f(X) \wedge \phi(f(X)) \in X]$, ϕ being a choice function for M. The condition can be explained as follows. Suppose $\mathbf{M}$ *is* well-ordered (and Zermelo's $\mathbf{M}$ is). Now put $\mathbf{M}' = \{\{x\}: x \in M\}$; then $f \upharpoonright \mathbf{M}'$ takes

$\mathbf{M}'$ one–one onto $\mathbf{M}$, i.e. $\mathbf{M} = \{f(\{x\}): x \in M \}$. Given this, the well-ordering of $\mathbf{M}$ is mirrored on $\mathbf{M}'$ by the function f^{-1}; the induced well-ordering on $\mathbf{M}'$ can then be mapped onto M itself by the function g which takes $\{x\}$ to x. Thus, it is not so much the map f which is important, but rather the existence of the map $f' = f \upharpoonright \mathbf{M}'$. That there is such an f' follows rather simply from the fact that $\mathbf{M}$ satisfies the conditions set out in (B) (or rather, the equivalent Kuratowski conditions in (B') explained below).

I do not propose to explain Zermelo's proof (for which see his [1908a], pp. 184–6) directly. Rather I shall explain why the $\mathbf{M}$ he defines does the required job by examining it through the work of Kuratowski's [1921] and [1922]. This will not distort Zermelo's proof, since Kuratowski's work simply makes clear why Zermelo and, following him, Hausdorff prove what they do prove about $\mathbf{M}$. Moreover, my explanation will make certain logical and historical points clear. In particular, it explains the precise role of the axiom of choice in Zermelo's proof, and how Kuratowski came to give his famous set definition of ordered pair.

Kuratowski [1921] radically simplifies the conditions of (A) and (B). Kuratowski proved:

(A')   A set $\mathbf{M}$ represents the set of all remainders of a total-ordering on a set M

if and only if

$\mathbf{M}$ is a maximal reverse inclusion chain in $P(M)$

and

(B')   A set $\mathbf{M}$ represents the set of all remainders of a well-ordering on a set M

if and only if

$\mathbf{M}$ is a maximal, well-ordered, reverse inclusion chain in $P(M)$.

(See Kuratowski [1921], pp. 161–8.) These results were highly important, as I shall explain. According to (B'), to explain why Zermelo's proof works it is enough to show that Zermelo's $\mathbf{M}$ is a maximal, well-ordered, reverse inclusion chain in $P(M)$. Zermelo's $\mathbf{M}$ is defined by the following steps:

(1)   Let M be the given set, and $\phi(X)$ a choice function on $P(M) - \{\emptyset\}$.
(2)   Call $\mathbf{Z}$ a Θ_M-chain in $P(M)$ if
    (i) $\mathbf{Z} \subseteq P(M)$
    (ii) $M \in \mathbf{Z}$
    (iii)   if $X \in \mathbf{Z}$ then $X - \{\phi(X)\} \in \mathbf{Z}$
    (iv)   if $\mathbf{Z}' \subseteq \mathbf{Z}$ then $\bigcap \mathbf{Z}' \in \mathbf{Z}$.
    There exist Θ_M-chains since $P(M)$ is one.
(3)   Define $\mathbf{M}$ as the interesection of all the Θ_M-chains. Since there are Θ_M-chains $\mathbf{M}$ is non-empty.

From this definition it is easy to show that **M** satisfies the (B′) conditions.

1. **M** *is well-ordered.* To explain this, we must turn to Kuratowski [1922]. In that paper, Kuratowski was looking for a way of eliminating the necessity for transfinite recursion over the ordinals as part of an attempt to justify Zermelian reductionism. Kuratowski openly acknowledged that transfinite numbers had been of enormous historical and heuristic importance. Applications of transfinite numbers, said Kuratowski ([1922], p. 76) '. . . have contributed time and time again to progress in various domains of mathematics; besides, it was in virtue of these applications that they were developed by Cantor.' But transfinite numbers were not available in Zermelo's system. Hence:

. . . in reasoning with transfinite numbers one implicitly uses the axiomatic assumption that these numbers exist; but, it is desirable both from the logical and mathematical point of view to reduce the system of axioms employed in proofs. In addition, this reduction will increase the aesthetic value of the arguments, for it will remove from them an essentially foreign element. (Kuratowski [1922], p. 77)

(Cf. Lindelöf [1905], p. 183, and §2.3(*b*).) Thus, Kuratowski took the Fraenkel-Zermelo view of transfinite numbers, namely that they are 'useful but unnecessary'. (This, one might add, was diametrically opposed to Hausdorff's view in his classic [1914] text. Hausdorff regarded Zermelo's axiom system as 'incomplete' ([1914], p. 2) and given the remarks from p. 275 quoted in §2.3(*b*) one of his reasons for this view, I suggest, was that the transfinite numbers were *not* available in Zermelo's system. Hausdorff preferred an informal approach with the cardinal and ordinal numbers postulated (see [1914], pp. 46-7, 73). Kuratowski then undertook to *prove* that the transfinite numbers are unnecessary, or at least that in their most important applications they can actually be dispensed with. Invariably, application of the ordinals in analysis, topology, etc. had focused on definitions by transfinite recursion over the ordinals. In pursuit of reductionism, Kuratowski succeeded in showing that in a large class of cases this kind of definition can be obviated by a pure set-theoretic method reproducible in Zermelo's system. As he noted:

From the viewpoint of Zermelo's axiomatic set theory one can say that the method explained here allows us to deduce theorems of a certain well-determined general type *directly* from Zermelo's axioms, that is to say, without the introduction of any independent, supplementary axiom about the existence of transfinite numbers. (Kuratowski [1922], p. 77)

Kuratowski's 'general method' is justified by his proof of the equivalence of the following two definitional procedures. Both start from a fixed set E, a fixed subset M of E, and a set function G defined on $P(E)$ with values in $P(E)$ and such that

$$\forall X \in P(E)[G(X) \subseteq X].$$

Procedure 1 is as follows. Define a Θ_E-chain to be any set **Z** satisfying the conditions

(i) $\mathbf{Z} \subseteq P(E)$
(ii) $M \in \mathbf{Z}$
(iii) if $X \in \mathbf{Z}$ then $G(X) \in \mathbf{Z}$
(iv) if $\mathbf{Z}' \subseteq \mathbf{Z}$ then $\bigcap \mathbf{Z}' \in \mathbf{Z}$.

Now let $\mathbf{M}(M)$ be the intersection of all Θ_E-chains; the set of Θ_E-chains is non-empty since $P(E)$ itself is one; hence $\mathbf{M}(M)$ (certainly non-empty) is itself a Θ_E-chain.

This, clearly, is a generalization of Zermelo's definition of his $\mathbf{M}$ given above.

Procedure 2 is as follows. Define the following sequence by transfinite recursion

(i) $M_0 = M$
(ii) $M_{\alpha+1} = G(M_\alpha)$
(iii) $M_\lambda = \bigcap_{\beta < \lambda} M_\beta$ for limit λ.

Then put

$$\mathbf{A}(M) = \{M_\alpha : \alpha \text{ an ordinal}\}$$

Kuratowski then shows that $\mathbf{M}(M)$ and $\mathbf{A}(M)$ represent precisely the same set. The argument is quite simple. Certainly $\mathbf{A}(M)$ must be a Θ_E-chain, so $\mathbf{M}(M) \subseteq \mathbf{A}(M)$ since $\mathbf{M}(M)$ is the smallest Θ_E-chain. Assume now that $\mathbf{A}(M) - \mathbf{M}(M) \neq \emptyset$ and that M_α is the set with the least ordinal index in $\mathbf{A}(M) - \mathbf{M}(M)$. If α is a limit ordinal, all M_β with $\beta < \alpha$ belong to $\mathbf{M}(M)$; hence by clause (iv) of Procedure 1, $M_\alpha \in \mathbf{M}(M)$. If α is a successor ordinal $\beta + 1$, then $M_\beta \in \mathbf{M}(M)$, and, by clause (iii) of Procedure 1, $M_\alpha \in \mathbf{M}(M)$. Thus, we have a contradiction, and so $\mathbf{A}(M) - \mathbf{M}(M) = \emptyset$.

Definitional Procedures 1 and 2 are therefore (informally) equivalent. The Zermelo-style Procedure 1 is a perfectly adequate replacement for the recursion 2. (Indeed, Kuratowski ([1922], p. 81) further proved an analogue of the transfinite induction theorem for the $\mathbf{M}(M)$, namely: Let ψ be a set property; if $\psi(M)$, $\forall X \subseteq E [\psi(X) \rightarrow \psi(G(X))]$, and $\forall \mathbf{Y} \subseteq P(E) [\forall X \in \mathbf{Y} [\psi(X)] \rightarrow \psi(\bigcap \mathbf{Y})]]$, then $\forall X \in \mathbf{M}(M)[\psi(X)]$.) More to the point, it shows immediately why Zermelo's $\mathbf{M}$, defined by Procedure 1, is well-ordered. To obtain Zermelo's $\mathbf{M}$, let M and E be the same, and put $G(X) = X - \{\phi(X)\}$ (ϕ being the choice function). By Kuratowski's result, $\mathbf{M}$ is equal to $\mathbf{A}$ defined by recursion, and $\mathbf{A}$ is well-ordered under the ordering of the indices. So, arguing informally, $\mathbf{M}$ must be well-ordered by the same ordering. This also gives us the clue as to the *final* ordering of $\mathbf{M}$ as it would look from inside Zermelo's set theory, deprived as it is of the presence of the ordinals. We know $M_\alpha \underset{\mathbf{A}(M)}{<} M_\beta$ if and only if $M_\alpha \supset M_\beta$, so 'inside' Zermelo's system we can say that **M** is well-ordered by reverse inclusion, which is just what is required according to Kuratowski's theorem (**B'**).

2. **M** *is maximal.* Maximality can be characterized as follows:

A well-ordered, reverse inclusion chain **M** in $P(M)$ is maximal

if and only if

the chain contains M and $\emptyset$, successors in the chain differ by only a single element, and any limit element is just the intersection of its predecessors.

Proof: It is clear that if the chain **M** does not contain M and $\emptyset$, or if two successors differ by *more* than one element, then it would be possible to *extend* **M** by 'squeezing in' another element, thus showing that **M** cannot be maximal. So the conditions specified are necessary for maximality. Let us show that they are sufficient. Assume that **M** is a chain in $P(M)$, well-ordered by reverse inclusion, containing $\emptyset$ and M, such that successors differ by only a single element, with any limit element equal to the intersection of its predecessors. Let **M'** be a well-ordered reverse inclusion chain in $P(M)$ with $\mathbf{M} \subset \mathbf{M'}$ and let B be the least element in **M'** which is not also in **M**. The initial segment of **M'** determined by B is isomorphic to some initial segment of **M** under identity. Let the element of **M** determined by this latter segment be A_0. But A_0 is also in **M'**. However, since it can neither be less than B in the **M'** ordering nor equal to it (since $B \notin \mathbf{M}$) then it must be greater than B, i.e. $A_0 \subset B$. Since the segments $\{X: X \in \mathbf{M'} \wedge B \subseteq X\}$ and $\{X: X \in \mathbf{M} \wedge A_0 \subseteq X\}$ are isomorhpic and B and A_0 are corresponding elements, either B and A_0 are both successor elements or both limit elements. Suppose both are limits. Then, being a limit in **M**, $A_0 = \bigcap\{X: X \in \mathbf{M} \wedge A_0 \subset X\}$. But then since $\{X: X \in \mathbf{M} \wedge A_0 \subset X\} = \{X: X \in \mathbf{M'} \wedge B \subset X\}$, A_0 must be the supremum of $\{X: X \in \mathbf{M'} \wedge B \subset X\}$ in **M'**, i.e. $B \subseteq A_0$. This contradicts $A_0 \subset B$. Suppose both A_0 and B are successors. Then by the definition of B, they must both have the *same* immediate predecessor, say A'. So we have $A_0 \subset B \subset A'$, but this contradicts the assumption that A_0 and A' differ by only a single element (A_0 being the immediate successor of A' in **M**). Thus, B can be neither a limit element nor a successor element, which is absurd. Consequently, $\mathbf{M} = \mathbf{M'}$ and **M** is maximal.

Given this equivalence it is now easy to show that Zermelo's **M** is maximal. Certainly it contains M (we shall deal with $\emptyset$ in a moment), and we saw from the analysis of the ordering on **M** that any limit element is the intersection of its predecessors. The interesting condition is that governing successors, for it is *precisely to guarantee this* that the axiom of choice is used. In the Kuratowski Procedures 1 and 2 above, the successor of any element X in $\mathbf{M}(M)$ is just $G(X)$. In the particular Zermelo case, $G(X)$ is defined by $G(X) = X$ if $X = \emptyset$, otherwise $G(X) = X - \{\phi(X)\}$, ϕ being the fixed Zermelo choice function on $P(M) - \{\emptyset\}$. Thus the shift from X to its successor $G(X)$ involves the 'wasting' of only a *single element*, the distinguished element $\phi(X)$ of X. Thus, successors differ by only a single element, the axiom of choice being needed to guarantee that $G(X)$ can be defined in just this way. Thus choice guarantees that nothing can be 'squeezed' in between two successors, and this is crucial to maximality.

At first sight it seems paradoxical that the axiom of choice is *not* needed to show that **M** is well-ordered, but only to show that it is maximal. But this is not really so strange. After all, it is easy to build up elementary well-ordered chains in $P(M)$, even when the conditions concerning M, $\emptyset$ and limit elements are imposed. But these simple chains will not generally enable us to define a well-ordering on M. For this we need maximality, and crucially the condition of successor elements. And as we have seen we need the axiom of choice to guarantee that this condition can be satisfied generally. Given maximality we can define a simple map f' which takes $\mathbf{M} - \{\emptyset\}$ one-to-one onto $\mathbf{M}' = \{\{x\} : x \in M\}$ namely by $f'(A) = A - A'$, A' being the successor of A in the reverse inclusion ordering of **M**. f' is one-to-one, for if $A, B \in \mathbf{M} - \{\emptyset\}$ and $A \neq B$, it is easy to show that $\phi(A) \neq \phi(B)$. For assume without loss of generality that $B \subset A$; then we know that $B \subseteq A'$, since $A' \subset B$ is ruled out by the single element difference between A and A'. But then $\phi(A) \notin B$, and hence $\phi(A) \neq \phi(B)$. Thus $A - A' \neq B - B'$, and f' is one-to-one. f' is *onto* $\mathbf{M}'$, for if $m \in M$ then $\{m\} = f'(A_0)$ where A_0 is the intersection of $\{X : X \in \mathbf{M} \wedge m \in X\}$, i.e. the greatest element of **M** containing m. Now the selected element $\phi(A_0)$ of A_0 must be m, for if not we would have $A_0' \subset A_0$ and $m \in A_0'$ since m would not be the unique element 'wasted' by the shift from A_0 to A_0'. But this means that there would be a *greater* element of **M** than A_0 containing m, and this is impossible. Thus $f'(A_0) = A_0 - A_0' = \{\phi(A_0)\} = \{m\}$. The well-ordering of $\mathbf{M} - \{\emptyset\}$ can then be imaged on $\mathbf{M}'$ and, by the simple map $g(\{x\}) = x$, consequently on M itself. (Hausdorff uses essentially these maps for imaging the ordering of **M** onto M; see his [1914], p. 138.) Note that the f' defined here using maximality (i.e. the singleton difference between successors) does exactly the job required of the f' I mentioned above after quoting Zermelo's central theorem.

The above analysis makes it quite clear that Zermelo's **M** is a maximal, well-ordered, reverse inclusion chain in $P(M)$. According to Kuratowski's (B') this guarantees that it represents a well-ordering on M. The map f' defined above shows explicitly how the well-ordering of **M** can be imaged on M. Close examination of Zermelo's proof, and particularly its much more perspicuous rendering by Hausdorff in his [1914] (pp. 136–8), shows that it follows more or less the steps set out above. It is proved that **M** is well-ordered by reverse inclusion, that successors differ by single elements (the condition that if $B \subset A$ then $B \subseteq A'$), and so on. (Note also that Zermelo refers to **M** as a set of 'remainders'.) The advantage of following Kuratowski's work is that it shows *why* the proof should take precisely these steps.

There are various points still to be cleared up concerning the details and structure of the Zermelo proof. First, why can the condition that **M** contain $\emptyset$ be dropped from the maximality conditions? The reason is that it actually follows from a result of Kuratowski's on the existence of fixed points for the function $G(X)$ in Procedures 1 and 2. Kuratowski proved $\exists M_0 \in \mathbf{M}(M)\ [G(M_0) = M_0]$. The proof is easy. Since $\mathbf{M}(M) \subseteq \mathbf{M}(M)$ then by definition $\cap \mathbf{M}(M) \in \mathbf{M}(M)$. Let $M_0 = \cap\mathbf{M}(M)$, then $M_0 \in \mathbf{M}(M)$ and so $G(M_0) \in \mathbf{M}(M)$. But $M_0 \subseteq X$ for all $X \in \mathbf{M}(M)$; in particular $M_0 \subseteq G(M_0)$. However, by definition of G, $G(M_0) \subseteq M_0$, so $G(M_0) = M_0$. Hence M_0 is the required fixed point. Now, in the Zermelo case $G(X) = X$ just in case $X = \emptyset$ and $X - \{\phi(X)\}$ otherwise. Kuratowski's theorem says there must be a fixed point in **M**; the definition of G says that this can only be the case if $\emptyset \in \mathbf{M}$. Thus, indeed, $\emptyset \in \mathbf{M}$. Consequently, in the Zermelo case, we need not trouble to prove independently that $\emptyset \in \mathbf{M}$. Of course, $M \in \mathbf{M}$ by definition. (Kuratowski's 'fixed-point theorem' is actually a special case of the Bourbaki fixed-point theorem given in Bourbaki [1968], Chapter 3, exercise 6, p. 222.)

The second point concerns the use of functions in Zermelo's proof, for example the choice function ϕ or the f' used to image the well-ordering of **M** onto M itself. The absence of any reductionist treatment of relations was precisely the reason for pursuing this rather complicated theory of chains. How then can he use *functions*? Actually Zermelo in his proof does not use functions explicitly. He contents himself with the notion of 'correspondence' without explaining what a correspondence is in set-theoretic terms. The axiom of choice, moreover, is framed in terms of the existence of a choice set, as it is in Zermelo [1908b]. Actually, Zermelo did not need the notion of function explicitly. For one could define a set M to be well-orderable if there exists a maximal, well-ordered, reverse inclusion chain **M**, and further, define **M** itself as the 'well-ordered set M'. This would then avoid having actually to map the set **M** onto M. This is Fraenkel's procedure. (See his [1927], pp. 137–8, and also Fraenkel and Bar-Hillel [1958], pp. 127–35.) It is also hinted at by Zermelo's reference to **M** as a set of remainders, and by his saying **M** well-orders M. (See Zermelo [1908a], p. 185.) In this sense, then, Zermelo's solution of the well-ordering problem *is* a reductionist one. But this is not completely satisfactory. For the theory of (cardinal) equivalence and (ordinal) similarity clearly *do* require a notion of function or correspondence. However, in his [1908b] Zermelo did define what he called 'a mapping of M onto N' quite independently of any general notion of relation or of ordered pair. The definition is not completely general but only defines the notion of a one-to-one onto mapping between disjoint sets M and N. It proceeds as follows.

First, Zermelo defines 'the product of M and N', denoted by $M.N$, which is the set of all unordered pairs $\{m, n\}$ having just one element in common with each of M and N. Then an $F \subseteq M.N$ is an equivalence mapping of M onto N if each element of $M \cup N$ occurs in one and only one element of F. The definition is extended to non-disjoint sets M', N' by showing that there is always another set disjoint from *both* but equivalent to one of them. This approach to equivalence was taken over by Fraenkel in his [1925] (see also Fraenkel and Bar-Hillel [1958], pp. 124-6). It succeeds in treating the notion of equivalence without a *general* notion of function or relation. But, of course, the lack of *these* notions is a crippling blow to any hope of rebuilding classical analysis within the Zermelo framework. And it is hard to see how these notions can be defined along the lines of Zermelo's treatment of equivalence.

The problem of reducing relations and functions to pure set theory was solved by Kuratowski's set definition of ordered pair. Hausdorff was the first to define relations and functions as sets of ordered pairs, and to reduce ordered pairs themselves to sets. But his definition of ordered pair was rather clumsy. He first introduces two objects 1 and 2 which are distinct. Then, for the elements a and b he defines the ordered pair (a, b) as $\{\{a, 1\}, \{b, 2\}\}$ (Hausdorff [1914], pp. 32-3). As Hausdorff states, the objects 1 and 2 must be different from a and b. But as Kuratowski remarks ([1921], p. 171) this means that the definition has to be modified if in fact a or b *are* 1 or 2, as they presumably could be if 1 and 2 are genuine objects inside the theory. Moreover, before forming $A \times B$ (as Hausdorff does) one must first show that 1, 2 $\notin A \cup B$. Wiener also gave a definition of ordered pair in his [1914]. This amounts to defining (x, y) as $\{\{\{x\}, \emptyset\}, \{\{y\}\}\}$ where x and y are assumed to be of the same type, and $\emptyset$ is the empty set for that type. (Wiener was attempting to show that one can dispense with the axiom of reducibility for propositional functions of two variables in the Whitehead–Russell treatment of relations.) But as with Hausdorff's definition, this suffers from the difficulty of involving an object different from x and y in the pair (x, y).

Kuratowski's definition gets round these difficulties very neatly; it follows immediately from his theorem (A$'$) on orderings. For if $M = \{a, b\}$ then the *only* maximal inclusion chains in $P(M)$ are $\{\{a\}, \{a, b\}, \emptyset\}$ and $\{\{b\}, \{a, b\}, \emptyset\}$. Using the definition (*) of the derived ordering $<$ given above, these chains must correspond to the orderings $a < b$ and $b < a$ on $\{a, b\}$ respectively. If $\emptyset$ is ignored, as Kuratowski does, the chain $\{\{a\}, \{a, b\}\}$ is associated with the ordered set (a, b). Kuratowski then, naturally, *defines* (a, b) as $\{\{a\}, \{a, b\}\}$.

There is one more key point to be explained, namely the relationship between Zermelo's 1904 proof and the 1908 proof. They are indeed very similar. Suppose we change Zermelo's 1908 definition to make **M** the intersection of all subsets **Z** of $P(M)$ satisfying

(i) $\emptyset \in \mathbf{Z}$
(ii) if $X \in \mathbf{Z}$ then $X \cup \{\phi(M - X)\} \in \mathbf{Z}$
(iii) if $\mathbf{Z}' \subseteq \mathbf{Z}$ then $\bigcup \mathbf{Z}' \in \mathbf{Z}$.

Suppose also that we replace (A), (A$'$), (B), and (B$'$) by their analogues for segments; then the proof that **M** well-orders M can be carried through in much the same way. Of course, we need analogues for Procedures 1 and 2 (for which see Kuratowski [1922]) and conditions on the maximality of a well-ordered inclusion chains (these are identical except that now any limit element must be the *union* of its predecessors). Kuratowski's fixed-point theorem also has an analogue. Hausdorff [1937] proves the well-ordering theorem in this way (pp. 65-8). However, it is not difficult to show that defined in this new way, **M** is just the collection of all segments which go to make up L_γ in the 1904 proof. In fact, $\mathbf{UM} = L_\gamma$.

7.4. The problem of definite properties

Although there is no great structural difference between the 1904 proof and the 1908 proof the context was considerably changed. Following the rigorous demands of his reductionism Zermelo purged his proof of any reference to ordinals, transfinite recursion, and rank ordering. Well-ordering relations on a set M were, in effect, treated as sets, namely certain members of $P^2(M)$, the double power-set of M. Even the questionable step of proving $L_\gamma = M$ was avoided by defining the required chain 'down' from M, instead of 'up' from $\emptyset$. Moreover, in the 1908 proof Zermelo makes clear which set existence principles he is relying on (principally choice, subset formation, and power-set). It is emphasized that these are the *only* principles used. In short, the new proof, unlike the old, is an axiomatic proof.

Zermelo founded his reply to his critics on this basis. Referring to the contradictions he stated:

But if in set theory we confine ourselves to a number of established principles such as those that constitute the basis of our proof—principles that enable us to form initial sets and to derive new sets from given ones—then all such contradictions can be avoided. (Zermelo [1908a], p. 195)

Such a claim is based, as I pointed out in §7.2(a), on the *assumption* that these 'established' principles are not responsible for the contradictions. Hence Zermelo's claim is manifestly too strong: it is *not* clear that 'such contradictions' are avoided by restriction to these principles. Nevertheless, the shift to the new proof both simplifies and clarifies Zermelo's position that the contradictions and contradictory 'sets' *cannot* (as some critics had feared) creep into the proof through the back door.

Zermelo now had the major part of an axiom system for set theory. Not surprisingly, the axioms he presented in his [1908b] are much like those used in the [1908a] proof: axiom 2 deals with 'elementary sets' (the empty set is a set, if a and b are objects, so are $\{a\}$, $\{a, b\}$); axioms 3, 4, and 5 are the central relative existence principles, separation, power-set, and union; axiom 6 is the axiom of choice; axiom 7 is the axiom of infinity. (Zermelo was apparently the first to recognize that 7 is not required for elementary set-theoretic arithmetic ([1909a], p. 192). But he also recognized ([1909a], p. 192) that it is indispensable if one is to go beyond arithmetic.) To these are added the axiom of 'definiteness' (axiom 1) which is a form of extensionality for sets: if two sets have the same members they are identical. Note that the power-set axiom appears here in a strong form, strong enough to yield Cantor's theorem and (some form of) the classical continuum. For, in order to satisfy the demands of the 1908 proof, the subset axiom must allow for the existence of impredicatively defined subsets. This is enough, say, to enable a proof of Cantor's theorem to be carried through. And Zermelo, of course, does allow the existence of impredicatively defined subsets, both in his [1908a] 'subset axiom' and in his [1908b] separation axiom. According to him, *any* 'definite property' will define a subset.

So much for the axioms, axioms which constitute the core of modern set theory. What I shall be concerned with now is the setting in which the axioms were placed, and particularly the formulation of the separation axiom. For it seems that here Zermelo allowed a non-reductionist element to creep into his system against the trend of the reductionism I have emphasized earlier.

The presentation of the axioms begins with the following statement: 'Set theory is concerned with a domain $\mathfrak{B}$ of individuals, which we shall call simply *objects* and among which are the *sets*.' The first important thing to notice is the realism which underlies this statement: the impression given is that the purpose of set theory is to *describe* objects within the fixed domain $\mathfrak{B}$. This realist impression is supported by the use of pure existence principles among the axioms, and by the earlier explicitly realist reply to Poincaré's objection to impredicative definitions ([1908a], p. 191). The second important aspect of this statement is the implicit admission that there may be objects which are not themselves sets. Of course, Zermelo does not make any specific use of 'elementless' non-set individuals (the empty set is the only set with no elements). But the redundancy of the notion seems to emphasize the conflict with reductionism, not excuse it.

There is, however, a more serious violation of reductionism in Zermelo's system, namely the use of so-called 'definite properties'. The second proof of the well-ordering theorem strongly emphasized the importance of the power-set axiom. This axiom seems important in the 1904 proof only for the purpose of framing the axiom of choice. In the 1908 proof it was clearly revealed as one of the two key hinges on which the whole proof swung. For it was *this axiom* which permitted the reductionist treatment of the well-ordering relations for arbitrary sets.

Nevertheless, the existence of most of the subsets of a given set is guaranteed by the principle of subset formation. But what yields the subsets? Zermelo's answer is clear: 'properties'. For example, the statement of the subset or separation axiom [*Aussonderungsaxiom*] in [1908a] is (p. 183): 'All elements of a set M that have a property $\mathfrak{E}$ well-defined for every single element are the elements of another set $M_{\mathfrak{E}}$, a "subset" of M.' This is now, clearly, the 'limitation of size' substitute for the comprehension principle discussed in §7.1. But if Zermelo is invoking properties as new entities then he is clearly violating his own reductionism. For *these* entities, unlike the non-set individuals (if there are any) do play a highly important, indeed crucial, mathematical role in Zermelo's system. But *are* the properties taken as new entities? There is some evidence to suggest that Zermelo does so take them, either indirectly *via* the use of classes, or directly as objects in their own right. The evidence, though, is not conclusive: Zermelo's intentions are obscured by his not building his system within a clear logical framework. (See his remarks on this on pp. 339–40 of his [1929].)

In his [1929], Zermelo attempts to explain his notion of 'definite property'. He proposes an axiomatization of definite properties, or, more strictly, an axiomatization of the concept of 'definiteness' for properties. This approach, he tells us, goes back to his [1908b]; it is '. . . the method which I myself had in view, though I did not expressly say so, and which was applied in the reasoning of the work mentioned' ([1929], p. 340). 'Definiteness' is now treated as a predicate under which properties (or relations etc.) fall, and, importantly, the axioms proposed for definiteness involve quantification over the properties. The permissibility of quantification over properties is confirmed in Zermelo [1930]. There he states what appears to be a second-order version of the axiom of separation: 'Every propositional function $f(x)$ separates from each set m a subset m_f containing all elements x for which $f(x)$ is true' ([1930], p. 30). It seems to me that this is a fairly clear admission that properties *can* be treated as objects. And Zermelo suggests that this theory goes back to his [1908b].

The evidence in this earlier paper itself is simply not clear. There, separation is not stated in a form which suggests second-order quantification, and Zermelo largely avoids the term 'property'. But this paper, if it does not invoke properties, does invoke classes, i.e. non-set collections. $\mathfrak{B}$ itself is one of them (Zermelo proves that it is not a set), and there are presumably others (Russell's collection, for instance). Their appearance here is puzzling; they serve no purpose which cannot be served either by sets or by properties. No doubt the classes are the extensional correlates of 'properties', here called by Zermelo '*Klassenaussagen*', literally 'class assertions' or 'class predicates'. Perhaps he thought that he could axiomatize the classes in a way which would obviate the properties they represent. Von Neumann, of course, attempted something like this, dispensing with properties and sets in favour of functions (in effect classes). And Zermelo later stated his qualified approval of von Neumann's method, likening it to his own intention of treating properties axiomatically.

If the status of properties and classes in [1908b] is unclear so is the explanation of definiteness. A 'class predicate' $\mathfrak{E}(x)$, with x ranging over a class $\mathfrak{K}$, is said to be definite for the class $\mathfrak{K}$ if its instantiation at each element of $\mathfrak{K}$ is a 'definite assertion'. And an assertion

. . . $\mathfrak{E}$ is said to be *definite* if the fundamental relations of the domain [$\in$ and identity], by means of the axioms and the universally valid laws of logic, determine without arbitrariness whether it holds or not. (Zermelo [1908*b*], p. 201)

To modern eyes, this is hopelessly confused. For example, there appears to be no separation between truth and proof. Is an assertion $\mathfrak{E}$ to be definite because either it or its negation are provable, or because either it or its negation is true (in domain $\mathfrak{B}$)?

Whether Zermelo treats properties as entities in their own right, or whether he intended to 'reduce' properties to classes, he clearly contravenes his own set-theoretic reductionism. (What now of the 'non-existence' of the 'set' of all ordinals?) In a sense, the 'class theory' interpretation of properties imposes less of a strain on reductionism than does the treatment of properties as objects in their own right. This is especially so if classes are taken to be 'like sets', only subject to rather more restrictions. And we know how to frame set-class theories in first-order logic without any further recourse to quantification over properties. Moreover, such theories (depending on the strength of the class axioms) can usually be recast in a pure set theory which yields exactly the same information about sets. However, all this comes only with a certain logical sophistication that was lacking to Zermelo and his contemporaries. What is more, the indications Zermelo gives in his [1929], unclear as they are, tend to point away from the class theory to the theory of properties. And the historical remarks there put the origins of this approach firmly back in 1908. Whatever the situation, the confusion over definite properties was undoubtedly a serious problem for Zermelo's successors.

There were various responses to this problem. Skolem proposed a first-order formulation, 'definite property' thus being replaced by 'predicate of the first-order language'. Fraenkel defined a recursive list of terms to take the place of the definite properties. And von Neumann proposed a new axiomatization of set theory altogether. Although a comparative study of the respective strengths of these approaches is highly important for a full-scale study of the development of set theory, there is no space to pursue it here. But, because of his shift to the highly important cardinal theory of limitation of size, I want to sketch von Neumann's approach. And because of the involvement of the axiom of replacement this will also require a statement, if not a discussion, of Fraenkel's proposals. (Skolem's proposals are in his [1922].)

VON NEUMANN'S REINSTATEMENT OF THE ORDINAL THEORY OF SIZE

Zermelo's conception of limitation of size (at least originally) was based on comprehensiveness, not on cardinal size. In this, Zermelo was followed by Fraenkel who, as we saw, distinguished between the 'all-too extensive' sets of the paradoxes and 'normal sets' which are presumably of restricted or reasonable extent. On the other hand, Cantor, Jourdain, and later Mirimanoff had used criteria based on cardinal size, though in Mirimanoff's case this was not the dominant theme. It was to such a cardinal conception that von Neumann returned and from which axiomatic set theory as we know it now stems.

Von Neumann not only reinstated the Cantor–Jourdain view of size based on the ordinals, but also a version of the Cantor–Jourdain proof of the aleph theorem. This involved two very important heuristic shifts with respect to Zermelo's programme. Zermelo had abandoned one important aspect of Cantorian finitism. For, in avoiding the use of ordinal number objects, he effectively jettisoned the counting theory of cardinality. Moreover his reductionism, in accord with the letter of the Cantorian principle of the non-mathematizability of the Absolute, dictated that the contradictory 'sets' must not be objects within the theory. Von Neumann reversed both these heuristic trends; that is to say, he restored Cantor's counting theory *and* allowed the contradictory collections as proper objects within his theory.

This heuristic shift is the subject of the present chapter. It involved two essential and quite separate steps: (1) showing how to avoid the reductionist objection to the use of ordinals as primitive objects; (2) disarming the objection that using *Ord* as part of a key proof, i.e. treating *Ord* or any other 'big' set as a legitimate mathematical object, must engender contradictions. Jourdain had stumbled because he could deal with neither of these problems successfully. Not so von Neumann. Von Neumann achieved (1) by a successful reduction of the full Cantorian theory of ordinals to a pure set theory. Thus he blazed a reductionist trail quite different from that mapped out by Zermelo. In effect, von Neumann showed for the first time how to harmonize full Cantorian finitism and set-theoretic reductionism. This achievement was quite independent of step (2), for von Neumann showed that it did not depend on his own views about how set theory should be axiomatized. Indeed, it became a corner-stone of what we now know as Zermelo-Fraenkel set theory. Von Neumann's step (2) came with his treatment of 'functions'. The reinstatement of the theory of ordinals involved essential use of 'Fraenkel's' replacement axiom. And any rigorous formulation of this required a successful handling of Zermelo's 'definite properties' or, more specifically, 'definite relations'. Von Neumann's solution to this problem reintroduced not only the cardinal limitation of size theory but also the treatment of absolute collections as legitimate objects.

8.1. The von Neumann theory of ordinals

By 1923 von Neumann had already largely carried through his programme for the axiomatization of set theory. This is testified to by Fraenkel (see Fraenkel's remarks quoted on p. 282), and by von Neumann's letter to Zermelo of 15 August 1923, published in Meschkowski [1967], pp. 271-3. A central part of this was his theory of ordinals. But in his [1923a] in which this theory was first presented he was at pains to stress that it by no means depended on the special features of his axiom system, and that it can be successfully transplanted into, say, the Zermelo-Fraenkel framework:

The aim of the present paper is to give unequivocal and concrete form to Cantor's notion of ordinal number.
    Ordinarily, following Cantor's procedure, we obtain this notion by 'abstracting' a common property from certain classes of sets. We wish to replace this somewhat vague procedure by one that rests upon unequivocal set operations. This procedure will be presented below in the language of naive set theory, but, unlike Cantor's procedure, it remains valid even in a 'formalistic' axiomatized set theory. Thus our conclusions retain their full validity even in the framework of Zermelo's axiomatization (if we add Fraenkel's axiom [replacement]). (Von Neumann [1923a], p. 347)

I shall consider the role of the axiom of replacement below. First let us consider the outlines of von Neumann's theory.
    Von Neumann's idea was quite simple and beautifully explained by von Neumann himself:

What we really wish to do is to take as the basis of our considerations the proposition: 'Every ordinal is the type of the set of all ordinals that precede it.' But in order to avoid the vague notion 'type', we express it in the form: 'Every ordinal is the set of the ordinals that precede it.' (Von Neumann [1923a], p. 347)

Not only is von Neumann's trick simple, but it is also very natural. In working with the 'naive' concept of ordinal it is often convenient to use instead of α the set of all ordinals less than α. One finds an example of this in, say, Hessenberg [1906], p. 592, and even in Cantor's original 'naive' treatment of ordinal it is the set of predecessors of the ordinal which really represents the well-ordered set (see §2.1). Given this inspired simplification, von Neumann's theory results from a direct translation into objective set-theoretical terms of Cantor's counting theory. Let $(X, <)$ be a well-ordered set (using modern notation, not von Neumann's). According to Cantor's idea, $(X, <)$ is assigned an ordinal by 'successively assigning' ordinal numbers to the elements of $(X, <)$ in order. The ordinal required is the ordinal type of the sequence of successive counting 'acts'. Von Neumann translated this in the following way. A 'numeration' of $(X, <)$ is not a 'succession' of counting 'acts', but a function f on X. Like the counting process, f must assign to each element of $(X, <)$ an ordinal matching its place in $(X, <)$. But since an ordinal is just the *set* of its predecessors, we must have $\forall x \in X[f(x) = \{f(y): y \in S_x\}]$ where S_x is the initial segment of $(X, <)$ determined by x. Then lastly, the ordinal number assigned to $(X, <)$ by the numeration f is the *set* of all 'counting acts' on $(X, <)$, namely $\{f(x): x \in X\}$.

This idea not only reduces Cantor's idea to set-theoretic operations, given that one can express what a function is set-theoretically, it actually simplifies it. In the Cantor process, to assign an ordinal to an element x of $(X, <)$ one *first* collects together the ordinals assigned to previous elements, *then* one 'calculates' the type of this collection, and lastly one assigns this type to x. In the von Neumann process the intermediate step is redundant: the collecting together of the previously assigned ordinals is *itself* the 'calculation' of the required ordinal. This is because in the Cantor process one is dealing with two different kinds of entity, the *set* of ordinals assigned to elements prior to x and the *ordinal type* of this set. Von Neumann simplifies the procedure by *identifying* the second entity with the first. Of course, the legitimacy of this depends on a proof of the existence of $\{f(y) : y \in S_x\}$ and $\{f(x) : x \in X\}$ as sets, a point we shall come to in a moment.

Von Neumann's procedure leads straight to the von Neumann ordinals. For:

If x_1, x_2, x_3, and x_4 are the 1st, 2nd, 3rd, and 4th elements of X [or $(X, <)$] then clearly for every numeration $f(x)$ of X we have

$f(x_1) = \emptyset$
$f(x_2) = \{\emptyset\}$
$f(x_3) = \{\emptyset, \{\emptyset\}\}$
$f(x_4) = \{\emptyset, \{\emptyset\}, \{\emptyset, \{\emptyset\}\}\};$

Consequently, if X has 0, 1, 2, or 3 elements, its ordinal is, respectively,

$\emptyset$
$\{\emptyset\}$
$\{\emptyset, \{\emptyset\}\}$
$\{\emptyset, \{\emptyset\}, \{\emptyset, \{\emptyset\}\}\}$.   (Von Neumann [1923a], p. 348)

(I have changed von Neumann's notation slightly.) Von Neumann shows that if a well-ordered set has a 'numeration' then it is unique; thus a well-ordered set can only have one ordinal. More importantly, he analysed the ordinals as generated above and arrived at a characterization of them which can then act as a definition, namely:

> P is an ordinal number if and only if
> (1) It is a set of sets that is capable of being ordered by inclusion;

(2) Its inclusion ordering is a well-ordering;
(3) For every element ξ of P, always $\xi = A(\xi, P)$ [i.e. $\xi = \{\beta : \beta \in S_\xi\}$]
(Von Neumann [1923a], p. 350)

It is interesting to look at this definition in the light of the Kuratowski work on ordering relations discussed in §7.3(*b*). For it is easy to see that if M is a von Neumann ordinal them $M \cup \{M\}$ is itself one of the 'ordering chains' in $P(M)$. It may be that this too had an influence on the structure of von Neumann's definition. For example, it may explain his insistence on the well-ordering of ordinals by inclusion rather than the equivalent ordering by '$\in$', an ordering which is preferable because '$\in$' is set-theoretically primitive and because it follows immediately from the von Neumann conception. (Von Neumann must have recognized that for his ordinals $\alpha \subset \beta$ just in case $\alpha \in \beta$. This is proved explicitly in his [1928a], p. 328.) Moreover, if M is a given set and $\mathbf{M}$ is a Kuratowski–Hessenberg well-ordering relation on M, then the von Neumann ordinal associated with $\mathbf{M}$ is a direct copy of $\mathbf{M}$. For we have

$\mathbf{M}$	$ord(M, \mathbf{M})$
$\emptyset$	$\emptyset$
$\{x\}$	$\{\emptyset\}$
$\{x, y\}$	$\{\emptyset, \{\emptyset\}\}$
$\{x, y, z\}$	$\{\emptyset, \{\emptyset\}, \{\emptyset, \{\emptyset\}\}\}$
etc.	etc.

Both are ordered by inclusion; the main differences are that $ord(M, \mathbf{M})$ has no element corresponding to the *last* element of $\mathbf{M}$, namely M itself, and the members of $ord(M, \mathbf{M})$ contain, not elements of M, but canonical representatives of them.

The connection between the Kuratowski–Hessenberg–Zermelo inclusion orderings and the von Neumann ordinals is shown beautfully in Mirimanoff's papers [1917a] and [1917b], which to a large extent anticipate the von Neumann definition. (Mirimanoff certainly knew the work of Zermelo and Hessenberg.) Mirimanoff uses the fact central to the Hessenberg work, that any well-ordered set E is similar to the set E' of its segments ordered under inclusion, provided that we add to E' a first element e which corresponds to the 'empty segment'. Now says Mirimanoff, if each element of E' (being well ordered) is replaced by the set of its segments, each member of this set by its segments, and so on, we end up with precisely the set

$$S = \{e, \{e\}, \{e, \{e\}\}, \ldots\}$$

which of course (putting $\emptyset$ for e) is the same as the von Neumann ordinal of E (see Mirmanoff [1971a], p. 46). And, of course, one arrives at the same S for any set similar to E. Thus, in Mirimanoff's conception, the inclusion ordering representing E is just a step on the way to the von Neumann ordinal of E.

Mirimanoff now follows the same route as von Neumann, for he shows how such sets S can be generally defined. In fact he says that they are those sets which are:

(a)   'ordinary' sets with the single node e;
(b)   connected by $\in$ (i.e. $\forall x, y \in S[x = y$ or $x \in y$ or $y \in x]$;
(c)   transitive (i.e. $\forall x \in S[y \in x \rightarrow y \in S]$).

By 'ordinary' Mirimanoff means that there are no infinite descending membership chains, and in addition here that the only indecomposable (or ur-) element one meets in descending chains is e. (See his [1917a], pp. 42 and 47, and §4.4 above.) Thus, since (a) amounts to assuming that the sets are well-founded (i.e. to assuming the axiom of foundation), the definition is virtually the same as the standard modern definition of the von Neumann ordinals. (The axiom of foundation (AF) says that $\forall y \exists u \in y[u \cap y = \emptyset]$. With the axiom of choice, the assumption that all sets are Mirimanoff 'ordinary' is equivalent to AF: see Fraenkel, Bar-Hillel, and Levy [1973], pp. 88-90. In the presence of AF, or the assumption that all sets are 'ordinary', the ordinals can be defined by conditions (b) and (c), (a) now being redundant. Without AF, the ordinals can be defined by (c) together with '$\in$ is a well-ordering of S'.)

Mirimanoff, unlike von Neumann, never quite took the step of identifying the ordinals with such sets. But he was quite clear that for every well-ordered set such a canonical representative S-set must exist (Mirimanoff [1917a], p. 49), and thus that these S-sets could do the job of the ordinals. This was shown quite clearly in his [1917b]. In this paper he approaches the S-sets rather differently, but in a way which is reminiscent of von Neumann's key step of replacing 'is similar to' by 'is'. Following Cantor, Mirimanoff says that an ordinal number is obtained by abstracting from the particularity of the objects in a given well-ordered set in such a way that the ordinal is a copy or replacement representative of it. Now, says Mirimanoff, suppose that the well-ordered set is represented by an ordinal whose elements are $A, B, C, \ldots$, what properties must these have 'so that their structure derives directly from the order relations of the elements of the given well-ordered set?' (Mirimanoff [1917b], p. 214). His answer is simple: 'It is sufficient to replace the words "precedes" or "is preceded by" by the words "is an element of" and "contains".' He then shows that one must again arrive at the sets S defined in [1917a], namely those satisfying conditions (a)-(c). He goes on to develop some of the properties of such sets. For example, he proves that given any two such sets, one will be an element of the other, adding:

The analogy between the S-sets and well-ordered sets is manifest, which was to be expected. In particular, the theorem we have just proved is analogous to the following theorem of Cantor: given two well-ordered sets, either they are similar or one is similar to a segment of the other. (Mirimanoff [1917b], p. 216)

But even having come this far, Mirimanoff still does *not* say: 'In a pure set theory, we can use these as ordinals.' Rather, he stops just short:

The analogy we have just underlined allows us to reduce the theory of well-ordered sets to that of the sets S. I do not know if this indirect method has any real advantages. In any case, the classical theory of Cantor appears now in a new light. The essential thing for us is that the order relations, instead of being artificial tags, are now in some way incorporated in the elements of the set, since the rank of each of them is determined by the structure of the elements. In this theory, the types of structure of the sets S correspond to the Cantorian ordinals. (Mirimanoff [1917b], p. 217)

(One should note that by 'rank' here Mirimanoff means something very like the notion of rank in the cumulative hierarchy; see §4.4.)

We must assume that von Neumann did not know of Mirimanoff's papers when he worked out his own theory; certainly he never mentions them in connection with the ordinals. But in any case von Neumann's achievement does not lie just in the discovery of a class of sets which can do the work of the ordinals. He went much further than this, for he showed precisely which principles of set existence are required to enable the new theory to flourish. Mirimanoff, too, gave a set of postulates for set theory in connection with his 'limitation of size theory', and thus he went much further than, for example, Jourdain did. (Arguably Jourdain needed a theory of set ordinals, and thus a collection of set existence postulates, to be able to make his limitation of size theory work (see §§4.2 and 4.3 and Hallett [1981]).) But he did not turn his attention to the question of which postulates are needed to give enough S-sets (or set ordinals) for the rebirth of the full Cantorian theory. The key theorem here is: every well-ordered set is similar to a set-ordinal. One of von Neumann's crucial achievements was to recognize that the proof of this requires some version of the axiom of replacement.

In proving that any well-ordered set $(M, <)$ is similar to an ordinal the key point is showing that if every proper segment of $(M, <)$ is similar to an ordinal then so is $(M, <)$. One can construct a functional relation with domain M taking each segment of M onto its ordinal. Replacement allows the range of this relation to be a set; its union is then an ordinal similar to M. (Von Neumann worked out the details in [1928a], pp. 327-8.)

But, as von Neumann points out, this was not the only way in which his theory of ordinals depended on replacement. To be able to define addition and multiplication operations for the ordinals in complete generality one needs to rely on definition by transfinite recursion over the whole ordinal sequence (see von Neumann [1923a], p. 353). Von Neumann was the first to undertake an axiomatic treatment of the legitimacy of such definitions, indeed even the first to give a rigorous *informal* treatment. Hausdorff used such definitions in his [1914] with no proof of their legitimacy; he contented himself with the remarks (p. 113): 'We can not only argue inductively . . . but also define inductively . . .' and the rather obvious '$f(\alpha)$ is defined for each α as soon as

$f(0)$ is defined and as soon as the definition of all $f(\xi)$ for $\xi < \alpha$ makes possible the definition of $f(\alpha)$.' As I explained in §7.3(b) Kuratowski tried to find ways of circumventing definitions by recursion. Actually he only showed that certain kinds of such definition can be avoided, and indeed can be replaced by definitional procedures legitimate in Zermelo set theory. What Kuratowski might have shown (but did not) was that definition by recursion over a well-ordered (Zermelo) set is quite legitimate in Zermelo set theory. (This is explained in §8.2.) But of course what von Neumann required was to be able to construct functional terms defined over well-ordered *classes*. And as von Neumann points out ([1928a], p. 321): 'The *possibility* of definition by transfinite induction is not at all obvious.'

The theorem von Neumann needed to prove was stated by him as follows:

Let $f(x)$ be a function that is defined for all sets of objects of a domain B and whose values are always objects of the domain B. Then there is one and only one function $\Phi(P)$ defined for all ordinals P and having values that are always objects of the domain B, such that for all ordinals P

$$\Phi(P) = f(\{\Phi(Q) : Q \ \overline{OZ}, Q \subset P\})$$
$$= f(\{\Phi(Q) : Q \in P\})$$

where $Q \ \overline{OZ}$ means that Q is an ordinal. (Von Neumann [1923a], pp. 353–4)

(Again I have changed von Neumann's notation slightly.) Here f and Φ are what would now be called functional terms which need not be restricted to set domains. Clearly, replacement is needed at each stage to guarantee that $\{\Phi(Q): Q \in P\}$ is a set.

Zermelo had shown that in his theory there are enough sets to prove (using Hessenberg–Zermelo well-orderings) that every set can be well ordered. Von Neumann's definition of ordinal can be used to supplement this by providing a notion of ordinal which interlocks completely with the Hessenberg–Zermelo theory. Adding the axiom of replacement, von Neumann could show that this theory is adequate, that is, one can develop the full theory of ordinal arithmetic and show that every set is equinumerous with an ordinal. Consequently, the Cantorian conception of the universe of sets is restored, in particular, the 'counting' or ordinal theory of cardinality. Indeed, von Neumann introduced the Cantorian alephs in the simplest possible way by defining them as initial ordinals. (See von Neumann [1928b], p. 401.) This was the way Cantor himself had indicated, and which Russell later independently discovered. (For this latter, see Hallett [1981], p. 395.) And all this can be done *within* a universe of sets in accordance with the Cantorian tendency in a fully reductionist way.

Von Neumann was not the only one, or even the first, to give a satisfactory set-theoretic definition of the ordinals, independent of any appeal to the general conception of well-ordered set and where the ordinals are sufficiently small to be 'safe'. As we have seen Mirimanoff was at least on the same trail (and size to him was also important). And Russell too, in his [1906a], had the beginnings of such an idea, since in showing that the 'Russell set' has a subclass equivalent to *Ord,* he produced a sequence very like the von Neumann ordinals. (See §4.3, and also Hallett [1981], pp. 397-8.) If Russell had wanted to give a new set-theoretic definition of ordinals in keeping with the limitation of size idea this would have been a good place to start. However, Russell by this time had lost interest in pure set theory, and was seeking a 'natural' logical solution to the paradoxes. (Nevertheless, it is curious that Jourdain did not follow up Russell's hint.) Interestingly, Zermelo too was pursuing such a definition before either von Neumann or Mirimanoff.

In his [1941] (p. 6), Bernays noted that Zermelo's idea dated back to around 1915. Bernays then states further:

The definition of an ordinal on which Zermelo based his independent theory of ordinals . . . was stated in terms of his axioms for set theory . . . It can, however, also be formulated with reference to our present system, as follows:

A set n is an ordinal if (1) either $0 = n$ or $0 \in n$, and (2) if $\alpha \in n$ then $\alpha' = n$ or $\alpha' \in n$ $[\alpha' = \alpha \cup \{\alpha\}]$, and (3) if S is a subset of n, the sum of the elements of S is represented by n or by an element of n [i.e. $\bigcup S \in n'$]. (Bernays [1941], p. 10)

(Bernays states that while Zermelo has only the condition $0 \in n$ in (1), his modification allows 0 also to be an ordinal (however, see below). In any case, as Quine [1969], p. 155, points out, Bernays's (1) obviously follows from (3).) This definition, says Bernays, can be proved equivalent to the usual definition using just the Zermelo axioms. There is no reference to this definition in Zermelo's published work, though there are some references in unpublished sources. For example, in a postcard to Hilbert of 5 May 1913, Zermelo writes:

Dear Professor,
There must be a mistake here. Recently I have done no work at all on logical calculuses. What Cara[theodory] means perhaps is an investigation I began in the winter into an independent *definition of the ordinal numbers,* thus the *W*-Reihe, in any case pure set theory.

(The card is in the Hilbert *Nachlass,* held in the *Handschriftenabteilung* of the Niedersächsische Staats- und Universitätsbibliothek, Göttingen.) And there is the following comment in a card from Bernays to Zermelo, 30 November 1920, referring presumably to a lecture in Göttingen:

Also your theory of ordinal numbers [*Wohlordnungszahlen*] was mentioned in the lecture, namely in connection with the Burali-Forti paradox, which thereby obtains a more precise formulation.

(Zermelo *Nachlass,* correspondence, envelope 'Bernays'. The *Nachlass* is held in the *Handschriftenabteilung* of the Universitätsbibliothek, Freiburg, F.R.G.) More intriguingly there are various scattered notes (all unfortunately rather scrappy and hard to read) in the Zermelo *Nachlass,* which give the definition Bernays cites, and trace some of its consequences. However, there is not much to indicate just how far Zermelo went with his definition, nor any solid confirmation of the date Bernays gives. The best evidence for the date comes from notes written on the backs of two business letters addressed to Zermelo in Switzerland, both dated 8 September 1915 (both are in *Kasten* 4 of the Zermelo *Nachlass*). Here Zermelo gives exactly the definition Bernays ascribes to him, the only differences being that he has as condition (1) $0 \in \alpha \cup \{\alpha\}$ instead of the $0 \in \alpha$ Bernays ascribes to him, and that he calls an ordinal 'a set of the second level' without explaining what this means. (The expression undoubtedly goes back to Hessenberg [1909] where it means that the elements of the set are also sets. If the elements are of the second level, Hessenberg calls the set 'of the third level'. As Mirimanoff points out ([1921], p. 30) this is an anticipation of the notion of rank.) It is also clear from the two notes that Zermelo had developed his definition somewhat, for he states and partially proves that his ordinals have various key properties. For example:

(a)   If λ is an ordinal, then $\forall \alpha \, [\alpha \in \lambda \cup \{\lambda\} \to \alpha \notin \alpha \wedge \alpha \notin \textbf{U} \, \alpha]$.
(b)   For any ordinal number λ, either $\lambda = \textbf{U} \, \lambda$ or $\lambda = \textbf{U} \, \lambda \cup \{\textbf{U} \, \lambda\}$.
(c)   If α, β are any two segments of an ordinal then $\alpha \subseteq \beta$ or $\beta \subseteq \alpha$.
(d)   If α, β are two segments of an ordinal with $\alpha \subset \beta$, then $\alpha \cup \{\alpha\} \subseteq \beta$.

As well as these two notes, there is a notebook (Zermelo *Nachlass*, Kasten 6) devoted to ordinal numbers and well-ordering where exactly the same definition is given together with the propositions stated above, though now always with explicit proofs. This time, though, Zermelo also proves:

(e)   The elements of an ordinal number are well-ordered by inclusion (p.10 of the notebook).
(f)   Every element β of an ordinal number contains all smaller segments as elements, and only these (p. 11).
(g)   Every element of an ordinal number is itself an ordinal number (p. 11).
(h)   If λ is an ordinal number, so is $\lambda' = \lambda \cup \{\lambda\}$ (p. 12).

Curiously one proof here is marked 'Bernays', namely that for proposition (d), and also one proposition itself, proposition (c). It could be, therefore, that these notes stem from the period when both Zermelo and Bernays were at the University of Zürich, that is 1912-1916.

There are other notes on ordinals in the Zermelo *Nachlass*, in particular some in an envelope postmarked 1915. (The envelope is in *Kasten* 3 and is marked '*Bankverein Göttingen*'; it is itself contained in a broken envelope marked (apparently, not in Zermelo's hand) '*Vermischte Notizen, z.T. Masstheorie*' ['Various notes, in part on measure theory'].) In these there is a treatment of well-ordered sets on the basis of inclusion orderings, followed by the definition:

An ordinal number α = well-ordered set in which each element = the set of all preceding elements: . . .

Again, this notion is developed somewhat, and in particular Zermelo asserts:

(i)   If A is an arbitrary sequence [i.e. well-ordered set] then it is similar to some ordinal number α

and he also begins a proof of this by induction on the segments of A. Moreover, we also find, based on the definition of $\xi < x$ by $\xi \in x$:

$x_0 = 0$ because [?] $\bigcap_\xi \overline{\xi < x_0}$ [i.e. $\forall \xi \neg [\xi < x_0]$]

$x_1 = \{0\} = 1, x_2 = \{0, \{0\}\}$.
$0 = 0, 1 = \{0\}, 2 = \{0, 1\}, 3 = \{0, 1, 2\}, \ldots,$
$\omega = \{0, 1, 2, 3, \ldots\}, \omega + 1 = \{0, 1, 2, \ldots, \omega\}$.

Although it looks from these details as if Zermelo had come somewhat further with ordinal numbers (after all (i) is the key theorem, and one where some kind of replacement argument is needed), the envelope dated 1915 may be very misleading. For one thing, the definition of ordinal given here—unlike the one Bernays cites—is very close to von Neumann's from 1923. Moreover, there seems a clear connection between these notes and Zermelo's [1930].

In this paper Zermelo gives the following definition of a *Grundfolge* [basic sequence], a term he uses quite often in discussions of inclusion orderings:

I call a 'basic sequence' a well-ordered set in which each element, with the exception of the first, which must be an 'urelement', is identical with the set of all preceding elements.

Thus, out of the urelement u there arise the basic sequences

$$g_0 = u, g_1 = \{u\}, g_2 = \{u, \{u\}\}, g_3 = \{u, \{u\}, \{u, \{u\}\}\}$$

and so on, according to the rule

$$g_{\alpha+1} = g_\alpha + \{g_\alpha\}, \text{ and } g_\alpha = \sum_{\beta < \alpha} g_\beta, \text{ if } \alpha \text{ is a limit number.}$$

In general, a basic sequence is a set which is ordered by the $\in$-relation and which, because of [the axiom of foundation], must be well-ordered so. (Zermelo [1930], pp. 31–2)

This definition is identical to that of ordinal number at the top of this page, and indeed Zermelo remarks ([1930], p. 29) that the 'basic sequences' serve as a tool in his present work because '. . . they are the simplest representatives of the different ordinal numbers present in each normal domain.' This connection, then, would seem to put these last-mentioned notes on ordinals rather later than 1915. Note too the close similarity with Mirimanoff [1917a] and [1917b]. Much of Zermelo's conception of ordinals is set out also in a manuscript called 'On closed and open domains [*Über Geschlossene und Offene Bereiche*]' in *Kasten* 4 of the *Nachlass*. It is dated September 1930 and is clearly connected with Zermelo [1930] (dated April 1930). There is one curious note to all this. In the Zermelo *Nachlass* there is a letter to Zermelo from Hessenberg of 17 March 1912 (Zermelo correspondence, envelope 'Hessenberg'), in which Hessenberg says '. . . we can perhaps define [the number 2] following Grelling, by the definition $2 = \{0, \{0\}\}$.' This is perhaps one more indication that the 'von Neumann' conception of ordinals was, so to speak, 'in the air'.

Despite these quite extensive anticipations it was nevertheless von Neumann who developed the theory. He himself was quite clear why, for example, Zermelo had not done this:

A treatment of ordinal number closely related to mine was known to Zermelo in 1916, as I learned subsequently from a personal communication. Nevertheless, the fundamental theorem, according to which to each well-ordered set there is a similar ordinal, could not be rigorously proved because the replacement axiom was unknown. (von Neumann [1928*a*], p. 321)

Von Neumann knew he had to adopt some version of the axiom of replacement. But *what* version and how is it to be formulated? Before considering von Neumann's solution to this problem I want to look briefly at the development of Zermelo's programme under Fraenkel's tutelage. Not only is this an integral part of the life history of the replacement axiom, but it is essential for understanding the strength of von Neumann's step (2).

8.2. The discovery of the replacement axiom

The discovery of the axiom of replacement is usually attributed to Fraenkel and Skolem, with appreciative nods in the direction of Cantor and Mirimanoff for providing 'hints'. (See e.g., Fraenkel, Bar-Hillel, and Levy [1973], p. 50, or Quine [1969], p. 88, n. 1.) Certainly 'axioms' of replacement were proposed simultaneously by Fraenkel in his [1922*b*] and Skolem in his [1922]. Both had noticed that the existence of sets of power $\geqslant \aleph_\omega$ is not provable in Zermelo set theory. For this one needs to take the union of sets a_n, $n \in \omega$, where each a_n has power $\geqslant \aleph_n$. But one cannot show using just Zermelo's axioms that $\{a_n : n \in \omega\}$ is a set. (Skolem's paper actually outlines how to build an 'inner model' of Zermelo's axioms which excludes such a set.) This was one reason why Fraenkel wrote in his [1921] (p. 97) that '. . . Zermelo's seven axioms are not sufficient for the foundation of legitimate set theory' and why Skolem wrote ([1922], p. 296) that: 'It is easy to show that Zermelo's axiom system is not sufficient to provide a complete foundation for the usual theory of sets.' Fraenkel noted that to get round this difficulty one needs, not, as one might at first suppose, a new axiom of infinity but rather ([1921], p. 97) 'an essentially new existence condition . . . which uses a suitable concept of function.' This remark was expanded in his [1922*b*]:

This hitherto unnoticed gap in Zermelo's foundation has to be filled by the addition of a new axiom or by the extension of one of the existing axioms. As far as the example given is concerned, one can get by with an extension—albeit very considerable—of axiom 7 [infinity]. However, one can form systematically more far reaching and more general counterexamples by much the same method as used in the examples above. Thus one can see that a *general* postulate of a new kind must be proposed. By such considerations one comes first to the following axiom:

Replacement axiom: If M is a set and if each element of M is replaced by a 'thing of the domain $\mathfrak{B}$' then M is transformed into a set. (Fraenkel [1922b], p. 231)

Skolem proposed that:

In order to remove this deficiency of the axiom system we could introduce the following axiom: Let U be a definite proposition that holds for certain pairs (a, b) in the domain B; assume, further, that for every a there exists at most one b such that U is true. Then as a ranges over the elements of a set M_a, b ranges over all elements of a set M_b. (Skolem [1922], p. 297)

But the 'discovery' of an axiom, like the discovery of a scientific 'law' or theory, must surely involve more than simply stating the axiom (or theory) for the first time. At the very least it should involve the development of some of its important consequences. Skolem's approach to set theory outlined in his [1922] had great merit. For example, he proposed framing Zermelo's axiom system in a first-order language with '$\in$' and '$=$' the only primitives, thus, as he showed, dispensing with any difficulty over Zermelo's 'definite properties'. Instead one can use 'one-place predicate of the given language' and formulate separation as a schema. All the shadowy entities which inhabit Zermelo's system (properties, classes, domains) and which threaten his reductionist achievements are now exorcised with one simple logical stroke. The same clarity can be brought to bear on the axiom of replacement: in this context a 'definite proposition' U is simply a 'definite relation' instantiated, and a 'definite relation' is simply a two-place predicate of the given language. Skolem's 'axiom' of replacement is thus very close to its modern ZF version. Nevertheless, so far as I am aware, Skolem did not attempt to develop Zermelo set theory with his replacement axiom added to it. Neither does he provide any convincing argument that sets provably of power $\geqslant \aleph_\omega$ are essential, supposedly the reason for the proposal of replacement.

Fraenkel has even less claim to be the discoverer (in the strong sense hinted at above) of the replacement axiom than Skolem. Fraenkel's 'axiom' is not really an axiom at all, only an intuitive sketch of a possible axiom. It lacks a precise explanation of the notion of 'replacing' one object by another, that is a precise statement of what he had earlier hinted at, 'a suitable concept of function'. Not only this, but Fraenkel did not begin to recognize the strength of the axiom, and neither did he do anything to develop its consequences. All this was the work of von Neumann. In his autobiography Fraenkel recalls the impact that von Neumann's paper on transfinite numbers had on him:

This paper held in store a special surprise for me. In 1921 in a lecture, published in 1922 [Fraenkel [1922*b*]] in volume 86 of the *Mathematische Annalen* at Hilbert's suggestion, I had added, among other things, a new axiom, the *Ersetzungsaxiom*, to the customary axioms of set theory. The importance of this axiom was now shown in a wholly unexpected way: von Neumann based the theory of transfinite numbers on my axiom in a way which showed that it is indispensable for this purpose. (Fraenkel [1967], p. 169)

Thus, the true importance of the replacement idea was a surprise to Fraenkel. He says here that in 1922 he 'added' replacement to the other axioms. But this is not correct. As I have noted, he did not formulate the 'axiom' precisely. But more importantly, he did not accept that replacement was necessary. And this remained the case even *after* von Neumann had shown how to derive the theory of ordinals with its help. Up to 1958, he continued to regard the theory of ordinals, and the theory of sets of power $\geqslant \aleph_\omega$ as 'special set theory'. I shall suggest a reason for this in §8.3. For the moment let me just quote von Neumann ([1928*a*], p. 322): 'The replacement axiom was first formulated by Fraenkel ([1922*b*]). However, he does not include it in his axiom system at the present time.' It seems to me that Fraenkel himself came much closer to the truth about the discovery of 'replacement' in his autobiography. He remarks, concerning his [1922*b*], that

I have to thank this paper primarily for the fact that for three or four decades an important axiom system for set theory has been called the 'Zermelo–Fraenkel' system. In two respects I had here undeserved good fortune.

The 'second respect' was the simultaneous and usually unacknowledged 'discovery' of replacement by Skolem. But more significantly, the 'first respect' is that:

I did not immediately recognise or take advantage of the full significance of my new axiom. Rather this was done by von Neumann a year earlier . . . (Fraenkel [1967], pp. 149–50)

I suggest that von Neumann, not Fraenkel or Skolem, was the true 'discoverer' of the replacement axiom. About a year after Fraenkel had 'proposed' replacement, von Neumann had already worked out his own axiomatization of set theory. This system contained a clearly formulated version of replacement (quite different from Skolem's axiom or Fraenkel's suggestion), and its consequences (in particular the theory of ordinals) were fully developed. This work was eventually published as von Neumann [1928*b*]. The delay was explained by Fraenkel:

Around 1922–3, being then Professor at Marburg University, I received from Professor Erhard Schmidt, Berlin (on behalf of the *Redaktion* of the *Mathematische Zeitschrift*) a long manuscript of an author unknown to me, Johann von Neumann, with the title *Die Axiomatisierung der Mengenlehre*, this being his eventual doctoral dissertation which appeared in the *Zeitschrift* only in 1928 (vol. 27) [von Neumann [1928*b*]]. I was asked to express my views since it seemed incomprehensible. I don't maintain that I understood everything, but enough to see that this was an outstanding work and to recognise *ex ungue leonem*. While answering in this sense, I invited the young scholar to vist me (in Marburg) and discussed things with him, strongly advising him to prepare the ground for the understanding of so technical an essay by a more informal essay which should stress the new access to the problem and its fundamental consequences. He wrote such an essay under the title 'Eine Axiomatisierung der Mengenlehre', and I published it in 1925 in the *Journal der Mathematik* (vol. 154) [von Neumann [1925]] of which I was then Associate Editor.

(From a letter of Fraenkel to Stanislaw Ulam, quoted in Ulam [1958], p. 10, n. 3. Fraenkel tells the same story in his [1967], pp. 168-9.)

I have mentioned that Fraenkel did not formulate his 'axiom' precisely in his [1922*b*], that he did not specify a 'suitable' concept of function. Interestingly, it may well be that he was not in a position to formulate a sufficiently strong version of replacement. Fraenkel *did* develop a highly important 'concept of function'. But it turned out to be not 'suitable', i.e. not wide enough for formulating a sufficiently strong version of replacement.

Fraenkel's most important work on Zermelo's axiom system concerned not the axiom of replacement nor any other possible addition to the axioms but rather an apparent weakening of the axioms. Fraenkel began with the question of the *necessity* of the existing Zermelo axioms, that is with the question of their mutual *independence*. (This problem is clearly stated in Fraenkel [1922*b*], p. 234.) But as Fraenkel states:

. . . in the proofs—particularly that of the independence of the axiom of choice—the lack of clarity in the concept 'definite', and consequently axiom 3 [separation], is very troublesome. (Fraenkel [1922*b*], p. 234)

The reason is (presumably) that in checking to see whether separation holds in a certain model (say one in which choice fails) one has to show that for any 'definite' ϕ, $\{x: x \in m \land \phi(x)\}$ is in the model if m is. And for this one needs some idea of what the 'definite' ϕ are. Fraenkel's proposal was to replace the notion of 'definite property' with a recursively defined, countable list of 'functions', all of which have a precise set-theoretic sense. These 'functions' are, in effect, what we would call set *terms*, but not all terms as we shall see.

Fraenkel's formulation of separation is: if m is a given set so is the collection of xs in m satisfying $\phi(x) \circ \psi(x)$ where '$\circ$' stands for any of '$\in$', '$\notin$', '$=$', and '$\neq$' and $\phi(x)$ and $\psi(x)$ are Fraenkel functions. The functions are defined recursively as follows:

(i) $x, a, P(x), \bigcup x$ are functions of x (a is any constant).

(ii)   If $m(x)$ is a function of x and $\phi(y), \psi(y)$ are given functions of y then the subset of $m(x)$ determined by $\phi(y) \circ \psi(y)$ and guaranteed to exist by the separation axiom is a function of x.

(iii)   If $\phi(x)$ and $\psi(x)$ are functions of x so are $\{\phi(x),\ \psi(x)\}$ and $\phi(\psi(x))$.

(See Fraenkel [1925], p. 254, or [1926], p. 132-3, or [1927], pp. 104-10.) The important thing to notice here is that Fraenkel's functions are just those terms which are legitimized by the Zermelo axioms or iterations of these axioms.

Using this quite precise notion of function and the precise version of separation, Fraenkel was able to prove a highly important independence result, namely the independence of the axiom of choice from the other Zermelo axioms (see Fraenkel [1922c]). But he achieved much more than this as far as the Zermelo programme was concerned. Such an independence proof was only any good if it could be shown that the new and apparently much more restricted Fraenkel version of separation had not emasculated Zermelo's theory. Fraenkel set out to show that Zermelo set theory, that is, principally the reductionist treatment of cardinal equivalence, Kuratowski ordering sets, and well-ordering sets, could be derived from the new 'Zermelo' axioms. This he succeeded in showing in the series of three papers [1925], [1926], and [1932a]. This was a remarkable victory for reductionism. For not only did Fraenkel do for ordering and well-ordering relations what Zermelo had himself done for the theory of equivalence, but he did it without Zermelo's 'definite properties'.

In its 1908 formulation, Zermelo's system had strayed somewhat from its reductionist path. Fraenkel, like Skolem, had developed a method of putting it firmly back on that path without sacrificing any of its promised mathematical content. Under Fraenkel's guidance, Zermelo set theory was in spirit reductionist again, fully in harmony with the heuristic doctrine which had such an important part in its development.

Prima facie it seems that Fraenkel *could* formulate 'the' axiom of replacement. For he had developed a precise concept of function, and therefore could frame replacement as follows

(*FR*): if $f(x)$ is any Fraenkel function, then $\{f(x) : x \in m\}$ is always a set for any set m.

Moreover, since Fraenkel's concept of function appears to go back to 1922, so could this Fraenkel version of replacement. That Fraenkel considered (*FR*) to be 'the' axiom of replacement is suggested in his [1926] (p. 134). He states a theorem: if $f(x)$ is a Fraenkel function then $\{f(x) : x \in m\}$ is a set *provided* that $\{f(x) : x \in m\}$ is *contained* in a set. Then he says that if the proviso is dropped '. . . one obtains an assertion (the replacement axiom) *which is independent of the axiom system* . . .' i.e. (*FR*). (Fraenkel was more cautious in the earlier [1925], p. 271.)

But it transpired that Fraenkel was quite wrong. Von Neumann showed that (*FR*) is not independent of the axioms of Zermelo set theory, but can be *derived* from them. Thus although Fraenkel's specification of function allows the formulation of *a* version of replacement, it is a redundant version. The consequence of von Neumann's analysis [1928a] was not that Zermelo set theory is sufficiently strong without replacement, but that Fraenkel's notion of function is too narrow.

Von Neumann's proof shows that if f is a Fraenkel function and m a Zermelo set, then it is always possible to find a Zermelo set $n(f, m)$ which includes $\{f(x) : x \in m\}$. Thus, separation will guarantee that the collection $\{f(x) : x \in m\}$ is itself a Zermelo set. The proof is rather simple (von Neumann [1928a], pp. 323-4), proceeding naturally enough by induction on the complexity of f. If f is identity $n(f, m) = m$; if f is a constant function with constant value a, $n(f, m) = \{a\}$. If f is $P(x)$, then $n(f, m) = P(P(\bigcup m))$. For if $x \in m$ then it is clear that $[y \subseteq x \to y \subseteq \bigcup m]$; but this is just to say $[y \in P(x) \to y \in P(\bigcup m)]$, or $P(x) \subseteq P(\bigcup m)$, i.e. $P(x) \in P(P(\bigcup m))$, which is what we require. If f is $\bigcup x$, then $n(f, m) = P(\bigcup \bigcup m)$. For if $x \in m$, $x \subseteq \bigcup m$, hence $\bigcup x \subseteq \bigcup \bigcup m$, and so $\bigcup x \in P(\bigcup \bigcup m)$. Assume now that the result holds for two Fraenkel functions f and g. It must also hold for the functions $\{f(x), g(x)\}$ and $f(g(x))$. In the latter case $n(f(g(x)), m) = n(f, n(g, m))$; for if $x \in m$, $g(x) \in n(g, m)$ by assumption, and hence $f(g(x)) \in n(f, n(g, m))$. In the former case, $n(\{f, g\}, m) = P(n(f, m) \cup n(g, m))$; for if $x \in m$, then $f(x) \in n(f, m)$ and $g(x) \in n(g, m)$, hence $\{f(x), g(x)\} \subseteq n(f, m) \cup n(g, m)$, and so $\{f(x), g(x)\} \in P(n(f, m) \cup n(g, m))$. Lastly, assume $a(x)$ is a function of x, and $\phi(x, y)$, $\psi(x, y)$ are functions of x with parameter y; assume that the result holds for $a(x)$, and let $a'(x) = a(x)_{\phi(x, y) \circ \psi(x, y)}$; then $n(a', m) = P(\bigcup n(a, m))$. For if $x \in m$, then $a'(x) \subseteq a(x) \subseteq \bigcup n(a, m)$, i.e. $a'(x) \in P(\bigcup n(a, m))$. This completes the proof.

The proof thus shows that replacement collections under Fraenkel functions are always bounded by Zermelo sets. But the kinds of replacement operation one requires in proving that every well-ordered set possesses an ordinal number and that definitions by transfinite recursion are legitimate give rise to (Zermelo) *unbounded* collections, as von Neumann pointed out in his [1928a] (p. 327, n. 18, and §3). This kind of unboundedness, of course, strongly violates the limitation of size idea underlying Zermelo's separation axiom (more of this later). To legitimize it, one has to *postulate* that these collections are indeed sets. Von Neumann showed how such a postulate can be formulated in the Fraenkel framework if one first extends the notion of Fraenkel-function to include the following:

(iv)   Let $\phi(x, y)$ and $\psi(x, y)$ be any two functions. If for each x, there is a set $f(x)$ containing just those y satisfying $\phi(x, y) \circ \psi(x, y)$, then $f(x)$ is a function of x. (Von Neumann [1928a], p. 323)

The corresponding replacement axiom is then just that replacement of a set under von Neumann functions always yields a set.

Thus there is *some* evidence (albeit inconclusive) to suggest that Fraenkel was not in a position in 1922 to formulate a substantially new version of replacement. There are some hints that Fraenkel became aware of the narrowness of his function concept before this was proved by von Neumann in print. For example, a need to extend the notion of function in framing replacement is hinted at in [1925] (p. 271), but this caution is retracted, as we have seen, in his [1926]. (The hint is restored in [1927], p. 114, n.8.) Apparently Fraenkel and von Neumann corresponded about this matter, but the only reference to this correspondence (von Neumann's [1928a], p. 323) is ambiguous: 'The necessity of this further extension [of the function concept] became clear to me in correspondence with Herr Fraenkel.' It does not attribute to Fraenkel prior knowledge of the 'necessity' in question, but neither does it rule it out.

8.3. Limitation of size revisited

Cantor's 1899 theory raised a question about the doctrine of the Absolute. According to this the Absolute is 'beyond mathematical determination', something which, although it embraces the whole of the finite and transfinite, and thus of mathematical activity, is not itself part of or subject to that activity. In so far as it can be regarded as a collection or a domain, it must be a collection or range of a different *type* from the transfinite collections. In the 1899 theory *the* Absolute is replaced by a notion of absolute collection which is not so summarily dismissed from the mathematical stage as the Absolute itself was. These collections do perform important mathematical tasks, both in the criterion of sethood and the Cantor–Jourdain proof of the aleph theorem. To this extent, the new theory violates the doctrine of the Absolute. Now, however, the previously implicit type distinction is made explicit. In particular, unlike 'normal' collections the absolute collections carry neither cardinal nor ordinal numbers. Thus they are 'beyond mathematical determination' in any traditional and specifically Cantorian sense. In *this* sense, then, the spirit of the Cantorian doctrine is preserved. Moreover one might say that while the absolute collections are certainly mathematically involved according to the 1899 theory, something of the same could be said of the old Absolute. For it was used as part of an explanation of the limitations of mathematics, something to reflect the transfinite against. The absolute collections (specifically *Ord*) perform this task in the 1899 theory, though, one might say, a little more openly. The 'extramathematical' nature of these collections is underlined by Cantor when he explicitly says that they cannot be single objects, they 'cannot be gathered together to a whole without contradiction'. This puts a further restriction on their entry into normal mathematical activity, for it means, says, Cantor, that they cannot be taken as *members* of any collections:

. . . an inconsistent multiplicity, because it cannot be understood as one *whole* and thus cannot be considered as *one thing*, cannot be taken as an element of a multiplicity. (Cantor, letter to Jourdain, 9 July 1904, in Grattan-Guinness [1971], p. 119)

In the two other limitation of size theories we have so far considered, those of Mirimanoff and Zermelo (Fraenkel), Cantor's absolute collections are decreed unnecessary, indeed as non-existent, totally outside mathematics as the original doctrine of the Absolute itself declares. Mirimanoff gets away with this view, but his theory is not strongly tested since he deals with sets only informally, and in any case more with the question of principle: which sets should/should not exist. Zermelo goes much further than a question of principle; he attempts to say explicitly which sets exist and to develop set-theoretic mathematics on this basis. But he does not get away completely without the absolute collections. They are deliberately not used directly in the proof of the well-ordering theorem. But the ghosts of these departed entities do haunt the central separation axiom. While Zermelo wants to do without the absolute collections, it seems, like Cantor, that he does not avoid them.

Skolem conjectured that the ghosts can be exorcised by replacing 'arbitrary property' by 'predicate of the set-theoretic language', thus by building them into the linguistic or logical framework of the system. This view was later confirmed, although it was only followed up later. Fraenkel tackled the problem instead by replacing the notion of 'definite property' with his notion of function. Skolem later showed that this idea can be reduced to his, and this Fraenkel accepted. (See Fraenkel and Bar-Hillel [1958], p. 40. Skolem [1929] showed that Fraenkel's $\phi(x) \circ \psi(x)$ correspond to just some among the predicates of the language of first-order set theory.) Nevertheless in the original conception the notion of function stood alongside the notion of set. Fraenkel's method had the highly desirable merit of replacing Zermelo's vague notion by something clear and precise. And it seemed, as with Skolem's method, that the ghosts of the absolute collections had been removed, for the functions used are specifically related to the basic set-theoretic operations. Thus, it seemed that now nothing was being used or referred to which was not an intrinsic part of Zermelo's systems of sets.

Thus something like the old or 'pure' view of the Absolute, or absolute collections, is restored. They are outside mathematics since they perform no mathematical function, and they are also outside the system, more like domains in which mathematical activity takes place, and thus more like the category of mathematical activity that Cantor's Absolute originally was. But although the absolute collections had been bypassed mathematically, it was at the cost of adding a new notion to the set-theoretic framework, the notion of function. And unlike Skolem's method, Fraenkel's functions were not explained as part of the linguistic, logical framework of the theory.

This is important when looking at the structure of von Neumann's system. For he proposed bringing the absolute collections back into the mathematical framework and asking of them as much mathematical work as we can consistently demand. He also chose to axiomatize a notion of function (which was quite different from Fraenkel's) and *only* this notion. Again, unlike Fraenkel, the main point for von Neumann was to get the axiom of replacement right. The upshot was a 'return' to the Cantorian theory of absolute collections. Von Neumann's decision to axiomatize the notion of function should not necessarily be seen as a preference for functions over collections, as he himself makes clear ([1925], p. 396). But behind it there *is* the idea that the system should be as unified as possible, *and* that the demands of the full Cantorian theory of ordinals must be taken seriously.

Since von Neumann's treatment does not essentially depend on using functions rather than collections, let us examine it first in terms of collections. This at least will make the connection with the Cantorian theory clear. Von Neumann's theory extends Zermelo's axiomatization: i.e., it has axioms corresponding to the infinity, union-set, and power-set axioms. But it replaces Zermelo's separation axiom by a new principle, explicitly a principle of limitation of *cardinal* size. Thus, we have a shift back in the direction of Cantor and Jourdain: a comprehension principle based on cardinal size. And, as with their theories, one of the collections which is 'too big' is used as a criterion of size, not *Ord* this time but the universe itself (we shall see shortly that this supplanting of *Ord* does not make too much difference). Von Neumann's central principle is beautifully simple:

(1)   A set is 'too big' if and only if it is equivalent to the set of all things.

(Quoted from von Neumann's letter to Zermelo, 15 August 1923 (henceforth [1923*b*]), p. 272 of Meschkowski [1967].) Indeed, it is not only simple but perfectly natural. For, whether one construes bigness as concerned with comprehensiveness or with cardinality, the 'set of all things' is clearly too big if anything is. (1) now says that a set is too big just in case it is as big as anything could possibly be. This is an extremely strong principle in von Neumann's hands. For it is formulated as an axiom in the system, and *everything in it is taken literally*. Reference to the 'set of all things' is not simply a manner of speaking. In von Neumann's system this set is an *object*. 'Equivalence' means literally 'cardinal equivalence'; for von Neumann, *a* is equivalent to *b* just in case there is a *function*, i.e. a certain object in the system, taking *a* one-to-one onto *b*. And this holds true, whatever *objects a* and *b* are, 'too big' or not. But this is not all: 'cardinal equivalence' can even be interpreted in its classical Cantorian sense. For von Neumann (1) is equivalent to:

(1′)   A set is too big if and only if it has *the same cardinal number* as the set of all things.

This must mean that the 'set of all things' does *have* a cardinal number.

But why *is* (1) incorporated as an axiom? This is not difficult to explain, for here we come immediately to the need for a full, broad axiom of replacement. Von Neumann himself is clear about this:

Fraenkel's 'replacement axiom' is added. This (among other things) is necessary for the setting up of the theory of ordinal numbers. Sets which are 'too big' are permitted (e.g. the set of all sets not containing themselves). I believe that this is necessary for the formulation of the 'replacement axiom'. (von Neumann [1923b], p. 271)

The heuristic pattern behind this remark can be reconstructed in the following way. Von Neumann clearly starts from the idea that Cantor's theory of ordinals in its fullest extent has to be restored to set theory. For him, the theory of ordinals is part of 'general' (i.e. essential) set theory (see von Neumann [1928a], p. 320), a thesis which Fraenkel for instance never accepted. (See e.g. Fraenkel and Bar-Hillel [1958], pp. 85-6.) But von Neumann's analysis of the theory of ordinal number had shown that some form of replacement axiom, or some principle yielding it, is essential. (One might add that this is true quite independently of von Neumann's particular definition of ordinal, which after all only uses the idea of a Zermelian subset ordering. Assignment of an ordinal to a well-ordered set must be a recursive process, i.e. it must depend on the *set* of unique assignments made to the proper segments of the set. But to be able always to form the *set* of 'previous assignments' one surely needs some form of replacement axiom.) Formulation of replacement requires a concept of 'function' and among other things von Neumann required 'functions' to include functional terms taking the whole ordinal sequence as domain. These functions cannot be sets in Zermelo's sense. For if (following Hausdorff) a function is a certain collection of ordered pairs (or just pairs in Zermelo's theory), and its domain is, say, the ordinal number sequence, then the collection must be 'large', 'too large' to be a Zermelo set. Zermelo's notion of 'definite property', or, what amounts to the same 'definite functional relation', is too vague to be a satisfactory basis for the notion of function. Consequently, if one does not rely on an underlying logical framework to supply the notion, the only reasonable course left is to axiomatize the notion of function itself. This was the course that von Neumann took.

Following this line a little further quickly shows the need for (1) as an axiom. Von Neumann was himself a reductionist to this extent: although non-set objects are permitted, von Neumann insists that they should be *as much like sets as possible*. This means, above all, that they must satisfy an axiom of extensionality, thus giving a clear-cut criterion of identity (conversely, of *individuation*) for them. (This was a major difference between von Neumann's approach and Zermelo's.) But if functions are treated as extensional collection-objects very much like sets, and if (on pain of contradiction) some functions are not full sets, there must be some formal means of distinguishing between them *within the system*. And it is for this reason that von Neumann invokes a limitation of size theory. The paradoxical sets are supposed to be 'too big'; hence those collection-objects which are 'too big', among them some of the important functions, are the collections we must treat cautiously. This, I think, fully explains von Neumann's remark that 'sets which are "too big" are permitted [and] necessary for the formulation of the replacement axiom'. Von Neumann's limitation of size principle is the principle (1); since he needs an internal principle he adopts (1) itself as an axiom.

I shall return to the subject of limitation of size in a moment. For the time being, let us concentrate on the contradictory sets. Von Neumann supposes that the sets which go wrong are 'too big'. But this does not tell us *why* they go wrong, nor how to prevent them from going wrong. Von Neumann's answer to *why* sets go wrong is a characteristically simple one. As Fraenkel says:

It was von Neumann's daring idea that it is not the existence of the overcomprehensive sets as such which leads to contradictions but *their being taken as members of other sets* (elementhood). (Fraenkel and Bar-Hillel [1958], p. 97)

This may already be suggested by Russell's paradox if one rejects Zermelo's conclusion that the Russell set does not exist. But I think the clearest way to arrive at von Neumann's answer is via the Burali-Forti antinomy. Von Neumann accepts that the collection *Ord* is an object, a set (of some kind). Starting from this, let us now trace the steps of the Burali-Forti argument. According to von Neumann's definition of ordinal, *Ord* must itself *be* an ordinal number. There are now two possible ways of generating the contradictions. First, *Ord* must be the greatest ordinal number. But there is surely a greater one, namely $Ord' = Ord \cup \{Ord\}$, hence contradiction. Second, *Ord* must belong to *Ord*; but no ordinal *can* belong it itself, hence contradiction. Let us now invoke the distinction between sets which are too big and those which are not; call the former sets$_1$ and the latter sets$_2$. A crucial step in the derivation of either form of the contradiction is the assumption that sets$_2$ can be members of sets. If this is *denied* then the formation of *Ord'* becomes impossible. Moreover, *Ord* must by definition be the collection of all set$_1$ ordinals, not the collection of all ordinals absolutely. Hence $\sim (Ord \in Ord)$, which is just as it should be. According to this though, *Ord* is still an ordinal; indeed the unique set$_2$ ordinal.

Denial of elementhood to sets$_2$ is just what von Neumann decrees. Thus, having characterized the sets which are 'too big' by principle (1), von Neumann adds ([1923*b*], p. 271): 'Although all ("definite") sets are permitted, in order to avoid the paradoxes those which are "too big" are declared to be impermissible as *elements*.' In the formal presentation of the theory this is actually built into the formulation of principle (1). Indeed, formally (1) becomes:

(2)   A set A is not an element of another set just in case A is cardinally equivalent to the set of all things.

With these doctrines von Neumann shows quite clearly and simply how to avoid the paradoxes while retaining the 'contradictory' sets as objects within the theory. This is the essential step (2) I mentioned at the beginning of Chapter 8.

If we glance back now to Cantor's doctrine of the Absolute, it seems that it has been further weakened. Absolute collections can and do have ordinal and cardinal numbers. *Ord* is the ordinal number of the collection of all (set$_1$) ordinals, and also the cardinal number of the universe (of all set$_1$ objects), indeed by (1) the cardinal number of every set$_2$ object. Thus it seems that absolute collections are no longer 'beyond mathematical determination'. But this really gives a false impression. For one thing, the most important and fundamental mathematical role of sets is denied to the 'big' collections, since they cannot be members. For another, although these collections have a number, very little can be done with this. It simply says that these are the biggest collections, that is, repeats what (1) already tells us. And one might say that this even fits the Cantorian concept of the Absolute as an 'Absolute maximum'.

The reason for introducing a limitation of size principle at all is to avoid the paradoxes. But there are many ways of doing this. The reason why (1) and (2) are adopted, instead of, say, the Cantor–Jourdain principle (3) below, is because of its heuristic power. Sets$_2$ were deemed necessary to allow the full Cantorian theory of ordinals to be derived. (1) and (2) not only provide the necessary discrimination between sets$_1$ and sets$_2$, they imply the two principles crucial to the Cantorian theory, the axiom of choice and a form of the axiom of replacement. The reason for this is that, as Gödel says, von Neumann's principle (1) (or (2)) is a maximum principle: it allows the Burali-Forti paradox to be put to constructive use. Gödel says (in a personal communication to Ulam quoted by Ulam [1958], p. 13, n. 5):

The great interest which this axiom [i.e. (1)] has lies in the fact that it is a maximum principle, somewhat similar to Hilbert's axiom of completeness in geometry. For, roughly speaking, it says that any set which does not, in a certain well-defined way, imply an inconsistency exists. Its being a maximum principle also explains the fact that this axiom implies the axiom of choice.

What Gödel must mean by the last remark is that according to von Neumann's principle there must be as many ordinals as possible, i.e. enough to well-order the whole universe. Indeed, this is precisely how von Neumann's proof of the axiom of choice (or well-ordering theorem) works. Informally, this proof uses the fundamental hypothesis 'all contradictory collections are too big' in a direct way. As von Neumann says (a similar informal proof can be found in [1925], p. 398):

The method of proof is this: the set of all ordinal numbers (which can be set up without further ado) would lead to the Burali-Forti antinomy, thus it is 'too big'. Thus it is equivalent to the set of all things. But this immediately gives rise to a well-ordering of the set of all things. (von Neumann [1923*b*], p. 272)

(And, one can add, since any set is a subset of the well-orderable set of all things it too must be well-orderable.) The strong step in this proof is supplied by principle (1). It is this which authorizes the argument from the contradictoriness (largeness) of *Ord* to the *completeness* of *Ord* in the sense of containing enough ordinals to well-order the universe.

The formal version of the proof is just as elegant. In this case one first uses the Burali-Forti argument to *prove* that *Ord* cannot be a set$_1$ (otherwise *Ord* $\in$ *Ord*, and contradiction, as above), i.e. *Ord* cannot be an element. Principle (2) then allows us to conclude that *Ord* must be equivalent to the universe.

Maximality is also responsible for the derivation of replacement from (1) or (2). Suppose X is a set equivalent to a set$_1$. Then X cannot be equivalent to the universe of sets$_1$. Hence X must be a set$_1$ and replacement follows. What is applied here is the principle: whatever is not too large can be taken as a set (on the grounds that it cannot be contradictory). This is the first aspect of maximality referred to by Gödel. In this respect (1) (or (2)) is much the same as the Cantor–Jourdain principle (9) in §4.1 which in this context would read:

(3)   A set A is not an element of another set just in case *Ord* is equivalent to a subset of A.

For (3) also yields replacement. Suppose a set X is equivalent to a set$_1$ a, then X cannot have a subset equivalent to *Ord* since this would then also be true of a. Hence X must be a set$_1$. But the crucial difference between (1) (or (2)) and (3) is that (3) does not yield the axiom of choice (see (15) in §4.1), though if one adds the local AC to (3) one can get (2). (3) (like (2)) is maximal in that it allows all non-large collections as sets. But unlike (2) it does not maximize the collection of ordinals; indeed, it characterizes 'too-bigness' by setting a *minimum* requirement (being bigger than *Ord*). (1) and (2), on the other hand, set a maximum requirement for bigness: a set X is too big if it is as big as it can possibly be.

Von Neumann's central axiom is thus extremely powerful as he himself recognized:

[It], to be sure, requires something more than what was up to now regarded as evident and reasonable for the notion 'not too big'. One might say that it somewhat overshoots the mark. But, in view of the confusion surrounding the notion 'not too big' as it is ordinarily used, on the one hand, and the extraordinary power of this axiom on the other, I believe I was not too crassly arbitrary in introducing it, especially since it enlarges rather than restricts the domain of set theory and nevertheless can hardly become a source of antinomies. (Von Neumann [1925], p. 402)

Talk of sets$_1$ and sets$_2$ is somewhat misleading. I mentioned that von Neumann axiomatized the notion of function. But this was *all* he axiomatized; he explains sets$_1$ and sets$_2$ in *terms* of the notion of function:

We prefer to axiomatise not 'set' but 'function'. The latter notion certainly includes the former. (More precisely, the two notions are completely equivalent, since a function can be regarded as a set of pairs, and a set as a function that can take two values.) The reason for this departure from the usual way of proceeding is that every axiomatisation of set theory uses the notion of function (axiom of separation, axiom of replacement, see pages 400 and 403), and thus it is formally simpler to base the notion of set on that of function than conversely. (Von Neumann [1925], p. 396)

However, the axiomatization is not as simple as von Neumann makes it seem here. He requires a fair amount of primitive machinery. For example, to replace the notion of set by that of its characteristic function, one needs two arguments A and B to be able to state '$x \in f$' $(f(x) = B)$ and '$x \notin f$' $(f(x) = A)$. Von Neumann introduces these objects A and B as primitives. Moreover, he needs two predicates (being a I-object, being a II-object) which distinguish between 'arguments' and 'functions' respectively. This distinction is crucial, because it is, in effect, the distinction between sets$_1$ and sets$_2$. A I-object is allowed to belong to (to be an argument of) a function, i.e. a II-object, and (roughly speaking) the sets, are those objects which are I–II, i.e. both arguments and functions. Proper II-objects cannot belong to (be arguments of) II-objects; this is just the denial of elementhood to sets$_2$. Von Neumann does not introduce a specific predicate standing for functions; rather he introduces a three-place relation '$[x, y] = z$' which is to be read 'z is the value of the II-object x for the I-object argument y'. In addition von Neumann introduces the notion of 'the ordered pair of x and y' as a primitive (denoted (x, y)), though in his later papers he dispenses with this in favour of the Kuratowski definition.

The axioms for functions are not so simple either. One of the complicating factors is that he has to introduce various 'logical' axioms (see, for example, von Neumann [1925], pp. 399–400). The reason for this is quite clear. What the system does is axiomatize an extensional version of Zermelo's 'definite property' (relation etc.); since general logical operations presumably apply to these, one needs to be able to capture logical operations for the functions also. This is shown up much more clearly in later class-set systems when one replaces the so-called predicative comprehension schema for classes by a finite list of axioms which explicitly cover the logical operations. (There is an excellent discussion of classes as an internalization of the notion of formula in Fraenkel, Bar-Hillel, and Levy [1973], pp. 119–20, 128–33; see particularly pp. 129–30.) Above all this is clear in Gödel's [1940] with his eight group B axioms for classes. There Gödel proves that using just these axioms, if $\phi(x_1, \ldots x_n)$ is any formula of the theory which involves no bound class variables then there is a class containing exactly the n-tuples $(x_1, \ldots x_n)$ which satisfy ϕ. (This is just what the predicative class comprehension principle says.) Thus what the general class axioms capture is precisely the notion of 'formula of the language of set theory', for in such ϕs containing no bound class variables all references to classes can be systematically removed. Gödel himself makes clear that essentially what is being formalized is Zermelo's vague idea of 'definite property'. (See Gödel [1940], p. 2.) In this same monograph Gödel introduces also his famous eight $\mathfrak{F}$-operations for generating constructible sets. As he says (p. 35), they are modifications of the eight group B class operations. While these suffice to capture the notion of 'formula of set theory', the constructible operations capture the notion of 'bounded formula of the language of set theory' where a bounded formula is one where all quantifiers are relativized to sets. (This is shown by a normal form theorem; see Jech [1978], pp. 92-6.) This is just what is required in defining the concept of constructible set (see Drake [1974], Chapter 5, §1).

In a sense, then, we have two-way traffic between the Skolem 'linguistic' approach to 'definite' properties and the Fraenkel–von Neumann approach. 'Formulae of the language of set theory' can be treated as classes, and classes are (normally) just extensional versions of these formulae.

There is no space here to go further into the details of von Neumann's axiomatization. He himself explains it very clearly in his [1925], and full details together with the derivation of the most important parts of set theory are given in his [1928b]. (There is also a clear account of the von Neumann system in Fraenkel and Bar-Hillel [1958], pp. 96-104.) Rather I want to return briefly to von Neumann's use of the idea of 'limitation of size'.

Von Neumann regarded his limitation of size theory as a clarification and extension of Zermelo's. For example, in his [1925] (p. 397), introducing his limitation of size axiom, he says he is 'going in the direction pointed out by Zermelo'. In discussing his axiom he suggests that part of its justification is that it makes the notion of being 'not too big' precise. (See the quote from von Neumann [1925] on p. 292 above.) For instance:

Whereas *Zermelo–Fraenkel* do not define precisely when the extent of a set is 'too big', but only put a lower limit on the notion through some postulates, we give this concept a precise definition: a set is 'not too big' if and only if it is of smaller power than the set of things generally . . . (Von Neumann [1929], p. 494)

But von Neumann's remarks require some further comment. In one important sense von Neumann does *clarify* Zermelo's and Fraenkel's position. But in other respects von Neumann shifts right away from it. Zermelo and Frankel use limitation of size metatheoretically, as Levy says, 'as a guide to the introduction of axioms' (Fraenkel, Bar-Hillel, and Levy [1973], p. 137); von Neumann, as we have seen, goes further, building his limitation of size idea *into* his system. This does lead to one important clarification of and improvement on Zermelo's position. For example, I suggested that Zermelo (and Fraenkel) assume the consistency of the basic existence principles (power-set, union-set, etc.) by taking for granted that these axioms assert the existence of *small* sets. This latter assumption is implicit and extra-theoretical. But von Neumann's use of his limitation of size principle as an axiom makes this assumption *explicit* and brings it *inside* the system. How is this done? Suppose an axiom asserts that a II-object b is also a I-II object. (This is tantamount to assuming that b is a set.) Then applying the limitation of size axiom (2) it can be concluded immediately that b is 'not too big', i.e. is *small*. Thus adoption of (2) has the effect of appending the assumption of smallness to any set existence assumption. Since von Neumann follows Zermelo in adopting infinity, union-set and power-set axioms (axioms V 1, 2, 3 respectively in von Neumann [1925]), he is also explicitly assuming that the sets asserted to exist by these axioms are all small. Indeed, he explains the content of these axioms in just these terms:

Their meaning (naively formulated) is approximately as follows:
There is an infinite set a that is not too big. If a is a set, not too big, of sets that are themselves not too big, then the set b of the elements of the elements of a is not too big either.
If a is a set that is not too big, then the set b of all subsets of a is not too big either. (Von Neumann [1925], p. 401)

This is an important clarification of Zermelo. For this reason, the following comment of Levy on von Neumann's central axiom (which Levy refers to as (*)) is quite misleading:

. . . contrary to von Neumann's intention, (*) does not embody the full limita-tion of size doctrine. Even though (*) establishes non-equinumerosity with the set V of all elements as a necessary and sufficient condition for elementhood, (*) in itself does not tell us when a given set is equinumerous to V. For instance, the axiom of power-set which clearly falls within the limitation of size doctrine, does not follow from (*) and the other axioms . . . (Fraenkel, Bar-Hillel, and Levy [1973], p. 137)

Levy's complaint here is difficult to understand. Presumably for Levy a principle embodying the 'full limitation of size doctrine' would be a limitation of size principle which by itself would be a substitute for the comprehension principle. Certainly von Neumann's main axiom is not this, since among things, as Levy says, it does not entail the power-set axiom. Von Neumann has to assume this axiom as well as his (*). But so do Zermelo and Fraenkel, and so does Levy. I cannot see that *this* is a criticism which can be directed specifically against von Neumann. Levy seems to imply that his own justification of the ZF axioms explains the adoption of power-set by its falling 'within the limitation of size doctrine'. But this is just what his justification has not done. I have argued that it is impossible to justify convincingly the smallness of power-sets. Certainly Levy himself gives no convincing argument, indeed no argument at all that the power-set axiom 'falls within the limitation of size doctrine'. As I have said, von Neumann explicitly assumes that 'all sets are small'. But since this appar-ently cannot be effectively argued for, von Neumann's explicit assumption seems preferable to Zermelo's, Fraenkel's, and Levy's obscurely hidden *implicit* assumption of the smallness of power-sets.

Von Neumann does make the notion of being not too big precise. But in doing so he moves away from the Zermelo–Fraenkel position in three respects. First he introduces non-set collections and uses them in an essential way. Secondly, while Zermelo and Fraenkel assume only that some key sets are small, von Neumann assumes in addition that all small collections are sets (maximality). Thirdly, and most importantly, von Neumann switches from a 'comprehensiveness' view of size to a *cardinal* view of size. And this seriously conflicted with the Zermelian position, certainly as this latter was characterized by Fraenkel.

As I suggested in §3.1 Cantor seemed to regard the assignment of cardinal number to a collection as a guarantee that the collection is of restricted extent. Von Neumann did not accept Cantor's doctrine in quite this form, since he allowed the assignment of cardinal number to the 'overcomprehensive' collections. But von Neumann's position certainly embraced the Cantorian view that cardinal equivalence with a set (or a set$_1$) is a sufficient condition for smallness, and therefore for safety. This, of course, does not *necessarily* conflict with the Zermelo comprehension axiom (separation). The latter simply imposes a stronger sufficiency condition, namely being bounded by (included in) a Zermelo set; any collection satisfying this automatically satisfies the cardinal condition. Thus a Zermelo-small set is von Neumann-small. But Fraenkel seems to have interpreted the Zermelo limitation of size doctrine much more strictly, namely as 'being bounded by an existing Zermelo set is a sufficient *and necessary* condition for being a set'. And this of course *does* conflict with the Cantor-von Neumann cardinal view (provided 'equivalence' is taken to include 'equivalence under functional relations' rather than, say, 'equivalence under Fraenkel functions').

Fraenkel's position is exemplified in his reluctance to accept the axiom of replacement as an axiom of 'general set theory'. For von Neumann, of course, it was a 'general' axiom, an axiom which stems quite naturally from the cardinal characterization of smallness. But Fraenkel regarded it as both *ad hoc* and unnecessary. For example, in his [1925] it is described as

. . . an axiom proposed by me as a *stopgap*, an axiom which nevertheless would seem to be too sweeping to be called upon without a painstaking investigation of its necessity. (Fraenkel [1925], p. 251)

Then in his [1926] and [1927] he claims that since replacement 'is not necessary for the general theory of ordering and well-ordering'

. . . *general* set theory can be derived in its full extent from the axioms I–VII, the Zermelo axioms. (Fraenkel [1927], p. 139)

Replacement is regarded as an 'unpleasant far reaching axiom' ([1926], pp. 130-1) needed 'only for guaranteeing special sets . . . but not for the foundation of general set theory' ([1927], p. 115). Fraenkel certainly seems to have regarded replacement as violating his own limitation of size doctrine, based on 'extent' or comprehension, not cardinality. For he frequently describes the 'special' sets guaranteed by replacement as 'very comprehensive' (for example [1925], p. 271 or [1926], p. 134), a designation which recalls his diagnosis of the contradictory collections as 'all too comprehensive'. It cannot be just the cardinality of these special sets which makes them go wrong, since some of them (the von Neumann ordinal $\omega + \omega$ for example) are *countable*. Fraenkel must be suspicious of the fact that they are Zermelo-unbounded.

As for Fraenkel's view that the axiom of replacement is unnecessary, he freely accepted that the replacement axiom adds much more content to set theory, and thus that it can accomplish much more than he had originally envisaged. What he seems to challenge is that any of this extra content which the axiom furnishes is set-theoretically important. This is clear in the following passage:

Whether ultimately the limits of set theory are extended in a mathematically valuable way with these two special existence axioms VII and VIII [infinity and replacement] is a question which has not yet been settled. (Fraenkel [1928b], p. 310)

The suggestion that the mathematical value of the axiom of infinity is not clear is hard to understand. As far as replacement is concerned, what Fraenkel probably means is that the ordinals do not add anything mathematically important which cannot also be done without them. Thus Fraenkel is still presumably following the reductionist programme set out by Zermelo and Hessenberg (§7.1), according to which number objects, while sometimes heuristically useful are not strictly necessary. This programme had indeed found some confirmatory support in the work of Zermelo himself (well-ordering theorem without using the ordinals), Lindelöf and the Youngs (see §2.3(b)), and Kuratowski (see §7.3). But there are two points to be made here.

First, despite this success, this kind of reductionism violates the basic Cantorian thesis of the 'arithmetic' nature of set theory, that is, that numbering by number objects is the fundamental operation in mathematics. Looked at from this point of view, one of the basic set-theoretic problems is to provide *enough* ordinals. And here replacement clearly is mathematically necessary, for it is required to provide enough ordinals to number sets which *are* mathematically necessary, the continuum for example. The desire to provide enough ordinals has had a continuing influence on the development of set theory. For this is precisely what gave the impetus to the study of large cardinal axioms. The motivation is simply to provide anough ordinals to be able to extract as much information as possible about 'ordinary' mathematical sets, in particular the continuum and its subsets (see §2.3(c)). In this sense then, von Neumann's system (ZF!) is Cantorian, while Zermelo's and therefore Fraenkel's is *not*.

Secondly, aside from the ordinals, recent work has thrown down another challenge to the thesis that replacement is mathematically dispensable. For in his [1975], Martin proved a theorem in ZF whose statement does not require the ordinal numbers, and yet which cannot be proved without the axiom of replacement. This is the statement: Every Borel game is determined. Let B be the Baire space ω^ω (which has the product topology, ω itself having the discrete topology). Then with each $A \subseteq B$ we can associate a two-person infinite game G_A, as follows. Player (1) chooses $a_0 \in \omega$, player (2) $b_0 \in \omega$, (1) $a_1 \in \omega$, (2) b_1, etc., to infinity. (1) wins if the complete sequence is in A; otherwise (2) wins. A game is said to be determined if (1) or (2) has a winning strategy, where a strategy is a rule which tells a player at each stage which move to make (given knowledge of what moves have been made previously), and a winning strategy is a strategy such that the person using it always wins. The statement of Borel determinacy is now just that G_A is determined for any Borel set A. Martin has made the following point about his theorem. Many of the most important (and 'nice') properties of Borel sets as a whole were first established for open and closed sets, and the proofs were then adapted so that they can be iterated using induction on the degree of complexity of definition to establish the result for the full hierarchy. The best example of this is the result discussed in §2.3(c), namely the extension of the Cantor-Bendixson theorem (or the perfect subset property) to all Borel sets. (For others, see Martin [1977], p. 798.) In terms of strength of theory required these proofs are relatively simple, as is the proof that all open and closed games are determined. But this does not apply to full Borel determinacy. As Martin says:

The proof of [Borel determinacy] is quite different from the proofs of the other regularity properties of Borel sets. In other cases the proofs can be carried out in the usual formal theory of ω and its power-set. Friedman [1971] shows that [Borel determinacy] cannot be proved in this theory and indeed cannot be proved in Zermelo set theory (set theory without the axiom of replacement). Hence although [Borel determinacy] is a statement about ω and [the Baire space] only, its proof involves in an essential way transfinite iterations of the power-set operation. (Martin [1977], p. 808)

(See also Kanamori and Magidor [1978], p. 246.) What Martin is suggesting is that replacement is necessary to extract *expected* information about relatively simple sets (ω and the Baire space), sets which hitherto had been quite adequately treated within a fragment of Zermelo set theory. And this is information about the sets themselves and *not* about their ordinal or cardinal numbers. There is, thus, at least *some* evidence that replacement is necessary for what Fraenkel would have regarded as 'general set theory'. It may be that in this sense not just replacement is necessary but other much stronger principles too (see §2.3(c)).

CONCLUSION

I hope that this book has done something to show that the set concept, although taken as elementary for more than 70 years, is not simple. It developed in and around a complicated network of ideas to a position of official mathematical pre-eminence, pre-eminence at least in a foundational sense. But in this progress it has brought with it two dense thickets of mystery, not unconnected, which should be taken more seriously than they perhaps have been hitherto. The first is centred around the notion of sethood itself and its explanatory role, and in particular around the 'oneness' or unity of a set: when does it make sense to say that a collection forms 'one thing' as opposed to a mere aggregate? The second surrounds the standard axiom system or systems: to a large extent these systems technically work and have given mathematics undreamt of freedom; yet, while they consist of only a few 'simple' axioms, there is no coherent and correspondingly simple explanation of *why* they work, or of why they should fit well together. By way of conclusion I want to stress the difficulties here. In doing so I do not mean to try to devalue set theory, but only to draw attention to its complexity and some of its fascination, and, by suggesting why these two pockets of mystery are connected, to try to draw together the material in the eight chapters above.

It was Cantor who first stressed the unity, the objecthood of sets, thereby marking them out from mere aggregations, and he was the first to insist that sets are the fundamental objects in mathematics. Of course the mathematical step from aggregation to objecthood had been taken before the official theoretical step, for example, with the Dedekind and Cantor definitions of real numbers and, even before that, with Dedekind's notion of an ideal (see Edwards [1980]). But when the theoretical step did come Cantor was unable to explain in any satisfactory way what actually *constitutes* unity (and therefore sethood). Bolzano before Cantor ([1851], p. 86) and Frege after ([1892], p. 165), seem to have regarded the wholeness or unity of an aggregate as explained by the concept or intension one uses to grasp or unify the elements. This is a perfectly plausible explanation, but not what Cantor was getting at. Cantor insisted on extensionality, and mathematics has largely followed him in this: the *elements* determine the set; the collecting intension is irrelevant, and thus *cannot* have anything to do with unity. Dedekind took this line too. Indeed, the characteristic mixture—collection, extensionality, and unity (objecthood, thingness)—is clearly displayed in a short passage at the beginning of Dedekind's [1888] (my translation):

It very frequently happens that different things *a, b, c, . . .* are for some reason comprehended under a common viewpoint, and put together in the intellect. One says then that they form a system *S*. One calls the things *a, b, c, . . .* the *elements* of the system *S,* and they are contained in *S*. Conversely *S consists* of these elements. Such a system (or aggregate, or manifold, or totality) is, as an object of our thought, likewise a thing [*Ding*]. (Dedekind [1888], p. 1)

Note the crucial unexplained jump from the extensional collection of *a, b, c, . . .* etc. to the thing *S*, the system, the set. (Note too that the last 'likewise a thing' represents type-collapse, and thus opens the way to iteration.)

The fact that the set concept is unexplained is a real difficulty since set theory was intended as a universal theory, something to which all other mathematical theories are reduced. The degree to which Cantor conceived it thus can be argued about, though I think it high, and certainly his successors, the axiomatizers, took this universality as basic, and this is what marks set theory apart from other theories, particularly axiomatic theories. Number theory and Euclidean geometry, for example, adapt reasonably well to axiomatization. By this I do not mean that they unproblematically capture *the* natural numbers, or all there is to the Euclidean plane; 'modern' metamathematical investigations have raised all kinds of problems with *this* idea. But they undoubtedly express a basic core of truth about numbers or planes; they list basic propositions about basic properties. Part of the reason why we accept the axiomatic standpoint here is that we feel we have some philosophical hold on the number concept or the concept of space *outside* the axiom systems which helps us to explain why the axioms are correct. In other words (even apart from the difficulties thrown up by Gödel's work) one need not accept that the axioms and their consequences represent *all* we can say about numbers or planes. But set theory is quite different. Part of the problem is that 'set' (unlike 'aggregate' perhaps) is not an ancient, well-understood concept which can easily be taken as an axiomatic primitive in the knowledge that it can be supported by extra-axiomatic explanation. (Unlike the case of natural number, this is largely *why* set theory is axiomatized, because we do *not* understand the set concept well.) But on top of this, because of the reductionist ambition, we demand that set theory genuinely explain all other mathematical concepts. When I say 'we demand', I mean that is what we should demand. Useful reductionism (philosophically speaking) cannot be just successful theoretical translation, though of course this would suffice in a relative consistency exercise, or as an original part of Hilbert's programme, say. Rather reduction must (in the best instance) be accompanied by a gain in conceptual clarity. Where set theory suffers as a foundation framework is that in general it does *not* bring this conceptual clarity with it. It is no good, philosophically speaking, reducing to set theory something so basic to human thought as the elementary theory of natural number if you cannot also explain why numbers are sets, and why the set concept is even more fundamental. But the set concept is too unclear for any such explanation to be given.

However, conceptual explanation, an attempt to gain conceptual clarity, was genuinely part of Cantor's aim in pursuing set theory. One finds this attitude both in his remarks on the continuum and in his abstractionist approach to cardinal and ordinal numbers. For example, he insisted that we should *not* attempt to derive a theory of the continuum from geometrical, spatial, or temporal concepts but rather build a theory of the continuum and continuity quite independently and then use this to clarify our concepts of time and space (see §1.1). Since he gave (as did Dedekind) a set-theoretic theory of the continuum he must have held, while not totally rejecting geometric intuitions or whatever, that set theory deals with something conceptually more fundamental. Unfortunately Cantor never gave an informal explanation of why the set-theoretic account of the continuum is conceptually (as opposed to technically or rigorously) adequate. But he did attempt an intuitive conceptual explanation of the nature of cardinal and ordinal numbers in set-theoretic terms through his theory of abstraction which I analysed rather exhaustively in Chapter 3. Let us look at this again for a moment.

Cantor's account says that we acquire our concept of number from counting, and counting must derive from our capacity to aggregate; that is, aggregation is both prior to and part of enumeration. This has a good deal of plausibility. According to Cantor's account we obtain a number from this by imagining the objects of the aggregate giving way to an aggregate of canonical representatives. And this 'canonical aggregate' is the origin of the number of the aggregate. Indeed, Cantor says it *is* the number. Obviously there is a need to say that this number is an object; thus the 'canonical aggregate' must be *one thing*, that is, a set. Thus, appeal to the notion of sethood fills out the explanation and enables us to explain how we get the notion of number from the rather plausible starting point of aggregation. However, despite the plausible way in which the analysis starts, the whole account may seem rather mysterious just *because* the primitive set concept is injected at the last gasp. Is this not really reduction without explanation, or at least does it not seem in the end that we are explaining the relatively well-understood (natural number) with the less well-understood (set)?

I said before that the set-concept was unexplained with Cantor. But Cantor often stressed (as does Dedekind in the passage cited above) *our* putting together the given elements to a set, as though a set is something which it is within our intellectual capacity to create as an object. It is not clear how much Cantor wanted to rely on such 'synthesis' in his concept of set. Some of his utterances suggest not at all, for we saw in §1.4 that he believed that it is really God who has put elements together to sets and not us, and certainly in remarks on his theory of number Cantor emphasized more than once that he did not rely on any capacity of ours. But there is, to say the least, a good deal of confusion in Cantor's theory (see §3.5), and it nevertheless depends on quasi-constructivist notions via the theory of counting and well-ordering. Perhaps, then, a quasi-constructivist 'synthesis' view of sets *would* be appropriate in Cantor's theory of numbers as sets. Thus we could say: the aggregation is finished, the marking of the elements is finished, and an object (the number) is introduced to represent this completion. And since for the purposes of seeing 'how many' elements the aggregate has only the marking objects are important, one identifies the object introduced with the collection of these; they *determine* the number. This is not only perfectly plausible, but, as we saw in §3.3, close in spirit to the very earliest notion of number. Thus here 'synthesis' does make some sense (the number is 'synthesized' out of the marking objects) and using it we obtain a rather full reductive explanation of numbers as sets, it being at the same time quite clear *what* explanatory role sets are playing and *why* the reduction is satisfactory. In this instance there *is* the connection to Kantian ideas which Gödel pointed to in his [1964] (see §6.2(*b*)):

> Note that there is a close relationship between the [iterative] concept of set . . . and the categories of pure understanding in Kant's sense. Namely the function of both is "synthesis", i.e. the generating of unities out of manifolds (e.g., in Kant, of the idea of *one* object out of its various aspects). (Gödel [1964], p. 272)

But the generality of Gödel's claim appears wrong. If synthesis is a reasonable account of what sethood represents it seems plausible that it must in turn be associated with process and completion, as in the account of number just given. (We saw in Chapter 6 that this is just what Kant in particular meant; and how else could we take it?) But having said this we come back to exactly the difficulties discussed in connection with completion. Maybe we could construe any well-ordered aggregate as representing a completed process of some kind (after all it has the right discreteness property). And perhaps one might say that the definition of individual real numbers, at least the definition using Cauchy sequences, is an idealization of a *process* of approximation. That is, if there were a real number at point *p*, then we could successively approximate it by rationals; this process, imagined completed, is then turned on its head and made a *definition* of the point *p*, *modulo* equi-convergent approximating sequences. Here, just as in the natural number case, the set concept is invoked to fill out the explanation, the set being again introduced to represent the full generated outcome of the process (or of all 'equivalent' processes). The set notion here just stands for the mind imagining done what it cannot (quite!) do itself. This perhaps would enable us to understand the notion of a real number in a given case; or, better put, the use of the set concept (as standing for 'completed totality' or something similar) would mean exactly that we *do* understand the given case. The problem, however, comes precisely with the extension of this idea to more or less arbitrary (and specifically impredicatively defined) collections. We might perhaps stretch the idea of introducing objects to stand for completions beyond the finite, but we simply cannot stretch it enough to include, say, the full classical continuum. In particular, the 'synthesis' (completion of a process of arrangement etc.) view of the cardinal number of the continuum makes no sense at all.

Now Cantor did make the mathematical transition to uncountable (and therefore impredicative) infinities and certainly intended his notion of number to apply to this new realm. But one consequence of this must be that we cannot take seriously any more the quite plausible and attractive view that sets are created or synthesized by us out of the aggregate of their elements, and we are therefore left without any plausible account of what the explanatory role of sets is (unless we want to make a clear distinction within the theory between those sets (finite?) which are synthesized and those which already exist, a division which would have been anathema to Cantor). In short Cantor showed how to develop a theory of higher infinities set theoretically, and how to begin to reduce other mathematical concepts to the set concept. But in doing so he (inadvertently?) shifted away from the aim of reductive conceptual explanation of other basic concepts towards 'formal' or theoretical reductionism instead. With this the attempt to understand the nature of the objects is gradually forgotten, and so is any attempt to understand the nature or role of sethood. What is left is rather the bald claim that sets exist.

Russell puzzled (in his [1903]) about how to explain what sets are, what the unity of an extensional collection could be, found no answer and gave up sets in favour of various intensional approaches. Cantor's successors on the other hand, particularly Zermelo and von Neumann, took the attitude that we have no call and no need to explain what sets are. Their philosophical transition (which Cantor only partly made himself) in a sense matches the bold mathematical steps Cantor had already taken, and certainly by removing all traces of Cantor's quasi-constructivist formulations, Zermelo and von Neumann relieved the Cantorian theory of much of its tension. But, to repeat, in this transition we must give up the idea that our use of sets is anything like a Kantian 'synthesis of manifolds into unities'. And it is precisely because set theory took this strong turn away from 'constructive explanation' that it is just no good going back to the axioms with attempts at quasi-constructivist justifications. This was a lesson we learned in Chapter 6.

Axiomatization went hand in hand with the divorce from any attempt to understand what sets are or what conceptual role they play. We cannot say with any kind of conviction what sort of things sets are, so we attempt a type of ostensive definition of them through axiomatization or 'listings'. This attitude to mathematical objects was stressed by Hilbert. For example in his [1900b] paper on the real number concept Hilbert writes:

In the theory of the number-concept the axiomatic method takes the following form:
We think of a system of things; we call these things numbers and denote

them by *a, b, c,* . . . We think of these numbers as standing in certain mutual relations, the exact and complete description of which is given by the following axioms: . . . etc. (Hilbert [1900*b*], pp. 257–8)

This was certainly Zermelo's position towards sets in his [1908*b*] : sets are just things in a domain, he says, about which the seven axioms he gives hold— there is no further explanation. (Von Neumann later took a similar attitude.) It is curious that this Hilbertian position is close to Cantor's description of what happens when a new concept is introduced into mathematics (see §1.1). In presenting this Cantor had in mind the concept of transfinite number, though it is remarkable that a similar approach was later taken to the set concept. The key difference with the latter, however, was that there was no possibility of relating this to 'concepts already present' as a means of establishing its co- herence, or even guiding the selection of axioms. What *are* the axioms governing this 'unknown' concept? Or in other words: which among all collections are the sets?

The first axiomatization was in effect due to Russell who tried to base everything on the comprehension principle: *all* collections are sets. This was simple, but it does not work. Not only this, but, as I tried to show, it violates Cantor's own (admittedly rather vague) notion of the set-theoretic universe. Now this latter view, a view about the structure of the transfinite and the way this fits together with the Absolute, at least gives rise to the limitation of size theory, and this *seemed* to give a plausible account of why Russell's comprehension principle does not work and how it might be modified. But the theory is difficult to formulate clearly, and indeed cannot be the complete answer. It suggests certain principles, and this certainly was a help in formulat- ing *some* of the Zermelo-Fraenkel axioms. But limitation of size, it seems, is quite incapable of 'supporting' the power-set axiom, an axiom of crucial importance in Zermelo's attempt to reduce the theory of well-ordering to pure set theory, and also to the framing of the Cantor-Dedekind theory of the continuum. Indeed (see Chapter 5) it appears that the power-set principle falls outside the limitation of size theory precisely because it is strong enough to render the classical theory of the continuum.

The upshot is that we have an axiom system to replace Russell's simple axiom; this system works both in that it seems to avoid contradiction and in that it gives mathematics enormous freedom. Yet we have no satisfactory simple heuristic explanation of *why* it works, why the axioms should hang together as a system. Our only plausible candidates, the limitation of size theory and the derivative iterative theory of set, do not work. So we have an axiom system about a concept we cannot easily understand, and yet apart from the demands of theoretical reduction to the theory of sets we find it hard to say why many of the key collections isolated by the axioms should be sets.

This situation is already foundationally unsatisfactory. As far as the lack of any explanatory account of what sets are, one *might* adopt a pragmatic view, insist on taking *carte blanche* and exploiting the unexplained axiomatic set notion for all it is worth. Indeed, despite the critical comments above, one has to face the fact that set theory provides an ideal background theory for large parts of modern mathematics. But really this is precisely because these parts (e.g. abstract algebra) have bypassed the ontological difficulties which stem from a genuine attempt to *explain* how all mathematical objects are built up as sets from a small number of initial elements. Thus for an algebraic structure what matters is the presence of a domain or 'object' and the nature of the constitutive elements is irrelevant. Set theory 'succeeds' for large segments of modern mathematics because it provides enough objects; but if set theory is to be the ultimate framework one is still left with the conceptual mystery of why all mathematical objects *should* be sets ('unities out of manifolds'). And in any case conceptual problems do not go away just because the theory they concern is successful, a lesson we should have learnt from the history of physics.

But the conceptual difficulties with set theory are thrown into sharper relief by theoretical difficulties. The theory just does not answer key questions about the continuum, and so far has not been extended in a way which does. It does not tell us the Cantorian size of the continuum, and it fails to answer 'purer' questions such as whether the continuum is characterized by the Souslin property, or whether every projective set is Lebesgue measurable. Now in a way it is natural to trace these theoretical difficulties back to the foundational difficulties. If there is no reasonable explanation of why the continuum should be cast as a classical set-theoretic structure, why should we *expect* a theory of sets to answer the deepest analytic questions about the continuum? In his [1967] (p. 52), Putnam ridicules the idea that the independence of *CH* from the axioms of *ZF* (and, one might add, the failure of any subsequent attempt to find axioms which decide *CH*) shows that the concept of set is unsound. Taken by itself this may be right. But that is not the point. The set concept was *always* unclear, at least after the 'Cantorian transition'; and it was axiomatized in large part because of that unclarity (the paradoxes made a bad situation worse). The mounting theoretical difficulties have simply forced us to go back and face that fact. Maybe use of sets is fundamental to abstract thought; but why should we expect that all of mathematics should be embraced by it?

As soon as we give utterance to something, we render it valueless in some strange way. We believe we have plunged to the uttermost depths, and yet when we come back to the upper surface the drops of water on our pale finger-tips no longer resemble the sea from which they came. We imagine we have discovered a hoard of wondrous treasure, but when we emerge again into the light of day we see we have brought only false stones and chips of glass, while the treasure shimmers on unchanged in the darkness beneath.

—Maeterlinck

BIBLIOGRAPHY

A The works of Georg Cantor

Most of Cantor's published works were republished in Zermelo's 1932 edition of the collected works (Cantor [1932]), and since this is the most easily accessible source I have referred to it rather than the original source wherever possible.

CANTOR, G. [1872]. Über die Ausdehnung eines Satzes aus der Theorie der trigonometrischen Reihen. *Mathematische Annalen* **5**, 123–32. (Cantor [1932], pp. 92–101.)

— [1874]. Über eine Eigenschaft des Inbegriffes aller reellen algebraischen Zahlen. *Journal für die reine und angewandte Mathematik* **77**, 258–62. (Cantor [1932], pp. 115–8.)

— [1878]. Ein Beitrag zur Mannigfaltigkeitslehre. *Journal für die reine und angewandte Mathematik* **84**, 242–58. (Cantor [1932], pp. 119–33.)

— [1879a]. Über einen Satz aus der Theorie der stetigen Mannigfaltigkeiten. *Nachrichten von der Königlichen Gesellschaft der Wissenschaften und der Georg-August Universität zu Göttingen*, pp. 127–35. (Cantor [1932], pp. 134–8.)

— [1879b]. Über unendliche, lineare Punktmannigfaltigkeiten, 1. *Mathematische Annalen* **15**, 1–7. (Cantor [1932], pp. 139–45.)

— [1880]. Über unendliche, lineare Punktmannigfaltigkeiten, 2. *Mathematische Annalen* **17**, 355–8. (Cantor [1932], pp. 145–8.)

— [1882], Über unendliche, lineare Punktmannigfaltgkeiten, 3. *Mathematische Annalen* **20**, 113–21. (Cantor [1932], pp. 149–57.)

— [1883a]. Über unendliche, lineare Punktmannigfaltigkeiten, 4. *Mathematische Annalen* **21**, 51–8. (Cantor [1932], pp. 157–64.)

— [1883b]. Über unendliche, lineare Punktmannigfaltigkeiten, 5. *Mathematische Annalen* **21**, 545–86. (Cantor [1932], pp. 165–209.)

— [1883c]. Surs divers théorèmes de la théorie de points situés dans un éspace continu à *n*-dimensions. *Acta Mathematica* **2**, 409–14.

— [1883d]. *Grundlagen einer allgemeinen Mannigfaltigkeitslehre. Ein mathematisch-philosophischer Versuch in der Lehre des Unendlichen.* Teubner, Leipzig. (Separate edition of [1883b] with its own *Vorwort* not in [1883b] and not reprinted in Cantor [1932].)

— [1884a]. Über unendliche, lineare Punktmannigfaltigkeiten, 6. *Mathematische Annalen* **23**, 453–88. (Cantor [1932], pp. 210–46.)

— [1884b], De la puissance des ensembles parfaits de points. *Acta Mathematica* **4**, 381–92. (Cantor [1932], pp. 252–60.)

— [1885a], Rezension der Schrift von G. Frege *Die Grundlagen der Arithmetik. Deutsche Literaturzeitung, 6. Jahrgang*, pp. 728–9. [1932], pp. 440–1.)

— [1885b]. Über verschiedene Theoreme aus der Theorie der Punktmengen in einem *n*-fach ausgedehnten stetigen Raume G_n. (Zweite Mittheilung.) *Acta Mathematica* 7, 105–204. (Cantor [1932], pp. 261–76.)

— [1886a]. Über die verschiedenen Ansichten in Bezug auf die actualunendlichen Zahlen. *Bihang Till Koniglen Svenska Vetenskaps Akademiens Handligar* 11 (19), 1–10. (Not in Cantor [1932].)

— [1886b]. Über die verschiedenen Standpunkte in bezug auf das aktuelle Unendliche. *Zeitschrift für Philosophie und philosophische Kritik* 88, 224–33. (Cantor [1932], pp. 370-77.)

— [1887-8]. Mitteilungen zur Lehre vom Transfiniten 1, II. *Zeitschrift für Philosophie und philosophische Kritik* 91, 81–125, 252–70; 92, 250–65. (Cantor [1932], pp. 378–439.)

— [1891]. Über eine elementare Frage der Mannigfaltigkeitslehre. *Jahresbericht der deutschen Mathematiker-Vereiningung* 1, 75–8. (Cantor [1932], pp. 278–80.)

— [1895]. Beiträge zur Begründung der transfiniten Mengenlehre, 1. *Mathematische Annalen* 46, 481–512. (Cantor [1932], pp. 282–311.)

— [1897]. Beiträge zur Begründung der transfiniten Mengenlehre, 2. *Mathematische Annalen* 49, 207–46. (Cantor [1932], pp. 312–56.)

— [1932]. *Gesammelte Abhandlungen mathematischen und philosophischen Inhalts.* (ed. E. Zermelo) Springer, Berlin (reprinted 1980).

In addition to these published works there are Cantor's unpublished writings, the Cantor *Nachlass*, kept, under the auspices of the Göttingen *Akademie der Wissenschaften*, in the *Niedersächsische Staats- und Universitätsbibliothek*, Göttingen. I have made frequent reference to three items from the *Nachlass*, the three surviving *Briefbücher* in which Cantor usually drafted his letters. In these references I have used the classification hitherto employed by the *Handschriftenabteilung* of the Göttingen *Universitätsbibliothek*.

CANTOR, G. *Nachlass VI* : *Briefbuch 1* (Letter book used by Cantor between 1884 and 1888).

— *Nachlass VII* : *Briefbuch 2* (Letter book used by Cantor between 1890 and 1895).

— *Nachlass VIII* : *Briefbuch 3* (Letter book used by Cantor between 1895 and 1896).

The *Nachlass* contains 17 items or sections altogether, though the three letter books constitute by far the most important part. For a list and brief account of the contents of the *Nachlass* see Grattan-Guinness, *Annals of Science* 27, (1971) 348–9. Just before going to press, I was informed by Dr Klaus Haenel, Director of the *Handschriftenabteilung* of the Göttingen *Universitätsbibliothek*, that a new classification system has been adopted by the library. According to this, Cantor *Nachlass VI-VIII* are now classified as *Cod. Ms. Cantor* 16–18 respectively.

B Other works

ALEXANDROFF, P. S. [1916]. Sur la puissance des ensembles mésurables B. *Comptes Rendus de l'Académie des Sciences de Paris* 162, 323–5.

ANNAS, J. [1976]. *Aristotle's Metaphysics: Books M and N*. Translated with an Introduction and Notes by Julia Annas. Oxford University Press, Oxford.

BAIRE, R., BOREL, E., HADAMARD, J., and LEBESGUE, H. [1905]. Cinq lettres sur la théorie des ensembles. *Bulletin de la Société Mathématique de France*. (References are to the reprinting in Borel [1950], pp. 150–8.)

BARWISE, J. (ed.) [1977]. *Handbook of mathematical logic*. North-Holland, Amsterdam.

BECKER, O. [1964]. *Grundlagen der Mathematik in geschichtlicher Entwicklung*. (2nd extended edn. 1974.) Suhrkamp, Frankfurt.

BELL, J. L. [1977]. *Boolean-valued models and independence results in set theory*. Oxford University Press, Oxford.

— [1979]. The infinite past regained: a reply to Whitrow. *British Journal for the Philosophy of Science* **30**, 161–5.

— [1981]. Category theory and the foundations of mathematics. *British Journal for the Philosphy of Science* **32**, 349–58.

— and MACHOVER, M. [1977]. *A course in mathematical logic*. North-Holland, Amsterdam.

BENACERRAF, P. and PUTNAM, H. (eds.) [1964]. *Philosophy of mathematics: selected readings*. Blackwell, Oxford.

BENDIXSON, I. [1883]. Quelques théorèmes de la théorie des ensembles de points. *Acta Mathematica* **2**, 415–29.

BENNETT, J. [1974]. *Kant's dialectic*. Cambridge University Press, Cambridge.

BERNAYS, P. [1941]. A system of axiomatic set theory, 2. *Journal of Symbolic Logic* **6**, 1–17.

— [1942]. A system of axiomatic set theory, 4. *Journal of Symbolic Logic* **7**, 133–45.

— [1946]. Review of Gödel [1944]. *Journal of Symbolic Logic* **11**, 75–9.

— [1948]. A system of axiomatic set theory, 6. *Journal of Symbolic Logic* **13**, 65–79.

— [1950]. Mathematische Existenz und Widerspruchsfreiheit. *Etudes de Philosophie des Sciences*, pp. 11–25. Griffon, Neuchâtel. (Page references in the text are to the reprinting in P. Bernays, *Abhandlungen zur Philosophie der Mathematik*, pp. 92–106. Wissenschaftliche Buchgesellschaft, Darmstadt, 1976.)

— [1961]. Zur Frage der Unendlichkeitsschemata in der axiomatischen Mengenlehre. In Y. Bar-Hillel (ed.) *Essays on the foundations of mathematics* pp. 3–49. Magnes, Jerusalem.

BERNSTEIN, F. [1905]. Über die Reihe der transfiniten Ordnungszahlen. *Mathematische Annalen* **60**, 187–93.

— [1905*b*]. Untersuchungen aus der Mengenlehre. *Mathematische Annalen* **61**, 117–55.

— [1908]. Zur Theorie der trigonometrischen Reihe. *Berichte über die Verhundlungen der Königlich-Sächsischen Gesellschaft der Wissenschaften zu Leipzig, mathematische-physische Klasse* **60**, 325–38.

BOCHENSKI, I. M. [1970]. *A history of formal logic*. Chelsea, New York.

BOLZANO, B. [1851]. *Paradoxien des Unendlichen*. (Translated by D. A. Steele as *Paradoxes of the infinite*, Routledge and Kegan Paul, London, 1959. Page numbers refer to this translation.)

BOOLOS, G. [1971]. The iterative concept of set. *Journal of Philosophy* **68**, 215–31.

BOREL, E. [1898]. *Leçons sur la théorie des fonctions*. Gauthier-Villars, Paris. (Page numbers refer to the reprinting in Borel [1950].)

— [1899]. A propos de l' "infini nouveau". *Revue Philosophique* 1899. (Page numbers refer to the reprinting in Borel [1950], pp. 135–41.)

— [1900]. L'antinomie du transfini. *Revue Philosophique,* 1900. (Page numbers refer to the reprinting in Borel [1950], pp. 142–6.)

— [1908]. Les "paradoxes" de la théorie des ensembles. *Annales de l'Ecole Normale* 1908. (Page numbers refer to the reprinting in Borel [1950], pp. 161–5.)

— [1950]. *Leçons sur la théorie des fonctions* (4th edn.). Gauthiers-Villars, Paris.

BOURBAKI, N. [1968]. *Elements of mathematics: theory of sets.* Addison-Wesley, Reading, MA.

BROWDER, F. (ed.) [1976]. *Mathematical developments arising from Hilbert's problems.* (*Proceedings of Symposia in Pure Mathematics*, Vol. 28, Parts 1 and 2.) American Mathematical Society, Providence, RI.

BUTTS, R. and HINTIKKA, J. (eds.) [1977]. *Logic, foundations of mathematics and computability theory.* Reidel, Dordrecht.

CAVAILLÈS, J. [1962]. *Philosophie mathématique.* Hermann, Paris.

CHIHARA, C. [1982]. A Gödelian thesis regarding mathematical objects: do they exist and can we perceive them? *Philosophical Review* **91**, 211–27.

COHEN, P. J. [1963–4]. The independence of the continuum hypothesis, I and II. *Proceedings of the National Academy of Sciences USA* **50**, 1143–8; **51**, 105–110.

— [1966]. *Set theory and the continuum hypothesis.* Benjamin, New York.

CRAIG, W. L. [1979]. Whitrow and Popper on the impossibility of an infinite past. *British Journal for the Philosophy of Science* **30**, 165–9.

VAN DANTSCHER, V. [1908]. *Vorlesungen über die Weierstrassche Theorie der irrationalen Zahlen.* Teubner, Leipzig.

DAUBEN, J. [1971]. The trigonometric background to Cantor's theory of sets. *Archive for the History of the Exact Sciences* **7**, 181–216.

— [1979]. *Georg Cantor: his mathematics and philosophy of the infinite.* Harvard University Press, Boston, MA.

DEDEKIND, R. [1872]. *Stetigkeit und irrationale Zahlen.* (Republished in 1969 by Vieweg, Braunschweig; English translation in Dedekind [1963] under the title *Continuity and Irrational Numbers*.)

— [1888]. *Was sind und was sollen die Zahlen?* (Republished in 1969 by Vieweg, Braunschweig; English translation in Dedekind [1963] under the title *The Nature and Meaning of Numbers*.)

— [1963]. *Essays on the theory of numbers.* (English translations by W. W. Beman of Dedekind [1872] and [1888].) Dover, New York.

DE MORGAN, A. [1858]. On the syllogism, III. In A. De Morgan, *On the syllogism and other logical writings,* pp. 74–146. Routledge and Kegan Paul, London, 1966.

DIEUDONNE, J. [1969]. *Foundations of modern analysis.* McGraw-Hill, New York.

DODD, A. J. [1984]. *Consistency proofs in set theory.* Oxford University Press.

DRAKE, F. [1974]. *Set theory.* North-Holland, Amsterdam.

DUMMETT, M. [1975]. The philosophical basis of intuitionist logic. In H. E. Rose and J. C. Shepherdson (eds.) *Logic colloquium '73.* North-Holland Amsterdam, pp. 5–50. (Reprinted in M. Dummett *Truth and other enigmas.* Duckworth, London (1978), pp. 215–47.)

DUMMETT, M. [1976]. Frege on the consistency of mathematical theories.

In M. Schirn (ed.) *Studien zu Frege,* Vol. 1, Frommann-Holzboog, Stuttgart, pp. 229–42.

— [1977]. *Elements of intuitionism.* Oxford University Press, Oxford.

EASTON, W. B. [1970]. Powers of regular cardinals. *Annals of Mathematical Logic* **1,** 139–78.

EDWARDS, H. M. [1977]. *Fermat's last theorem.* Springer, Berlin.

— [1980]. Genesis of ideal theory. *Archive for the History of the Exact Sciences* **23,** 321–78.

FEFERMAN, S. [1964]. *The number systems.* Addison–Wesley, Reading, MA.

FELGNER, U. [1976]. Choice functions on sets and classes. In G. Müller (ed.) *Sets and classes: on the work of Paul Bernays,* pp. 217–55, North-Holland, Amsterdam.

— [1979a]. Bericht über die Cantorsche Kontinuumshypothese. In Felgner (ed.) [1979b], pp. 166–205.

— [1979b]. *Mengenlehre.* Wissenschaftliche Buchgesellschaft, Darmstadt.

FRAENKEL, A. [1919]. *Einleitung in die Mengenlehre.* Springer, Berlin.

— [1921]. Über die Zermelosche Begründung der Mengenlehre. *Jahresbericht der deutschen Mathematiker-Vereinigung* **30,** 97–8.

— [1922b]. Zu den Grundlagen der Cantor–Zermeloschen Mengenlehre. *Mathematische Annalen* **86,** 230–7.

— [1922c]. Der Begriff "definit" und die Unabhängigheit des Auswahisaxiom. *Sitzungsberichte der Preussischen Akademie der Wissenschaften, physikalische-mathematische Klasse,* pp. 253–7. (Translated in van Heijenoort [1967], pp. 284–9.)

— [1923]. *Einleitung in die Mengenlehre.* (2nd expanded edn.) Springer, Berlin.

— [1924]. Die neueren Ideen zur Grundlegung der Analysis und Mengenlehre. *Jahresbericht der deutschen Mathematiker-Vereinigung* **33,** 97–103.

— [1925]. Untersuchungen über die Grundlagen der Mengenlehre. *Mathematische Zeitschrift* **22,** 250–73.

— [1926]. Axiomatische Theorie der geordneten Mengen (Untersuchungen über die Grundlagen der Mengenlehre II). *Joural für die reine und angewandte Mathematik* **155,** 129–58.

— [1927]. *Zehn Vorlesungen über die Grundlegung der Mengenlehre.* Teubner, Leipzig.

— [1928a]. Zusatz zu vorstehendem Aufsatz Herrn von Neumanns. *Mathematische Annalen* **99,** 392–3.

— [1928b]. *Einleitung in die Mengenlehre.* (3rd further expanded edn.) Springer, Berlin.

— [1930]. Georg Cantor. *Jahresbericht der deutschen Mathematiker-Vereinigung* **39,** 189–266.

— [1932a]. Axiomatische Theorie der Wohlordnung (Untersuchungen über die Grundlagen der Mengenlehre III). *Journal für die reine und angewandte Mathematik* **167,** 1–11.

— [1932b]. Das Leben Georg Cantors. (Abridged version of Fraenkel [1930] in Cantor [1932], pp. 452–83.)

— [1967]. *Lebenskreise: aus den Erinnerungen eines jüdischen Mathematikers.* Deutsche Verlags-Anstalt, Stuttgart.

— and BAR-HILLEL, Y. [1958]. *Foundations of set theory.* North-Holland, Amsterdam.

— — and LEVY, A. [1973]. *Foundations of set theory.* North-Holland, Amsterdam.

FREGE, G. [1884]. *Die Grundlagen der Arithmetik*. (English translation by
J. L. Austin as *The foundations of arithmetic*. Blackwell, Oxford, 1974.
Page numbers refer to this edition and to both the German and English
texts which appear side by side; quotations are from the translation.)
— [1890-2]. Entwurf zu einer Besprechung von Cantors Gesammelte Abhandlungen zur Lehre vom Transfiniten. (In Frege [1969], pp. 76-86; page
numbers in the text refer to the translation in Frege [1979], pp. 68-71.)
— [1891-2]. Über den Zahlbegriff. In Frege [1969] or Frege [1979],
pp. 72-86.
— [1892]. Rezension von: Georg Cantor, *Zur Lehre vom Transfiniten.
Gesammelte Abhandlungen aus der Zeitschrift für Philosophie und philosophische Kritik. Zeitschrift für Philosophie und philosophische Kritik* **100**,
269-72. (Page numbers in the text refer to the reprinting in Frege [1967],
pp. 163-6.)
— [1893]. *Grundgesetze der Arithmetik*. Vol. 1, Olms, Hildesheim, 1962.
— [1903]. *Grundgesetze der Arithmetik*, Vol. 2. Olms, Hildesheim, 1962.
— [1906]. Reply to Thomae. *Jahresbericht der deutschen Mathematiker-
Vereiningung* **15**, 586-90. (English translation from Frege [1971], pp. 121-7.)
— [1967]. *Kleine Schriften*. Olms, Hildesheim.
— [1969]. *Gottlob Frege: Nachgelassene Schriften* (ed. H. Hermes, F. Kambartel
and F. Kaulbach). Meiner, Hamburg.
— [1971]. *On the foundations of geometry and formal theories of arithmetic:
a selection* (ed. E-H. W. Kluge). Yale University Press, New Haven, CO.
— [1979]. *Gottlob Frege: posthumous writings*. Blackwell, Oxford. (English
translation by P. Long and R. White of Frege [1969].)
FRIEDMAN, H. [1971]. Higher set theory and mathematical practice. *Annals
of Mathematical Logic* **2**, 326-57.
GÖDEL, K. [1931]. Über formal unentscheidbare Sätze der *Principia Mathematica* und verwandter Systeme, I. *Monatshefte für Mathematik und
Physik* **38**, 173-98. (English translation in van Heijenoort [1967], pp. 596-
616.)
— [1938]. The consistency of the axiom of choice and of the generalized
continuum hypothesis. *Proceedings of the National Academy of Sciences,
U.S.A.* **24**, 556-7.
— [1940]. *The consistency of the axiom of choice and of the generalized
continuum hypothesis*. Annals of Mathematics Studies, Vol. 3. Princeton
University Press, N.J.
— [1944]. Russell's Mathematical Logic. In P. Schilpp (ed.) *The philosophy
of Bertrand Russell*, pp. 125-53. (Reprinted in Benacerraf and Putnam
[1964], pp. 211-32.)
— [1946]. Remarks before the Princeton bicentennial conference on problems
in mathematics. In M. Davis (ed.) *The undecidable*, Raven Press, New York,
1965.
— [1947]. What is Cantor's continuum problem? *American Mathematical
Monthly* **54**, 515-25.
— [1964]. What is Cantor's continuum problem? In Benacerraf and Putnam
[1964], pp. 258-73. (Revised and slightly expanded version of Gödel
[1947].)
GRATTAN-GUINNESS, I. [1970]. An unpublished paper by Georg Cantor:
*Principien einer Theorie der Ordnungstypen. Erste Mittheilung. Acta Mathe-
matica* **124**, 65-107.

— [1971]. The correspondence between Georg Cantor and Philip Jourdain, *Jahresbericht der deutschen Mathematiker-Vereinigung* 73, 111–30.

— [1972]. A mathematical union: William Henry and Grace Chisholm Young. *Annals of Science* 29, 105–86.

— [1974]. The rediscovery of the Cantor–Dedekind correspondence. *Jahresbericht der deutschen Mathematike-Vereinigung,* 76, 104–39.

— [1977]. *Dear Russell–Dear Jourdain.* Duckworth, London.

GUTBERLET, C. [1878]. *Das unendliche metaphysisch und mathematisch betrachtet.* Faber, Mainz.

HALLETT, M. F. [1979a]. Towards a theory of mathematical research programmes I and II. *British Journal for the Philosophy of Science* 30, 1–25, 135–59.

— [1979b]. *The nature of progress in mathematics.* Ph.D. Thesis, University of London.

— [1981]. Russell, Jourdain and 'Limitation of Size'. *British Journal for the Philosophy of Science* 32, 381–99.

— [1984]. Russell's rejection of classes. In preparation.

HALMOS, P. [1960]. *Naive set theory.* Van Nostrand, Princeton, NJ.

HARDY, G. H. [1904]. A theorem concerning the infinite cardinal numbers. *The Quarterly Journal of Pure and Applied Mathematics* 35, 87–94.

HARTOGS, F. [1915]. Über das Problem der Wohlordnung. *Mathematische Annalen* 76, 438–443.

HAUSDORFF, F. [1904]. Der Potenzbegriff in der Mengenlehre. *Jahresbericht der deutschen Mathematiker-Vereinigung* 13, 569–71.

— [1906]. Untersuchungen über Ordnungstypen. *Berichten der Königlich-Sächsischen Gesellschaft der Wissenschaften zu Leipzig, mathematisch-physische Klasse* 58, 106–69.

— [1907]. Untersuchungen über Ordnungstypen. *Berichten der Königlich-Sächsischen Gesellschaft der Wissenschaften zur Leipzig, mathematisch-physische Klasse* 59, 84–159.

— [1908]. Grundzüge einer Theorie der geordneten Mengen. *Mathematische Annalen* 65, 435–505.

— [1914]. *Grundzüge der Mengenlehre.* Von Veit, Leipzig. (Reprinted by Chelsea, New York, 1965.)

— [1916]. Die Mächtigkeit der Borelschen Mengen. *Mathematische Annalen* 77, 430–7.

— [1937]. *Mengenlehre* (3rd revised edn. of Hausdorff [1914]). Berlin and Leipzig. (Page numbers in the text refer to the English translation *Set theory* (2nd edn.), Chelsea, New York, 1962.)

HAWKINS, T. [1970]. *Lebesgue's theory of integration.* Chelsea, New York, 1979.

HEATH, T. L. [1921]. *Greek Mathematics,* Vols. 1 and 2. Oxford University Press, Oxford.

— [1925]. *The Thirteen Books of Euclid's Elements* (2nd edn.). Cambridge University Press, Cambridge.

VAN HEIJENOORT, J. (ed.) [1967]. *From Frege to Gödel.* Harvard University Press, Boston, MA.

HESSENBERG, G. [1906]. Grundbegriffe der Mengenlehre. *Abhandlungen der Fries'schen Schule, Neue Folge* 1, 479–706.

— [1909]. Über Kettentheorie und Wohlordnung. *Journal für die reine und angewandte Mathematik* 135, 81–133.

HILBERT, D. [1900*a*]. Mathematische Probleme. *Nachrichten von der königlichen Gessellschaft der Wissenschaften zu Göttingen*, 1900, 253–97. (Page numbers in the text refer to the English translation *Mathematical problems*, in Browder [1976], pp. 1–34 (originally from *Bulletin of the American Mathematical Society* **8**, 437–79.)

—— [1900*b*]. Über den Zahlbegriff. *Jahresbericht der deutschen Mathematiker-Vereinigung* **8**, 180–4. (Page numbers in the text refer to the reprinting in D. Hilbert, *Grundlagen der Geometrie* (3rd end.), pp. 156–32, Teubner, Leipzig, 1909.)

—— [1923]. Die logischen Grundlagen der Mathematik. *Mathematische Annalen* **88**, 151–65. (Page numbers in the text refer to the reprinting in Hilbert [1935], pp. 178–91.)

—— [1935]. *Gesammelte Abhandlungen*, Vol. 3. Springer, Berlin. (Reprinted by Chelsea, New York, 1965.)

JECH, T. [1978]. *Set theory*. Academic Press, New York.

JOHNSON, D. [1979]. The problem of the invariance of dimension in the growth of modern topology, Part 1. *Archive for the History of the Exact Sciences* **20**, 97–188.

JOURDAIN, P. E. B. [1904*a*]. On the transfinite cardinal numbers of well-ordered aggregates. *Philosophical Magazine* **7** (6), 61–75.

—— [1904*b*]. On the transfinite numbers of number-classes in general. *Philosophical Magazine* **7** (6), 294–303.

—— [1905]. On transfinite cardinal numbers of the exponential form. *Philosophical Magazine* **9** (6), 42–56.

—— [1906]. The definition of a series similarly ordered to the series of all ordinal numbers. *Messenger of Mathematics* **35**, 56–8.

—— [1907]. On the question of the existence of transfinite numbers. *Proceedings of the London Mathematical Society* **4** (2), 266–83.

—— [1908]. The multiplication of alephs. *Mathematische Annalen* **65**, 506–12.

—— [1909]. The development of the theory of transfinite numbers, 2. *Archiv der Mathematik und Physik* **14**, 289–311.

—— [1910]. Transfinite numbers and the principles of mathematics. *The Monist* **20**, 93–118.

—— [1917]. Existents and entities. *The Monist* **27**, 142–51.

—— [1922]. A proof that every aggregate can be well-ordered. *Acta Mathematica* **43**, 239–61.

KAMKE, E. [1950]. *Theory of sets*. Dover, New York.

KANAMORI, A. and MAGIDOR, M. [1978]. The evolution of large cardinal axioms in set theory. In G. H. Müller and D. S. Scott (eds.) *Higher set theory*. *Springer Lecture Notes in Mathematics*, Vol. **669**, Springer, Berlin.

KANT, I. [1770]. *Inaugural dissertation: on the form and principles of the sensible and intelligible world*. (Translated by G. B. Kerferd and D. E. Walford in *Kant: Selected Pre-Critical Writings*, Manchester University Press, Manchester, pp. 43–92.)

—— [1783]. *Prolegomena to any future metaphysics*. (Translated by P. G. Lucas, Manchester University Press, Manchester 1971.)

—— [1787]. *Critique of pure reason*. (Translated by N. Kemp Smith, MacMillan, London, 1973.)

KELLEY, J. L. [1955]. *General topology*. Van Nostrand, Princeton, NJ.

KLEIN, J. [1968]. *Greek mathematics and the origin of algebra*. MIT Press, Cambridge, MA.

KLINE, M. [1972]. *Mathematical thought from ancient to modern times.* Oxford University Press, Oxford.
— [1980]. *Mathematics: the loss of certainty.* Oxford University Press, New York.
KÖNIG, J. [1904]. Zum Kontinuum-Problem. In A. Krazer (ed.) *Verhandlungen des dritten internationalen Mathematiker-Kongresses in Heidelberg vom 8 bis 13 August 1904,* pp. 144-7. Teubner, Leipzig. (Reprinted in *Mathematische Annalen* **60** (1905) 177-80.)
KOWALEWSKI, G. [1950]. *Bestand und Wandel.* Oldenbourg, Munich.
KREISEL, G. [1973]. Bertrand Arthur William Russell, Earl Russell, 1872–1970. *Biographical Memoirs of Fellows of the Royal Society* **19**, 583-620.
— [1976]. What have we learnt from Hilbert's second problem? In Browder [1976], pp. 93-130.
— [1980]. Kurt Gödel. *Biographical Memoirs of Fellows of the Royal Society* **26**, 1-76.
KURATOWSKI, C. [1921]. Sur la notion de l'ordre dans la théorie des ensembles. *Fundamenta Mathematicae* **2**, 161-71.
— [1922]. Une méthode d'élimination des nombres transfinis des raisonnements mathématiques. *Fundamenta Mathematicae* **3**, 76-108.
— [1980]. *A half-century of Polish mathematics.* Pergamon, Oxford.
KUNEN, K. [1980]. *Set theory: an introduction to independence proofs.* North-Holland, Amsterdam.
LEISENRING, A. C. [1969]. *Mathematical logic and Hilbert's ε-symbol.* Macdonald, London.
LEVY, A. [1960]. Axiom schemata of strong infinity in axiomatic set theory. *Pacific Journal of Mathematics* **10**, 223-38.
— [1969]. The definability of cardinal numbers. In J. Buloff *et al.* (eds.) *Foundations of mathematics: symposium papers commenorating the sixtieth birthday of Kurt Gödel,* pp. 15-38. Springer, Berlin.
— [1979]. *Basic set theory.* Springer, Berlin.
LINDELÖF, E. [1906]. Remarques sur un théorème fondamental de la théorie des ensembles. *Acta Mathematica* **29**, 183-90.
LUSIN, N. [1917]. Sur la classification de M. Baire. *Comptes Rendus de l'Académie des Sciences de Paris* **164**, 91-4.
— [1925]. Sur les ensembles projectifs de M. Henri Lebesgue. *Comptes Rendus de l'Académie des Sciences de Paris* **180**, 1572-4.
— SIERPINSKI, W. [1923]. Sur un ensemble non mesurable B. *Journal de Mathématique* **2**, 52.
MARTIN, D. A. [1975]. Borel determinacy. *Annals of Mathematics* **102**, 363-71.
— [1976]. Hilbert's first problem: the continuum hypothesis. In Browder [1976], pp. 81-92.
— [1977]. Descriptive set theory. In Barwise [1977], pp. 783-815.
— [?]. Projective sets and cardinal numbers, to be published.
— and SOLOVAY, R. [1970]. Internal Cohen extensions. *Annals of Mathematical Logic* **2**, 143-78.
MARTIN, G. [1955]. *Kant's metaphysics and theory of science.* Manchester University Press, Manchester. (Translated by P. G. Lucas from the German: *Immanuel Kant, Ontologie und Wissenschaftstheorie,* Cologne 1951.)
— [1968]. *General metaphysics: its problems and its method.* Allen and Unwin, London. (Translated by D. O'Connor from the German: *Allgemeine Metaphysik: ihre Probleme und ihre Methode,* Berlin, 1965.)

MESCHKOWSKI, H. [1961]. *Denkweisen grosser Mathematiker.* Vieweg,
Braunschweig.
MESCHKOWSKI, H. [1965]. Aus den Briefbüchern. Georg Cantors. *Archive
for the History of the Exact Sciences* 2, 503-19.
— [1967]. *Probleme des Unendlichen: Werk und Leben Georg Cantors.* Vieweg,
Braunschweig.
MIRIMANOFF, D. [1917a]. Les antinomies de Russell et de Burali-Forti et
le problème fondamental de la théorie des ensembles. *L'Enseignement Mathématique* 19, 37-52.
— [1917b]. Remarques sur la théorie des ensembles et les antinomies Can-
toriennes (I). *L'Enseignement Mathematique* 19, 208-17.
— [1921]. Remarques sur la théorie des ensembles et les antinomies Can-
toriennes (II). *L'Enseignement Mathematique* 21, 29-52.
MITTAG-LEFFLER, G. [1927]. Zusätzliche Bemerkungen zu Schönflies [1927],
Acta Mathematica 50, 25-9.
MOORE, G. H. [1982]. *Zermelo's axiom of choice: its origins, development
and influence.* Springer, Berlin.
MOSCHKOVAKIS, Y. N. [1980]. *Descriptive set theory.* North-Holland,
Amsterdam.
VON NEUMANN, J. [1923a]. Zur Einführung der transfiniten Zahlen. *Acta
Litterarum ac Scientiarum Regiae Universitatis Hungaricae Francisco-
Josephinae. Sectio Sci-Math.*, 1, 199-208. (References are to the translation
in van Heijenoort [1967], pp. 346-54.)
— [1923b]. Letter to Ernst Zermelo, 15 August 1923. In Meschkowski [1967],
pp. 271-3.
— [1925] Eine Axiomatisierung der Mengenlehre. *Journal für die reine und
angewandte Mathematik* 154, 219-40. (References are to the English translation in van Heijenoort [1967], pp. 393-413.)
— [1928a]. Über die Definition durch transfinite Induktion und verwandte
Fragen der allgemeinen Mengenlehre. *Mathematische Annalen* 99, 373-91.
(References are to the reprinting in von Neumann [1961], pp. 320-38.)
— [1928b]. Die Axiomatisierung der Mengenlehre. *Mathematische Zeitschrift*
27, 669-752. (References are to the reprinting in von Neumann [1961],
pp. 339-422.)
— [1929]. Über eine Widerspruchsfreitheitsfrage in der axiomatischen Mengenlehre. *Journal für die reine und angewandte Mathematik* 160, 227-41. (References are to the reprinting in von Neumann [1961], pp. 494-508.)
— [1961]. *John von Neumann: collected works,* Vol. 1. Pergamon, Oxford.
NOETHER, E. and CAVAILLÈS, J. (eds.) [1937]. *Briefwechsel Cantor-
Dedekind.* Hermann, Paris.
PARSONS, C. [1965]. Frege's theory of number. In M. Black (ed.) *Philosophy
in America,* pp. 180-203. George Allen and Unwin, London.
— [1977]. What is the iterative concept of set? In Butts and Hintikka [1977],
pp. 335-67.
POPPER, K. [1978]. On the possibility of an infinite past: a reply to Whitrow.
British Journal for the Philosophy of Science 29, 47-8.
PUTNAM, H. [1967]. Mathematics without foundations. *Journal of Philosophy*
64, 5-22. (References are to the reprinting in H. Putnam, *Mathematics,
matter and method. Philosophical papers,* Vol. 1, pp. 43-9, Cambridge
University Press, Cambridge 1978.)
QUINE, W. V. O. [1969]. *Set theory and its logic.* Harvard University Press,

Boston, MA.

REINHARDT, W. [1974]. Remarks on reflection principles, large cardinals and elementary embeddings. In T. Jech (ed.) *Axiomatic set theory* (*Proceedings of Symposia in Pure Mathematics,* Vol. 13, Part 2), pp. 189-205. American Mathematical Society, Providence, RI.

ROWBOTTOM, F. [1971]. Some strong axioms of infinity incompatible with the axiom of constructibility. *Annals of Mathematical Logic* 3, 1-44.

RUSSELL, B. A. W. [1902]. Letter to Frege. In van Heijenoort [1967], pp. 124-5.

— [1903]. *The principles of mathematics.* George Allen and Unwin, London.

— [1906a]. On some difficulties in the theory of transfinite numbers and order types. *Proceedings of the London Mathematical Society* 4 (2) 29-53. (References are to the reprinting in Russell [1973], pp. 135-64.)

— [1906b]. Les paradoxes de la logique. *Revue de Métaphsique et de Morale* 14, 627-50. (Translated by Russell as 'On "insolubilia" and their solution by symbolic logic' in Russell [1973], pp. 190-214.)

— [1911]. Sur les axiomes de l'infini et du transfini. *Comptes Rendus des Séances de la Société Mathématique de France* 2, 22-35. (References are to the English translation by Grattan-Guinness in his [1977], pp. 161-74.)

— [1914]. *Our knowledge of the external world.* George Allen and Unwin, London, 1972.

— [1973]. *Essays in analysis.* George Allen and Unwin, London.

SCHOENFLIES, A. [1899]. Die Entwicklung der Lehre von den Punktmannigfaltigkeiten. *Jahresbericht der deutschen Mathematiker-Vereinigung* 8, 1-251.

— [1905]. Über wohlgeorgnete Mengen. *Mathematische Annalen* 60, 181-86.

— [1908]. *Die Entwicklung der Lehre von den Punktmannigfaltigkeiten, Zweiter Teil.* Teubner, Leipzig.

— [1922]. Zur Erinnerung an Georg Cantor. *Jahresbericht der deutschen Mathematiker-Vereinigung* 31, 97-106.

— [1927]. Die Krisis in Cantors mathematischem Schaffen. *Acta Mathematica* 50, 1-23.

— and HAHN, H. [1913]. *Entwicklung der Mengenlehre und ihrer Andwendungen.* Teubner, Leipzig.

SCHRÖDER, E. [1898]. Über zwei Definitionen der Endlichkeit und G. Cantorsche Sätze. *Nova Acta, Abhandlungen der kaiserlichen Leopold-Carolinschen deutschen Akademie der Naturforscher* 71, 301-62.

SCOTT, D. [1961]. Measurable cardinals and constructible sets. *Bulletin de l'Academie Polonaise des Sciences, Séries des Sciences Mathématiques, Astronomiques et Physiques* 9, 521-4.

— [1974]. Axiomatising set theory. In T. Jech (ed.) *Axiomatic set theory* (*Proceedings of Symposia in Pure Mathematics,* Vol. 13, Part 2), pp. 207-14. American Mathematical Society, Providence, RI, 1974.

— [1977]. Foreword to Bell [1977].

SHOENFIELD, J. R. [1967]. *Mathematical logic.* Addison-Wesley, Reading, MA.

— [1977]. Axioms of set theory. In Barwise [1977], pp. 321-44.

SIERPINSKI, W. [1926]. Sur une propriété des ensembles (A). *Fundamenta Mathematicae* 8, 362-9.

SILVER, J. [1971]. Measurable cardinals and Δ_3^1 well-orderings. *Annals of Mathematics* 94, 414-46.

SKOLEM, T. [1922]. Einige Bemerkungen zur axiomatischen Begründung der

Mengenlehre. *Matematikerkongressen i Helsingfors den 4–7 Juli 1922, Den femte skandinaviska matematikerkongressen, Redogörelse.* (References are to the translation in van Heijenoort [1967], pp. 290–301.)

— [1929]. Über die Grundlagendiskussionen in der Mathematik. *Proceedings of the Seventh Scandinavian Mathematical Congress, Oslo, 1929,* pp. 3–21. (References are to the reprinting in J. E. Fenstad (ed.) *Thoralf Skolem: selected works in logic* pp. 207–25. Scandinavian University Books. Oslo.)

SOLOVAY, R. [1969]. On the cardinality of Σ_2^1 sets of reals. In J. Bulloff *et al.* (eds.) *Foundations of mathematics: symposium papers commemorating the sixtieth birthday of Kurt Gödel,* pp. 58–73. Springer, Berlin.

— [1970]. A model of set theory in which every set of reals is Lebesgue measurable. *Annals of Mathematics* **92**, 1–56.

SOUSLIN, M. [1917]. Sur une definition des ensembles mesurables *B* sans nombres transfinis. *Comptes Rendus de l'Académie des Sciences de Paris,* **164**, 88–91.

STRAWSON, P. F. [1966]. *The bounds of sense.* Methuen, London.

SZABO, A. [1978]. *The origins of Greek mathematics.* Reidel, Dordrecht.

TANNERY, P. [1884]. Note sur la théorie des ensembles. *Bulletin de la Société Mathématique de France* **12**, 90–6.

ULAM, S. [1958]. John von Neumann, 1903–1957. *Bulletin of the American Mathematical Society* **64**, 1–49.

WANG HAO [1954]. The formalisation of mathematics. *Journal of Symbolic Logic* **19**, 241–66. (References are to the reprinting in Wang [1962], pp. 559–84.)

— [1962]. *A survey of mathematical logic.* (Reprinted as *Logic, computers and sets,* Chelsea, New York, 1970.)

— [1974]. *From mathematics to philosophy.* George Allen and Unwin, London.

— [1977]. Large sets. In Butts and Hintikka [1977], pp. 309–33.

— [1981]. Some facts about Kurt Gödel. *Journal of Symbolic Logic* **46**, 653–9.

WEYL, H. [1931]. *Die Stufen des Unendlichen.* Fischer, Jena.

— [1932]. *The open world.* Yale University Press, New Haven, CO.

— [1940]. The mathematical way of thinking. *Science* **92**, 437–46. (Reprinted in H. Weyl, *Gesammelte Abhandlungen,* Vol. 3, pp. 710–18, Springer, Berlin.)

— [1946]. Mathematics and logic. A brief survey serving as a preface to a review of *The Philosophy of Bertrand Russell. American Mathematical Monthly* **53**, 2–13. (Reprinted in H. Weyl, *Gesammelte Abhandlungen,* Vol. 4, pp. 268–79, Springer, Berlin, 1968.)

— [1949]. *Philosophy of mathematics and natural science,* Princeton University Press, Princeton, NJ.

— [1951]. A half-century of mathematics. *American Mathematical Monthly* **58**, 523–53. (Reprinted in H. Weyl, *Gesammelte Abhandlungen,* Vol. 4, pp. 464–94, Springer, Berlin, 1968.)

WHITEHEAD, A. N. [1902]. On cardinal numbers. *American Journal of Mathematics* **24**, 367–94.

— and RUSSELL, B. A. W. [1910–12]. *Principia Mathematica,* Vols. 1–3. Cambridge University Press, Cambridge.

WHITROW, G. [1978]. On the impossibility of an infinite past. *British Journal for the Philosophy of Science* **29**, 39–45.

WIENER, N. [1914]. A simplification of the logic of relations. *Proceedings of the Cambridge Philosophical Society* **17**, pp. 387–90. (Reprinted in van Heijenoort [1967], pp. 224–7.)

YOUNG, W. H. and YOUNG, G. C. [1906]. *The theory of sets of points.*
Cambridge University Press, Cambridge.
ZERMELO, E. [1904]. Beweis, dass jede Menge wohlgeordnet werden kann.
Mathematische Annalen **59**, 514–16. (References are to the translation in
van Heijenoort [1967], pp. 139–41.)
ZERMELO [1908a]. Neuer Beweis für die Möglichkeit einer Wohlordnung.
Mathematische Annalen **65**, 107–28. (References are to the translation
in van Heijenoort [1967], pp. 183–98.)
— [1908*b*]. Untersuchungen über die Grundlagen der Megenlehre, I. *Mathe-
matische Annalen* **65**, 261–81. (References are to the translation in van
Heijenoort [1967], pp. 200–15.)
— [1909*a*]. Sur les ensembles finis et le principe de l'induction complète.
Acta Mathematica **32**, 185–93.
— [1909*b*]. Über die Grundlagen der Arithmetik. In G. Castelnuovo (ed.)
Atti dei IV Congresso Internazionale dei Matematici, Vol. 2, pp. 8–11.
Accademia dei Lincei, Rome.
— [1929]. Über den Begriff der Definitheit in der Axiomatik. *Fundamenta
Mathematicae* **14**, 339–44.
— [1930]. Über Grenzzahlen und Mengenbereiche: neue Untersuchungen
über die Grundlagen der Mengenlehre. *Fundamenta Mathematicae* **16**, 29–47.
— [1932]. Editor's comments in Cantor [1932].

NAME INDEX

SUBJECT INDEX

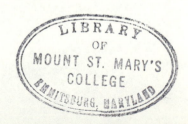